The Mechanics of Threaded Fasteners and Bolted Joints for Engineering and Design

The Mechanics of Threaded Fasteners and Bolted Joints for Engineering and Design

Toshimichi Fukuoka
Professor Emeritus at Kobe University,
Kobe, Japan

ELSEVIER

Elsevier
Radarweg 29, PO Box 211, 1000 AE Amsterdam, Netherlands
The Boulevard, Langford Lane, Kidlington, Oxford OX5 1GB, United Kingdom
50 Hampshire Street, 5th Floor, Cambridge, MA 02139, United States

ISBN: 978-0-323-95357-3

Publisher: Matthew Deans
Acquisitions Editor: Dennis McGonagle
Editorial Project Manager: Sara Greco
Production Project Manager: Fizza Fathima
Cover Designer: Christian J. Bilbow

Typeset by STRAIVE, India

Contents

Preface

Threaded fasteners, typified by bolts and nuts, are the most widely and commonly used machine elements, hence, their existence is often forgotten. However, once a failure or loosening occurs in threaded fasteners, the function of machines and structures clamped by them may be completely lost. It can be said, therefore, that the primary task imposed on the engineers engaged in jobs related to threaded fasteners is the design and assembly of bolted joints, so that they are not broken or loosened in the tightening process or in service conditions.

This book intends to provide the wide variety of knowledge and information required for engineers in charge of threaded fasteners and bolted joints to attain secure tightening, sufficient strength, and no loosening; and these three major issues are explained in an easy-to-understand manner for practical use. At the same time, to maintain mechanical stringency, core chapters and sections are written largely based on reliable research results published by prominent academic societies as journal papers or conference proceedings.

The contents of this book cover a variety of job levels, ranging from the basic design of bolted joints, such as the determination of the joint assembly procedure, to the sophisticated design and strength evaluation of bolted joints by use of advanced CAE technologies. The problems that this book is written to solve for readers are as follows:

- Selection of an appropriate tightening method and establishment of a reliable tightening procedure.
- Understanding what the static and fatigue strengths of bolted joints are and how they behave.
- Evaluation methods of static and fatigue strengths of bolted joints by solid mechanics and FEM.
- Establishment of prevention methods of fatigue failures by solid mechanics and FEM.
- Effect of high and low temperature thermal loads on the strength of bolted joints by taking account of thermal contact resistance at the interface.
- Understanding the loosening mechanism of threaded fasteners by classifying into rotation loosening and nonrotation loosening.
- Establishment of the method to prevent or mitigate the loosening of threaded fasteners.
- How to apply advanced CAE technologies to design bolted joints with high safety and evaluate their thermal and mechanical behaviors.

In Chapter 1, a brief history of screw threads and the progress in the relevant research areas are introduced, and the basics of threaded fasteners and bolted joints are explained from the geometrical and mechanical points of view. The idea of the pressure cone is discussed in detail, because the contact pressure distribution at the interface significantly affects the interface separation that leads to the fatigue failure of bolted joints. The last section provides an outline of the strength and the thermal and mechanical properties of fastener materials, where various properties of engineering materials commonly used for threaded fasteners are tabulated with their temperature dependency.

In Chapter 2, in order to understand the fundamentals of threaded fasteners, the essential characteristics of their strength and stiffness are discussed in detail. It is shown how the true profile of the cross section of screw threads can be expressed by mathematical formulas, by which finite element models of bolted joints with perfect geometry can be constructed in order to perform numerical analysis with high accuracy. In addition, a summary of the interface stiffness and thermal contact resistance, both of which are primarily due to the microprojections existing on contact surfaces, is given to conduct finite element analysis by taking account of the conditions of contact surfaces existing in bolted joints.

In Chapter 3, the tightening methods of threaded fasteners are discussed—which represents the first problem of bolted joints. The four commonly used methods, i.e., torque control method, elastic angle control method, direct tension method, and thermal expansion method, are explained in great detail from the practical point of view. The chapter also covers the tightening principle, application range, and tightening guidelines of each method. A new tightening method that combines torque control method with elastic angle control method is proposed. This can evaluate nut factor, which relates tightening torque to bolt force, during the tightening operation with high accuracy. The problem of elastic interaction, which occurs in the case of tightening multiple bolts one by one, is solved with help of FEM, and it is shown how to estimate the bolt force scatter and obtain the optimal tightening sequence.

In Chapter 4, the static and fatigue strengths inherent to threaded fasteners and bolted joints are dealt with—which is the second problem of bolted joints. The distributions of stress and stress amplitude along the thread root are evaluated by FE analysis using helical thread models. It is pointed out that bolted joint diagrams, which have widely been used for bolted joint design, have some essential defects and are effective only in limited conditions. Another important point is that the fatigue failure of bolted joints is greatly affected by the interface separation between fastened plates. The fatigue strength can primarily be improved by raising the bolt force, which is substantiated by comprehensive FE analysis. In addition, various practical measures are introduced to enhance the fatigue strength of threaded fasteners and bolted joints.

In Chapter 5, thermal and mechanical behaviors of bolted joints are described based on the theory of thermal stress and the results of FE analysis. High and low temperature thermal loads often cause the failure or loosening of threaded fasteners and bolted joints. In order to analyze the phenomena with high accuracy, it is explained how to take account of the effect of the heat transferred through contact surfaces and small gaps using the thermal contact coefficient and apparent thermal contact coefficient. The mechanism of seizure occurrence in threaded fasteners, which is a nasty problem sometimes encountered during the tightening operation, is clarified by using the theory of heat conduction, and its prevention methods are explained from the theoretical point of view.

In Chapter 6, loosening of threaded fasteners, which is the third problem of bolted joints, is discussed by classifying the loosening mechanism into rotation loosening and nonrotation loosening. As for the former, the loosening mechanism is analyzed in detail by finite element simulation. Hence, it is shown that the slip occurrence on the nut- or bolt head-bearing surface is a decisive factor. In contrast, nonrotation loosening inevitably occurs to some extent. Therefore, it is explained how to estimate the bolt force reduction by use of a bolted joint diagram. In addition, practical countermeasures are introduced that are effective in preventing rotation loosening and mitigating nonrotation loosening for various types of bolted joints. In the last section, the inspection methods of bolted joint integrity by torque measurement are discussed.

In Chapter 7, thermal and mechanical behaviors of pipe flange connections are expounded by conducting experiments and finite element analysis, aiming at preventing the leakage of contained fluids at high or low temperature. Numerical analysis by FEM and experiments quantitatively clarify how much bolt force is reduced in the running condition and during the shutdown operation of pipelines. It is shown that the mechanical behavior of gaskets, the stiffness of which is much lower than that of metal gaskets, is the primary influencing factor on leakage. One of the most important issues stated here is that the stiffness of sheet gaskets and spiral wound gaskets is significantly reduced at elevated temperature and this causes bolt force reduction leading to the leakage of contained fluids.

In Chapter 8, aiming at providing effective hints to solve various on-site problems concerning threaded fasteners and bolted joints, several problems and accidents that actually happened in the past are discussed based on rigorous research results. This covers wheel falling-off accidents of large vehicles, the rupture of the threaded portion of a roller coaster axle, fatigue failures of reamer bolts used for rigid flanged shaft couplings, etc. Also treated is how much the natural frequencies of bolted joints are increased with increasing bolt force. In the last section, in order to perform finite element analysis effectively, some concrete mesh patterns and simplified models of engaged thread portions are shown, and a few practical strategies to improve the computational efficiency are presented by utilizing the geometric symmetry of bolted joints.

The author would like to express his sincere appreciation to three distinguished professors, i.e., the late Prof. Minoru Hamada, Prof. Hiroshi Kitagawa, and Prof. Noboru Kikuchi. Prof. Hamada enthusiastically supervised the author's PhD thesis on the mechanical behavior of bolted joints and Prof. Kitagawa led the author to the world of numerical analysis at Osaka University. Prof. Kikuchi motivated the author to apply the latest theories of computational mechanics to bolted joints at The University of Michigan. The author would also like to express special thanks to his many ex-students for their cooperation in achieving the research results forming the basis of this book. Finally, the author would like to give the deepest appreciation to his wife Noriko for her providing a comfortable environment over 40 years to support his research activity.

Toshimichi Fukuoka

List of symbols

A	bolt cylindrical area, other cross-sectional areas
A_1, A_2, A_3	true cross-sectional areas of screw thread
A_I, A_{II}	cross-sectional areas of straight bar
A_{ap}, A_{re}	apparent contact area and real contact area
A_{cn}	contact area
A_e	true cross-sectional area of male thread
A_{ex}, A_{in}	cross-sectional areas of male and female threads
A_f	cross-sectional area of fastened plates
A_{fk}	contact area on thread surface
A_{ins}	true cross-sectional area inside female thread
A_n	nut-bearing surface area
A_r	cross-sectional area at thread root
A_s	stress area of screw thread
a, b	inner and outer radii of hollow bolt, hollow cylinder and hollow shaft
a_1, a_2	constants for representing load distributions
a_p	radius of contact area of microprojection
B	mean diameter of hexagon bolt head or hexagon nut
B_p	diameter of bolt pitch circle
b_1, b_2	constants for determining approach of interface
C_1–C_5, C_{1n}, C_{2n}	constants for representing true cross-sectional areas
c, c_1, c_2, c_3	constants
c_o, m_o	constants in the Ostrovsky equation
c_{sp}	specific heat
D, D_1, D_2	major, minor, and pitch diameters of female thread
D_o	outer diameter of fastened plates
d	nominal diameter, major diameter of male thread, diameter of round bar
d_1, d_2	minor and pitch diameters of male thread
d_{bt}	equivalent diameter of bottoming torque
d_c	diameter of projection part of bottoming stud
d_e	true pitch diameter
d_h	bolt hole diameter
d_{hd}	equivalent friction diameter of bolt head-bearing surface
d_{nu}	equivalent friction diameter of nut-bearing surface
d_r	thread root diameter
d_{rm}	diameter of reamer bolt
d_s	diameter of stress area
E, E_I, E_{II}	Young's modulus
E_b, E_f	Young's moduli of bolt-nut and fastened plates
E_{ex}, E_{in}	Young's moduli of male and female threads
e	width across corner of hexagon
F	axial force acting along bolt axis

F_{02}	bolt force corresponding to 0.2% proof stress
F_b	axial bolt force, bolt preload
F_{b0}	axial bolt force for the case without inclination at bearing surface
F_{b1}, F_{b2}, F_{b3}	axial bolt forces of multibolted joints
F_{bi}	initial bolt force
F_{bt}	compression force generated by bottoming torque
F_c	reduction of compression force in fastened plates due to external force
F_{cn}	compression force acting on bottom surface of bolt hole
F_f	objective bolt force in tightening operation by hydraulic tensioner
F_i	initial bolt force of bolt i, normal force acting on bolt i
F_i^N, F_i^μ	bearing force and friction force sustained by bolt i
$F_{inn}\ F_{out}$	bolt forces generated in inner-wheel and outer-wheel tightening processes
F_n	normal force acting on contact surface
F_{nu}, F_{hd}, F_{pl}	friction forces acting on joint contact surfaces
F_o, f_1, f_2	axial forces generated when tightening outer nut
F_{rm1}, F_{rm2}	normal forces acting on reamer surface
F_s	shear force, shear load
F_{slp}	critical shear load corresponding to whole slip
F_t	initial tension applied by hydraulic tensioner
F_{th}	axial force generated in engaged threads of bottoming stud
F_z	bolt force reduction due to embedment
f	natural frequency of bending vibration
H	height of fundamental triangle
H_1	thread overlap
H_{nu}	thickness of nut
H_V, H_{V1}, H_{V2}	Vickers hardness
h	plate thickness, beam height
h_c	thermal contact coefficient
h_{cv}	coefficient of convection heat transfer
h_e	apparent thermal contact coefficient
h_r	coefficient of radiation heat transfer
I	second moment of area
i	number of threads, bolt number, order of vibration mode
K	nut factor
$\overline{K}$	restraint coefficient
$[K], [M]$	stiffness matrix and mass matrix
$K_{cn}, K_{th}, K_{nu}, K_{hd}, K_f$	spring constants representing interface stiffness
K_{inn}, K_{out}	nut factors in inner-wheel and outer-wheel tightening processes
K_{tru}	true nut factor
k	spring constant, spring constant of straight bar
k_A, k_B	spring constants of plate and thin hollow cylinder
k_b	spring constant of bolt-nut connection
k_{bm}, k_{plc}, k_{ple}	spring constants of jointed components under shear load
k_{cyl}	spring constant of bolt cylindrical portion
k_{ex}, k_{in}	spring constants per one ridge of male thread and female thread
k_f, k_f^*	spring constants of fastened plates
k_n, k_t	spring constants representing interface stiffness

k_s	spring constant of unengaged threads
k_{pt}	spring constant of fastened plates under tension
k_{th}	spring constant of engaged threads
k_{total}	spring constant of entire bolted joint
L	lead of screw thread, length of straight bar
L_1, L_2, L_a	lengths of jointed parts under eccentric external force
L_{cyl}	length of bolt cylindrical portion
L_{ex}	length of protrusion added to joint end portion
L_f	grip length
L_{fk}	length of helix
L_{hd}	equivalent length of bolt head
L_s	length of unengaged threads
L_{th}	equivalent length of engaged threads
m	parallel row number of multibolted joint
m_h, n_h	constants in equation for evaluating thermal contact coefficient
N	number of repeated external force, number of microprojections
Nu	Nusselt number
n	number of bolts, perpendicular row number of multibolted joint, mating node numbers at interface
n_p	division number of one-pitch helical model in axial direction
P	thread pitch
p_m	plastic flow stress of material
p_n	contact pressure at interface
p_{nu}	contact pressure on nut-bearing surface
Pr	Prandtl number
p_t	stress in tangential direction
$Q_{th}, Q_{shk}, Q_{hd}, Q_f$	amounts of heat flowing through each part of bolted joint
Q_{total}	total amount of heat flowing through bolted joint
q	heat flux
R, S	constants in Bach's equation
$\{R\}$	load vector
R_{a1}, R_{a2}	arithmetic mean roughness of mating surfaces
R_{at}	sum of arithmetic mean roughness of mating surfaces
R_{cn}	thermal contact resistance
Re	Reynolds number
R_i	shear load-bearing ratio of bolt i
R_i^N, R_i^μ	shear load-bearing ratios by bearing force and friction force
R_{rm}, R_μ	shear force transfer ratios by reamer surface and friction force
Rz, Rz_1, Rz_2	maximum height roughness
Rzt	sum of maximum height roughness of mating surfaces
r, r_1, r_2	radial coordinates
r_i	radial coordinate of node
s	width across flat of hexagon
T	temperature, torque
T_1, T_2	thread torque and nut-bearing torque
T_b, T_f	bolt temperature and temperature of fastened plates
T_{bt}	bottoming torque
T_e	temperature difference between mating surfaces

T_{inn}, T_{out}	tightening torques of inner nut and outer nut
T_l	loosening torque
T_{l1}, T_{lbt}	loosening torques of bottoming stud
T_{rt}	re-tightening torque
T_{sng}	snug torque in elastic angle control method
T_{st}	second tightening torque
T_t	tightening torque
U	strain energy
U_f	circumferential force acting on thread surface
u	displacement, axial displacement of straight bar
$\{u\}, \{\ddot{u}\}$	displacement vector, acceleration vector
$\overline{u}$	uniform displacement applied to bolt cylindrical portion
$\overline{u}_{hd}, \overline{u}_{nu}$	mean displacements under bolt head- and nut-bearing surfaces
u_i	nodal displacement
u_{slp}	critical displacement corresponding to whole slip
$\overline{u}_{up}, \overline{u}_{dw}$	uniform displacements applied to bolt cylindrical portion
V	amount of volume pushed away by nut-bearing surface
v	sliding speed
W	external force
W_{trq}, W_{th}, W_{nu}	tightening energy
w_{fk}	contact width between male and female threads
Z	embedment factor
z	axial coordinate
α	thread angle, stress concentration factor
α_1	flank angle
α_1'	flank angle in the plane perpendicular to helix
α_I–α_{IV}	stress concentration factors of plates with semicircular notch
α_b, α_f	coefficients of linear expansion of bolt-nut and fastened plates
α_{ex}, α_{exI}, α_{exII}	coefficients of linear expansion
β	lead angle, notch factor
γ	effective tensile coefficient
γ_c	contact ratio
ΔF_b	variation of bolt force
Δr_i	average width between nodes
ΔT, ΔT_I, ΔT_{II}	temperature change or temperature difference
ΔT_b, ΔT_f	temperature changes of bolt and fastened plates
ΔT_e	temperature difference between mating surfaces
ΔT_m	mean temperature increase of microprojection
$\Delta\sigma_b$	variation of axial bolt stress
$\Delta\sigma_f$	variation of compression stress of fastened plates
$\Delta\sigma_{rm}$	bolt stress reduction per unit temperature difference
$\Delta\phi$	return rotation angle of nut
δ	bolt elongation, amount of thermal expansion or shrinkage
δ_{air}	gap size of air layer
δ_b	elongation of bolt-nut connection
δ_c	fit of reamer bolt
δ_f	shrinkage of fastened plates
δ_{pl}	clearance at bearing surface of oil seal plug
δ_z	amount of embedment
ε^p	plastic strain

ζ	approach of interface
$\zeta_{\max}$	maximum value of approach of interface
ζ_t	displacement in tangential direction due to interface stiffness
ζ_{th}, ζ_{nu}, ζ_{hd}, ζ_f	approaches of interface at each joint interface
η	bolt force generated per unit nut rotation angle, notch sensitivity factor
η_3, η_4	efficiencies of triangular and square screw threads
θ	angle, circumferential coordinate
θ_1–θ_6	angles specifying thread geometry
θ_{cn}	pressure cone angle
θ_{hd}	inclination angle of bolt head-bearing surface
θ_{nu}, θ_{tp}	inclination angle of nut-bearing surface
λ, λ_1, λ_2	thermal conductivity
λ_{air}	thermal conductivity of air
λ_i	Eigen value of ith mode
μ, μ_r, μ_θ	coefficient of friction
μ_{hd}	coefficient of friction on bolt head-bearing surface
μ_{ith}, μ_{inu}	coefficients of friction in inner-wheel tightening process
μ_{oth}, μ_{onu}	coefficients of friction in outer-wheel tightening process
μ_{th}, μ_{nu}	coefficients of friction on thread- and nut-bearing surfaces
ν	Poisson's ratio
ρ	root radius, thread root radius
ρ_{dn}	density
$\rho_{\max}$, $\rho_{n\max}$	upper limits of ρ and ρ_n
ρ_n	radius of female thread root
ρ_{th}, ρ_{th}'	friction angles of thread surface for square and triangular threads
σ	stress, true stress, thermal stress, standard deviation
$\overline{\sigma}$	von Mises stress
σ_1, σ_2, σ_3	three components of principal stress
σ_I, σ_{II}	thermal stresses generated in bar I and bar II
σ_a	stress amplitude
σ_B	tensile strength
σ_b	axial bolt stress
σ_{bi}	initial bolt stress
σ_{bnd}	bending stress generated in bolt
σ_{IN}, σ_{OUT}	axial stresses on bolt cylindrical surface in radial direction
σ_L, σ_R	axial stresses on bolt cylindrical surface in circumferential direction
$\sigma_{\max}$	maximum stress occurring at notch root
$\overline{\sigma}_{\max}$	maximum von Mises stress
σ_n	mean stress at notched cross section
σ_r	mean stress at thread root
σ_{th}	tensile stress of bolt threaded portion
σ_Y	yield stress
σ_w	fatigue limit
σ_{wo}	fatigue limit of smoothed test specimen
σ_z	axial stress
τ	shear stress
τ_{th}	shear stress of bolt threaded portion due to thread torque
τ_w	fatigue limit of shear stress
ϕ	nut rotation angle
ϕ_u	load factor

Thread standards and forms

1

1.1 Brief history of screw threads and relevant research activities

It is sometimes suggested that the origin of screw threads traces back to the "screw pump" invented by Aristotle before the Common Era, after which, large-sized wooden power screws, such as the screw-type mechanical press, came to be used. Then, metal screws appeared in the 16th century. It was in 1543 that screw threads were first introduced into Japan along with the arrival of guns. It has been reported that gun smiths of the day struggled to process the screw threads of the tail plugs.

In the Industrial Revolution era in the 19th century, Maudslay in the United Kingdom succeeded in the manufacturing of screw threads with a modern processing method using lathes. Clement, who worked in Maudslay's factory, devised a tap that could form female threads by a cutting operation. Whitworth, who was once an apprentice of Maudslay, is called the father of modern screw threads because he proposed a screw standard for the first time in 1841, the early stage of the Industrial Revolution. Thread cutting lathes, invented in those days, had almost the same structure as the present ones, which enabled the mass production of power screws and fastening screws. A thread rolling device was designed in 1851 by Broomann in the United Kingdom, and then the PW Corporation in Germany manufactured thread rolling machines in 1938 that had a similar structure to the present ones. Detailed history on screw threads is available in [1,2], for instance.

The primary purpose of this book is to explain the mechanics of threaded fasteners as simply as possible, aiming at the provision of an efficient design method for bolted joints with a high level of safety. In the following, a brief review is made of how research on screw threads progressed, which built the foundation of the current design method.

The main cause of the failure of screw threads is metal fatigue. The study of metal fatigue was started around 1860 by Wöhler, who is famous for developing the S-N curve. Research on the fatigue strength of fastening screws was begun by Thum in the 1930s. The idea of a bolted joint diagram, which is widely used today to evaluate the fatigue strength of threaded fasteners, first appeared in the book written by Rötcher [3]. He proposed the idea of a pressure cone with respect to the area at the interface composed of clamped plates, over which contact pressure is exerted due to bolt force. Goodier, who is the coauthor of the historical masterpiece the "Theory of Elasticity" written by Timoshenko, evaluated the particular load distribution pattern along the engaged threads of a bolt-nut connection. He succeeded in the evaluation by applying the theory of elasticity to the displacements measured on the outer surface of the nut [4]. Sopwith further developed the study of screw threads by means of the theory of elasticity, and he proposed nut shapes that could improve the load distribution pattern along the engaged threads [5].

The Mechanics of Threaded Fasteners and Bolted Joints for Engineering and Design.
https://doi.org/10.1016/B978-0-323-95357-3.00002-2

In Japan, some noteworthy studies were published in the 1970s. Sawa et al. clarified various mechanical behaviors of bolted joints, such as the amounts of equivalent lengths of engaged threads and bolt head that represent the stiffness of a bolt-nut connection, by applying the three-dimensional theory of elasticity in a sophisticated manner [6]. Using complex stress functions, Otaki analyzed how geometric factors, such as nominal diameter and thread pitch, affect the stress concentration at the thread root [7] and the fatigue strength [8]. As a pioneering study by numerical analysis, Maruyama evaluated the stress concentration factor at the thread root [9], in which a point-matching method was incorporated into the FEM (finite element method) when modeling the geometry of engaged threads. Bretl et al. calculated the load distribution along engaged threads by replacing the engaged threads with a stack of layered finite elements [10]. Miller et al. proposed a simple method using spring elements, which are discussed later in this book, for calculating the load distribution along engaged threads [11]. Tanaka et al. proposed a finite element approach that could solve contact problems and analyzed various mechanical properties of bolted joints [12], and they also calculated the load distribution along engaged threads in an ingenious way by means of spring elements [13].

As for experimental studies that evaluated the stress distribution in a bolted joint, a series of researches by Hetenyi through photo-elastic experiments [14] are worthy of special mention, in which the researcher revealed the specific load distributions along engaged threads in special nuts with various configurations. It can be said that his research achievements were the starting point of a lot of contrivances found thereafter in various-shaped nuts, aiming at smoothing the load distribution along engaged threads and mitigating the stress concentration at the thread root. Stress concentrations at the thread root of actual bolt-nut connections were successfully measured by the copper plating method. Maruyama compared the measured results to the stress concentration factors obtained by FEM [9,15]. Seika et al. obtained the magnitude and location of the maximum stress [16] using the same experimental method. They concluded that it occurred at the male thread root 2/3 pitch apart from the nut-bearing surface.

The jointed portions fastened by use of screw threads are generally called a bolted joint. In this book, therefore, the components clamped by various kinds of threaded fasteners are referred to as bolted joints. Basic forms and the usages of screw threads have been fundamentally unchanged since the Industrial Revolution. Nevertheless, threaded fasteners and bolted joints have continuously been important research objects in the mechanical engineering field. The biggest reason is that the external loads exerted on bolted joints are continuously being increased!

1.2 Geometry and application purposes of screw threads

Screw threads are roughly classified into two types, which are used for clamping and movement transmission, respectively. Most threaded fasteners such as bolts and nuts are basically used for clamping a number of parts composing machines and structures. On the other hand, screw threads used for movement transmission deliver various movements by utilizing their helical shape. They are further categorized into two

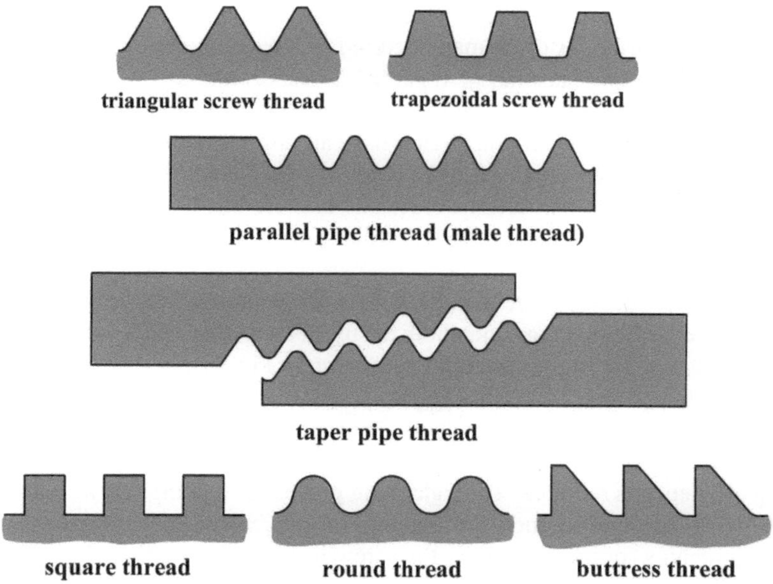

Fig. 1.1 Thread ridge forms of various types of screw threads.

types. One is used for generating large forces by making use of the action of the thread as an inclined plane, e.g., a jack for lifting heavy parts; the other facilitates the conversion between linear and rotational motions, e.g., feed screws built into machine tools. As a similar example, a micrometer provides a dimensional accuracy of 10 μm, in which a large rotation angle is converted into a small linear movement. A certain type of micrometer equipped with a digital display device provides higher accuracy of 1 μm.

Fig. 1.1 shows screw threads of various forms. Triangular screw threads are the most commonly used when clamping multiple parts. On the other hand, trapezoidal screw threads are used for motion and power transmission. These types of screw threads are often called power screws [17]. The relationship between thread form and application purpose is explained in detail in section 3.2.4. To explain it briefly for now, triangular screw threads are primarily used for clamping because of their high antiloosening performance. Trapezoidal and square screw threads are employed for motion and power transmission due to high thread efficiency. Pipe threads are classified into parallel pipe threads and taper pipe threads. The former is used for mechanical joining of structures with hollow cylindrical shapes. The latter is applied to connecting pipelines that require leakage prevention performance to contain fluids. Taper pipe threads are processed on the target surface along one-sixteenth of taper. The mating thread surfaces come into close contact as male threads are progressively screwed into female threads. The thread angle of triangular screw threads is 60 degrees, while the angle of pipe threads is 55 degrees.

Besides the aforementioned types of screw threads, square threads, round threads, and buttress threads are sometimes used. The thread efficiency of square threads is higher than that of trapezoidal screw threads; however, due to the processing difficulty, the thread surface is often manufactured to have an inclined angle of about 5 degrees [17]. Round threads can be regarded as trapezoidal screw threads to which large radii are added to the crest and thread root. Although the stress concentration at the thread root is lowered, round threads have an essential drawback of small thread overlap. Buttress threads are shaped like a saw with asymmetric geometry to support large forces acting from one definite side. The buttress thread's surface often has an inclined angle of about 7 degrees [17] for the same reason as in the case of square threads.

Power screws, such as acme threads, whose shape is similar to trapezoidal screw threads, and buttress threads are standardized by ANSI (American National Standards Institute). The tapping screw, although not shown in Fig. 1.1, is a kind of triangular-shaped screw thread used for clamping parts. It is a male thread screwed into the main body, thus producing a female thread hole. The parts to be clamped are commonly thin metal plates or wooden plates, where female threads are easily processed and high working efficiency is required. The tightening process of tapping screws can be evaluated in terms of three torques, i.e., tapping torque required for processing female threads, tightening torque for seating the screw head and generating clamping force, and breakage torque causing the rupture of the tapping screw [18]. From the point of view of the tightening operation, it is necessary that the breakage torque be larger than the tightening torque, to the extent that the rupture of screw threads does not occur even if being somewhat over-tightened.

In this book, the focus is placed on bolts and nuts with triangular thread geometry, which are most widely used for clamping all kinds of parts.

1.3 Standards of screw threads

1.3.1 Standard specifications of screw threads

In Japan, for instance, standard specifications of screw threads are specified in detail in JIS (Japanese Industrial Standards). Since screw threads are commonly used for all sorts of machines, structures, electric devices, etc., the highest level of international consistency is required in the standard specifications. Accordingly, the standards on screw threads specified in JIS almost completely agree with ISO (International Organization for Standardization) standards, as well as the national standards established in many countries. However, perfect consistency between JIS and ISO is practically impossible. To cope with those cases, therefore, the degree of inconsistency between ISO and JIS is explained for each specification in JIS. As a result of considering the above situation, the screws and threaded fasteners treated in this book are basically standardized in JIS. The correspondence between JIS and ISO, described in this book, is presented in Table 1.1, in which the standards specific to JIS are shown separately.

According to the unit of length, screw threads are classified into metric thread and inch thread. The most commonly used screw threads are metric. Pipe threads employed in piping systems are inch threads. Unified threads, which are another type

Table 1.1 Correspondence between JIS and ISO in order of appearance.

JIS ISO	B 0206 263	B 0208 263	B 0205 68-1, 261, 262	B 0209-1 965-1	B 0216 2901, 2902, 2904	B 0202 228-1	B 0203 7-1
JIS ISO	B 1180 4014–4018, 8676, 8765	B 1181 4032–4036, 8673–8675	B 1001 273	B 1052 898-2, 898-6	B 1051 898-1	B 1083 16,047	B 1056 2320
JIS specific	B 1082	B 1173	B 2238	B 2251	B 1089	B 1090	B 1451

of inch thread, are extensively used in aircraft, and are specified in JIS B 0206 and JIS B 0208.

1.3.2 Basic profile of screw threads

Fig. 1.2 represents the basic profile of metric triangular screw threads for general use, specified in JIS B 0205. The external or "male" thread is processed on the outer surface of cylinders or cones, while the internal or "female" thread is processed on the inner surface of hollow cylinders or hollow cones. Major diameter of external thread d is equal to major diameter of internal thread D. The two major diameters are the basic dimensions representing the size of screw threads and threaded fasteners, which are denoted as nominal diameter. The pitch diameters of male and female threads, d_2 and D_2, are virtual diameters, at which the width of the thread ridge is equal to that of the thread groove. They are used when evaluating thread strength and calculating the helix angle of the screw thread, termed the lead angle. Minor diameter of external thread d_1 is equal to minor diameter of internal thread D_1. The dimensions of each part

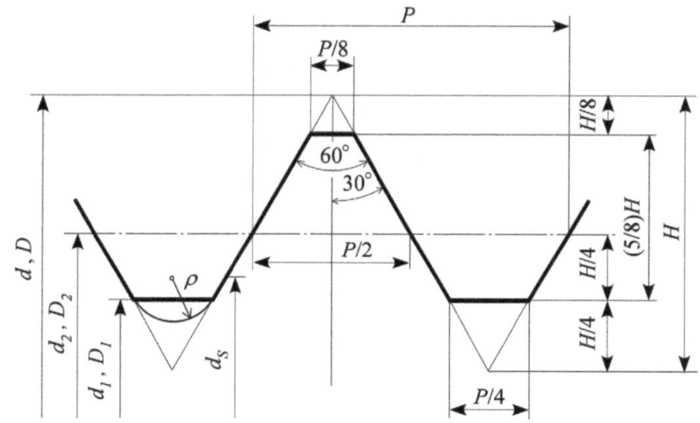

Fig. 1.2 Basic profile of metric triangular screw thread.

of triangular screw threads are expressed as follows, by means of nominal diameter d, thread pitch P and fundamental triangle height H:

$$H = \frac{\sqrt{3}}{2}P = 0.8660P \tag{1.1}$$

$$d_2 = D_2 = d - \frac{3}{4}H = d - 0.6495P \tag{1.2}$$

$$d_1 = D_1 = d - \frac{5}{4}H = d - 1.083P \tag{1.3}$$

$$H_1 = \frac{5}{8}H \tag{1.4}$$

Thread pitch P is the distance between the corresponding points on neighboring ridges. H_1 denotes basic thread overlap, which is the engaged length of male and female threads in the radial direction. However, it is difficult to derive a rigorous equation that can precisely evaluate the strength of screw threads taking into account the complex helical thread geometry. In JIS B 1082, therefore, a concept of stress area A_s proposed to evaluate the stress and stiffness of high-strength bolts, in which the threaded portion is replaced by a cylinder of cross-sectional area A_s:

$$A_S = \frac{\pi}{4}\left(\frac{d_2 + d_3}{2}\right)^2 = \frac{\pi}{4}d_S^2 = 0.7854(d - 0.9382P)^2 \tag{1.5}$$

In Eq. (1.5), d_3 corresponds to the diameter at thread root when root radius is equal to $H/6$. d_s is an important dimension for strength evaluation and designates the diameter of stress area. The amount of d_s can be calculated in terms of nominal diameter d and thread pitch P:

$$d_S = d - 0.9382P \tag{1.6}$$

Comparing Eq. (1.6) to Eqs. (1.2) and (1.3), it is found that d_s is located between minor diameter d_1 and pitch diameter d_2, as shown in Fig. 1.2.

High stress concentration occurs at the thread root due to its notched shape. To lower the stress concentration, therefore, an appropriate size of radius is processed there. As for the root radius ρ of the male thread, it is recommended in JIS B 209-1 that ρ should be larger than $0.125P$ for bolts of strength classification of 8.8 and higher. The definition of strength classification is explained in Section 1.6.1.

$$\rho \geq 0.125P \tag{1.7}$$

Fig. 1.3 shows the basic profile of metric trapezoidal screw threads, as specified in JIS B 0216. The thread angle is 30 degrees, which is half that of triangular screw threads. Fig. 1.4 depicts the basic profiles of parallel and taper pipe threads, specified

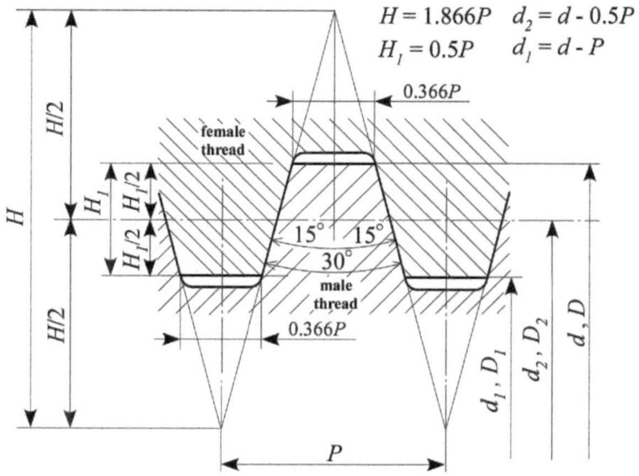

Fig. 1.3 Basic profile of metric trapezoidal screw thread.

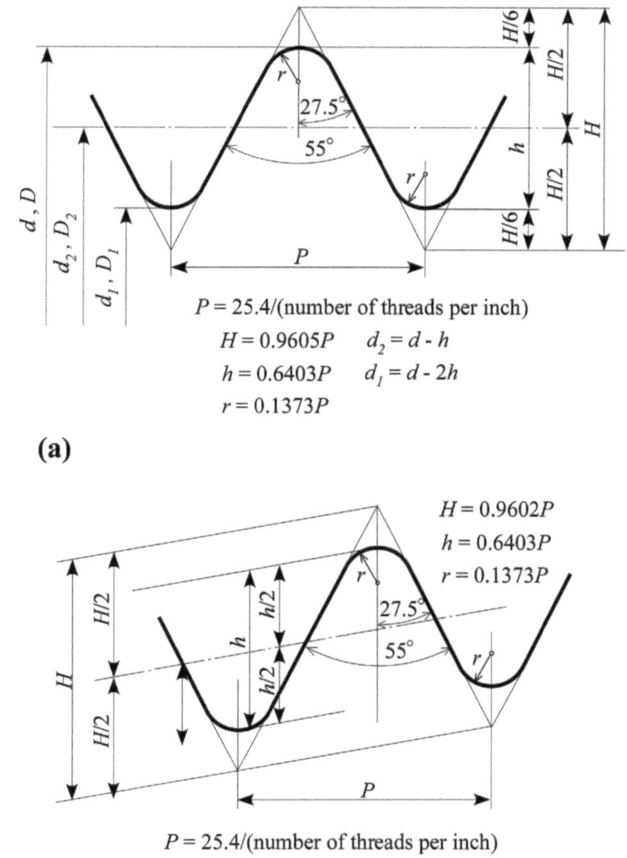

Fig. 1.4 Basic profiles of parallel and taper pipe threads. (A) Parallel pipe thread and (B) taper pipe thread.

in JIS B 0202 and 0203, respectively. In the case of pipe threads, the thread angle is 55 degrees, which is somewhat smaller than that of triangular screw threads. Taper pipe threads are processed on the target surface along one-sixteenth of taper.

1.4 Thread pitch and number of threads

1.4.1 Mathematical expression of Helix

The shape of *helix* is expressed by the following equation in terms of helix diameter d [19]:

$$x = \frac{d}{2}\cos\theta, \quad y = \frac{d}{2}\sin\theta, \quad z = \frac{d}{2}\theta\tan\beta \tag{1.8}$$

Fig. 1.5 shows a helix drawn by use of Eq. (1.8). Winding a right triangle with sides of πd and L on a cylinder of diameter d, the hypotenuse of the triangle becomes a helix. The distance of linear motion with one turn of the screw thread is termed lead L, and the angle made by the helix and the horizontal surface is called lead angle β. The relationship between d, L, and β is given in a simple equation:

$$\tan\beta = \frac{L}{\pi d} \tag{1.9}$$

1.4.2 Coarse screw threads and fine screw threads

A threaded fastener is a kind of machine element in which fundamental shapes such as triangle, trapezoid, and square are wound spirally around the outer or inner surface of a cylinder or cone. In the case of a male taper pipe thread, for example, a triangle is wound on the outer surface of a hollow cone. Screw threads are classified, in addition to the thread forms shown in Fig. 1.1, by the size of thread pitch P and the number of threads. With regard to thread pitch P, screw threads having the same nominal

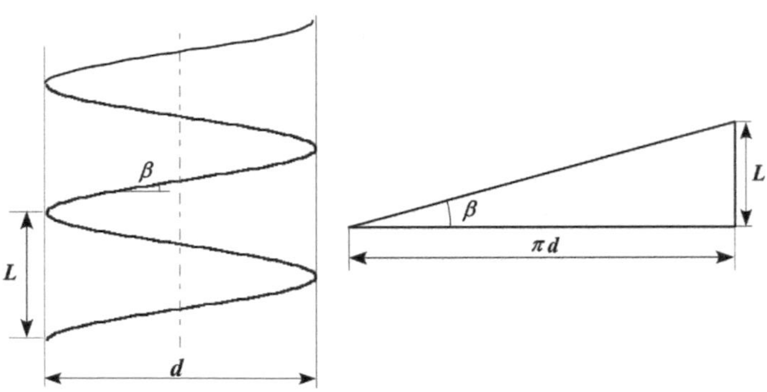

Fig. 1.5 Lead and lead angle of helix.

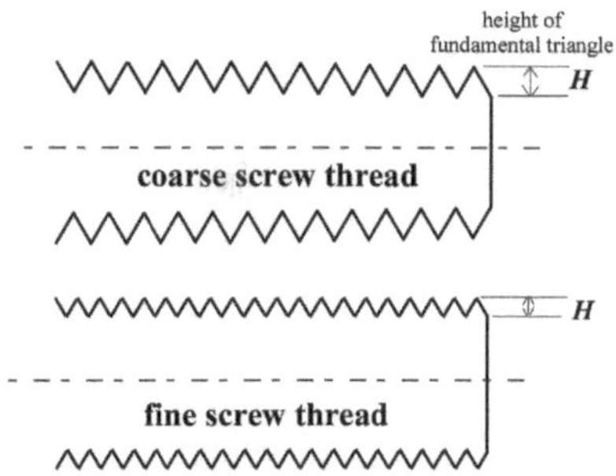

Fig. 1.6 Coarse screw thread and fine screw thread.

diameter d are categorized into coarse screw thread and fine screw thread. As shown in Fig. 1.6, coarse screw threads have a standard size of thread pitch and fine screw threads have smaller pitches for the same nominal diameter. In the case of M16, for example, thread pitch of coarse screw thread is 2 mm, and those of fine screw thread are 1.5 and 1 mm. Since fine screw threads have a relatively small lead angle, they are considered to be more resistant to loosening. Therefore, fine screw threads are often used for joints under severe service conditions. Basic profiles, shown in Fig. 1.2, are identical in coarse and fine screw threads; the only difference is that they have different thread pitches. Accordingly, the dimensions of each portion of screw thread, such as fundamental triangle height H, etc., are calculated by using Eqs. (1.1)–(1.4). As for thread strength, fine screw threads tend to give lower shear strength because basic thread overlap H_1 is smaller due to small thread pitch. On the other hand, there is no significant difference in the total area of pressure flank between coarse and fine screw threads. This is explained in Section 1.4.4 using mathematical expressions.

1.4.3 Number of threads and lead angle

Lead L is the distance of linear motion with one turn of the screw thread, which corresponds to the symbol L in Fig. 1.5. In commonly used screw threads, lead L is equal to thread pitch P. These screw threads are termed single-thread screw. In contrast, screw threads moving two pitches and three pitches with one turn are termed double-thread screw and triple-thread screw, respectively. They are collectively called multiple-thread screws.

Fig. 1.7 shows the configuration of male threads with single-thread and double-thread screws. Although both threads have an identical pitch size, the double-thread screw has a larger lead angle β because doubling the number of threads doubles the lead L. The inclined angle of thread surface is termed lead angle β, as shown in Fig. 1.5. Since the lead L is equal to the product of thread pitch P times number of threads i, lead angle β can be calculated by an equation of the same form as Eq. (1.9):

$$\tan\beta = \frac{iP}{\pi d_2} \tag{1.10}$$

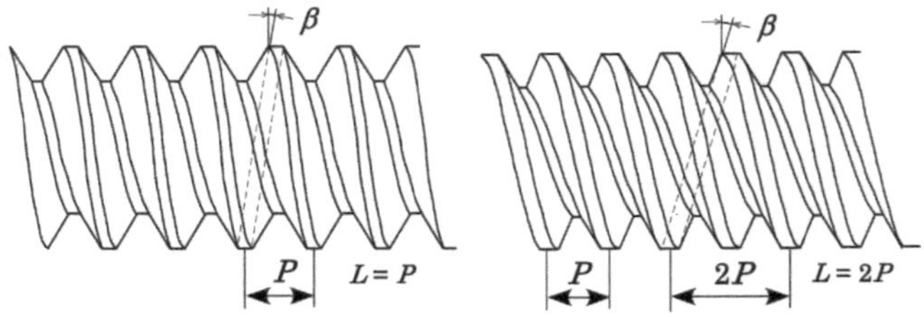

Fig. 1.7 Single-thread screw and double-thread screw.

As shown above, lead angle β is commonly calculated in terms of pitch diameter d_2. The equation also suggests that lead angle increases radially inwards, although slightly, from nominal diameter d to minor diameter d_1. In order to rigorously consider the thread geometry, therefore, it should be noted that lead angle β somewhat varies in the radial direction. For example, in the case of M16 with coarse thread, thread pitch P is 2mm. Therefore, the lead angles of a single-thread screw of M16, calculated on nominal diameter $d=16$mm, pitch diameter $d_2=14.701$mm, and minor diameter $d_1=13.835$mm, are 2.28, 2.48, and 2.63 degrees, respectively. The assumption of $\tan\beta \approx \beta$ holds when β is small. Accordingly, lead angles of multiple-thread screw, β, are almost proportional to number of threads i, which means that multiple-thread screws can provide large axial movements with small rotation of screw threads. By utilizing these characteristics, multiple-thread screws are often used as feed screws in the emergency shutdown valves of various plants, in which the quick stop of liquid flow is demanded. They are also ingeniously applied to the bottle caps of cosmetics because they provide quick opening and closing.

As explained above, multiple-thread screws have various specific features owing to their large lead angle. Triangular screw threads are unlikely to loosen and have lower thread efficiency compared to square and trapezoidal screw threads, the reason being that the thread surface is inclined by 30 degrees to the plane perpendicular to the bolt axis. This is discussed further in section 3.2.4. Regarding thread efficiency, it becomes higher with increasing lead angle. Making use of the high thread efficiency and moderate antiloosening performance of multiple-thread triangular screws, it is expected to be possible to develop a unique power transmission mechanism equipped with both high efficiency and stable stationary performance.

– Learning by numerical example 1.1: Dimensions of each part of screw thread and lead angle

Table 1.2 gives examples of the dimensions of each part of a triangular screw thread, i.e., nominal diameter d, thread pitch P, fundamental triangle height H, pitch diameter d_2, minor diameter d_1, diameter of stress area d_s, and lead angle β. Tabulated are the cases of M6, M10, M12, and M16 with coarse and fine threads, and M30 and M64 with coarse thread.

– Learning by numerical example 1.2: Lead angle of multiple-thread screws

Table 1.3 shows lead angles of M10 and M12 with coarse thread, ranging from single-thread to septuple-thread screws, at the position of pitch diameter d_2.

Table 1.2 Relationship between nominal diameter and dimensions of each part of the screw thread (unit: mm, degree only for β).

	M6	**M10**	**M12**	**M16**	**M30**	**M64**
d	6	10	12	16	30	64
P (coarse)	1	1.5	1.75	2	3.5	6
H	0.866	1.299	1.516	1.732	3.031	5.196
d_2	5.350	9.026	10.863	14.701	27.727	60.103
d_1	4.917	8.376	10.106	13.835	26.211	57.505
d_S	5.062	8.593	10.358	14.124	26.716	58.371
β	3.405	3.028	2.935	2.480	2.301	1.820

	M6	**M10**		**M12**		**M16**	
d	6	10	10	12	12	16	16
P (fine)	0.75	1.25	1	1.5	1.25	1.5	1
H	0.650	1.083	0.866	1.299	1.083	1.299	0.866
d_2	5.513	9.188	9.350	11.026	11.188	15.026	15.350
d_1	5.188	8.647	8.917	10.376	10.647	14.376	14.917
d_S	5.296	8.827	9.062	10.593	10.827	14.593	15.062
β	2.480	2.480	1.950	2.480	2.037	1.820	1.188

Table 1.3 Relationship between number of threads and lead angle (unit: degree).

Number of threads i	**1**	**2**	**3**	**4**	**5**	**6**	**7**
M10	3.028	6.039	9.018	11.948	14.816	17.610	20.320
M16	2.480	4.950	7.402	9.827	12.217	14.565	16.864

1.4.4 Contact area between male and female threads

The two sides composing the thread triangle are termed the pressure flank and clearance flank. Contact between male and female threads occurs only on the pressure flank in the normal clamping state. There is no contact on the clearance flank, the other side of thread triangle. Derived in this section is the equation to calculate the total contact length between male and female threads. The contact area of mating threads, which is equal to the total area of pressure flank, is also given in a mathematical formula, aimed at considering the effect of thread pitch P and number of threads i on the thread contact area.

Contact width w_{fk} between male and female threads is calculated in terms of basic thread overlap H_1, given in Eq. (1.4), and flank angle α_1, which is half the thread angle α:

$$w_{fk} = \frac{H_1}{\cos\alpha_1} = \frac{5}{8}P \tag{1.11}$$

The length of helix L_{fk} on the cylindrical surface of diameter d_2 is expressed in the following form using thickness of nut H_{nu}:

$$L_{fk} = \frac{H_{nu}}{iP} \times \sqrt{(iP)^2 + (\pi d_2)^2} \times i = \frac{H_{nu}}{P} \sqrt{(iP)^2 + (\pi d_2)^2} \tag{1.12}$$

For simplicity, thread contact area A_{fk} is calculated as the product of w_{fk} times L_{fk}:

$$A_{fk} = w_{fk} \times L_{fk} = \frac{5}{8} H_{nu} \sqrt{(iP)^2 + (\pi d_2)^2} \tag{1.13}$$

– Learning by numerical example 1.3: Contact area between male and female threads

Thread contact areas are compared between M16 with $P = 2$ mm (coarse thread) and M16 with $P = 1.5$ and 1 mm (fine thread). Assuming $H_{nu} = 13$ mm and using Eq. (1.12), the length of helix L_{fk} is calculated to be 300.5, 409.3, and 627.0 mm, respectively. This suggests that L_{fk} increases almost inversely with P. Contact width w_{fk} increases in proportion to P, as found from Eq. (1.11). As a result, thread contact areas A_{fk} are calculated to be 375.6, 383.7, and 391.9 mm^2, which means that A_{fk} slightly increases with decreasing P. Furthermore, thread contact areas of multiple-thread screws are calculated to be 376.7 mm^2 for double-thread screw and 378.4 mm^2 for triple-thread screw. Even for septuple-thread screw, the amount of A_{fk} is 392.1 mm^2, which indicates an insignificant effect of number of threads i.

It is found from the above calculation results that no significant difference is found in thread contact area A_{fk} between coarse and fine threads and between single- and multiple-thread screws.

1.4.5 Nonsimilarity of threaded fasteners

Basic profiles of triangular screw threads are all similar regardless of nominal diameters, as shown in Fig. 1.2. On the other hand, threaded fasteners, such as bolts and nuts, are not similar in most cases for different nominal diameters. In Fig. 1.8, thread profiles of male threads are compared for varying nominal diameters from M6 to M64 with coarse thread, in which thread pitches are drawn with the same dimension for easy comparison. The ratio of thread pitch to nominal diameter P/d decreases with larger nominal diameter. Accordingly, as shown in Table 1.2, lead angle β gets smaller as nominal diameter d increases. In addition, the lead angle of the fine screw thread is smaller, due to its small pitch, than that of the coarse screw thread of the same nominal diameter. Therefore, comparing among coarse screw threads, it can be stated that screw threads tend to have relatively finer thread pitch as nominal diameter increases.

In some cases, however, P/d is identical for different nominal diameters, e.g., the value of P/d is 0.125 for M16, M20, and M24 with coarse thread. Fig. 1.8 shows the complete similarity among threaded fasteners with these nominal diameters. Meanwhile, in the case of threaded fastener with screw thread of M12, its configuration is fairly different from the foregoing three threaded fasteners, because P/d is 0.146.

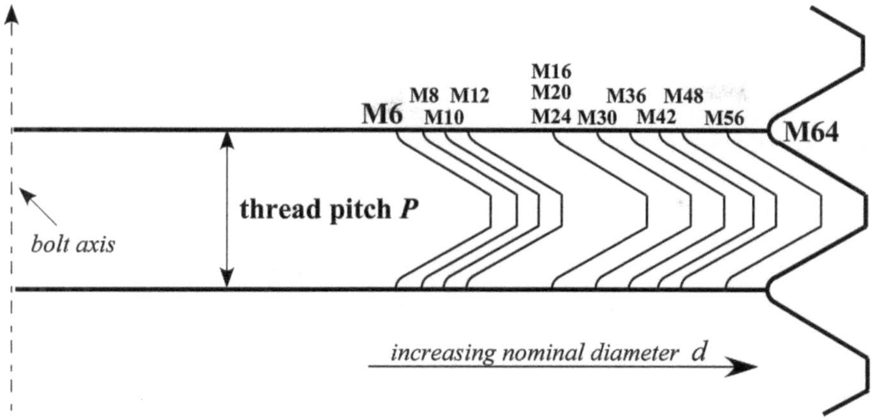

Fig. 1.8 Nonsimilarity among threaded fasteners with different nominal diameters.

For example, the effect appears as the difference in the expansion of the plastic region in the bolt tightening process, which is explained in section 4.2.5.

1.5 Clamping configuration and various threaded fasteners

1.5.1 Bolt-nut connection and stud

Bolts and nuts are the most widely used threaded fasteners. They are used in pairs in order to clamp multiple parts inserted between the nut-bearing surface and bolt head-bearing surface. The aggregate of the parts clamped by threaded fasteners is termed the fastened plates, herein. The total thickness of the fastened plates is called the grip length. Fig. 1.9 illustrates typical clamping configurations. When bolts and nuts are used in pairs, as shown in Fig. 1.9A, the usage is called through bolt. Meanwhile, when clamping thick plates within the body of large machines or structures, a combination of bolt and nut is often inapplicable, due to various limitations of joint configuration. In these cases, threaded fasteners called studs are used, which are geometrically cylindrical bars threaded at both ends. When using studs, the longer threaded portion is screwed into the female thread processed in the body, and the shorter one is combined with a nut to tighten by the torque control method. In some cases, an ordinary hexagon

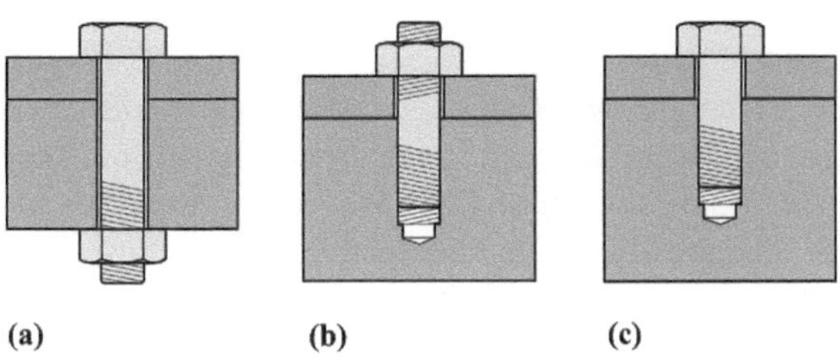

Fig. 1.9 Typical clamping configurations of threaded fasteners. (A) Through bolt, (B) stud, and (C) tap bolt.

head bolt is simply screwed into the female thread instead of studs, aiming at enhancing the workability. This type of clamping configuration is called tap bolt.

In this book, female threads processed in the main body are called body-side female threads, and the engaged portion consisting of the stud and the body-side female thread is termed body-side engaged threads. Body-side female threads involve a processing difficulty because of high stiffness, which may lead to lower dimensional accuracy and lower strength, compared to the female thread of a nut. As a countermeasure, body-side engaged thread length is designed to be longer than that of a bolt-nut connection. The thickness of nut H_{nu} is usually about 0.8 times the nominal dimeter d. On the other hand, in the case of studs specified in JIS B 1173, the body-side engaged length is recommended to be considerably longer than $0.8d$. When the body is made of carbon steel, cast steel, forged steel, stainless steel, or similar materials, the recommended engaged length is about $1.25d$. For cast iron and similar materials, the engaged length is selected to be about $1.5d$. For light alloys and other soft materials, the engaged length is set to be about $2d$. Studs and tap bolts are basically identical in their applications. The selection between them should be determined considering their advantages and disadvantages, as listed below.

1) Stud

Advantage: In the tightening process by applying torque to nut, the body-side engaged threads are unlikely to be damaged because no slip occurs in the circumferential direction.
Disadvantage: A stud must be lightly screwed into the main body prior to being tensioned. Accordingly, the workability is lower than the case of tap bolt. In addition, to avoid stud rotation in the disassembly process, the incomplete threaded portion of the stud is sometimes screwed into the body-side female threads, which may produce high stress concentration in the surrounding area. The use of bottoming studs, which is explained in section 3.2.8, is an effective countermeasure to avoid this problem.

2) *Tap bolt*

Advantage: Workability is high because the tightening operation is carried out by applying torque to bolt head using ordinary hexagon head bolts.
Disadvantage: The surface of body-side female threads may possibly be damaged due to the relative slip occurring there.

1.5.2 Geometric factors affecting mechanical behavior of threaded fasteners

Dimensions and shapes of hexagon head bolts and hexagon nuts are specified in JIS B 1180 and JIS B 1181, respectively. There are two types of hexagon nuts, i.e., Style 1 and Style 2, for both coarse and fine screw threads. Hexagon thin nuts having smaller nut thickness are also specified. Regarding the bearing surface shape, two types as illustrated in Fig. 1.10A and B are specified, i.e., double-chamfered type and raised-face type. Comparing the characteristics of these two nuts to that of single-chamfered type, shown in Fig. 1.10C, which was specified in the old JIS, it has been

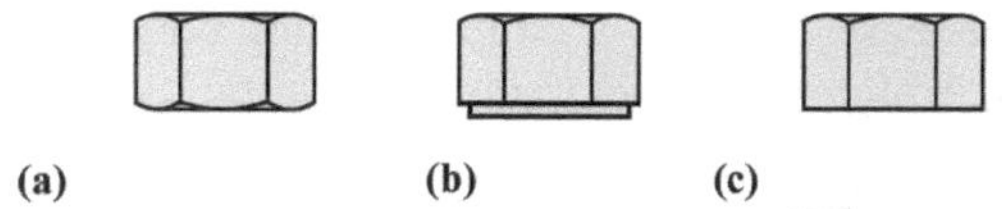

Fig. 1.10 Shapes of nut-bearing surfaces. (A) Double chamfered, (B) raised-face, and (C) single-chamfered.

reported that the latter type tends to generate lower axial bolt force when applying the same amount of torque. This is because imperfect contact inevitably occurs along the outer edge of the nut-bearing surface. When performing the strength calculation or numerical analysis of hexagon head bolts and hexagon nuts, the hexagonal geometry can be expediently converted into an equivalent circle of dimeter B, which is the mean value of width across flat s and width across corner e of the hexagon:

$$B = \frac{s + e}{2} \tag{1.14}$$

In the actual tightening operation using bolts and nuts, the mechanical properties of the bolted joint are influenced by the degree of parallelism of nut- and bolt head-bearing surfaces, the degree of perpendicularity of the side faces of nut and bolt head, bolt hole diameters, etc., as illustrated in Fig. 1.11. Geometric errors of bolts and nuts are specified in the aforementioned JIS specifications, and bolt hole dimensions are standardized in JIS B 1001. The errors in parallelism and perpendicularity are stipulated to be less than 1 or 2 degrees, depending on the grade. Too large geometric errors cause partial contact at the bearing surface, which may lead to surface damage and excessive bending of the bolt body. Meanwhile, from the plastic working point of view, a small taper is sometimes processed radially inwards on the nut-bearing surface. The existence of a taper changes the axial stiffness of the engaged thread portion, hence, the stress concentration at the thread root is also changed. When the tightening operation is done by the torque control method, the center of friction force acting on the nut-bearing surface tends to move radially outwards, because a small gap exists around the inner part of the contact surface. Accordingly, the equivalent friction diameter of nut-bearing surface is necessarily increased, which produces an unfavorable result of smaller axial bolt force regardless of applying the same torque. Additionally, since the actual contact area of the nut-bearing surface is increased with increasing bolt force, the above phenomenon is nonlinearly related to axial bolt force. This problem is discussed in detail in section 3.2.3.

Regarding bolt hole diameter d_h, there are usually three choices from first grade to third grade for a pair of bolt and nut with identical nominal diameter. In the case of

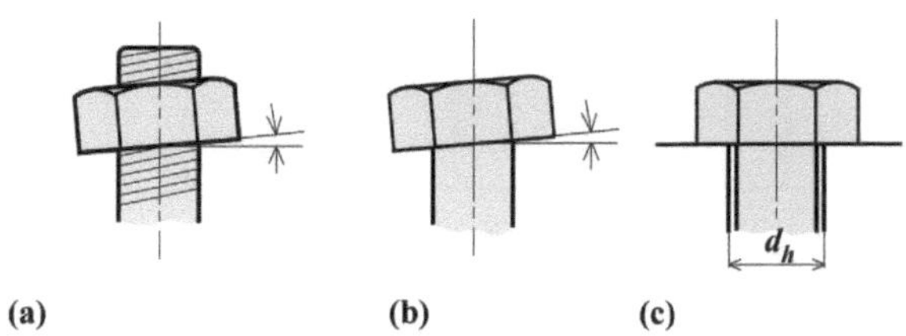

Fig. 1.11 Parallelism of nut- and bolt head-bearing surfaces and bolt hole diameter. (A) Nut bearing surface, (B) bolt head bearing surface, and (C) bolt hole idameter.

third grade having a large bolt hole, it is easy to install a target bolt, and the stress concentration at the first bolt thread root, which is adjacent to the nut-bearing surface, becomes smaller. This is because the engaged thread portion deforms easily in the axial direction due to its small stiffness. In the case of first grade, it provides precise bolt positioning, while it produces higher stress concentration at the first bolt thread root. In addition, when the joint is subjected to thermal load, the heat applied to fastened plates is easily transferred to bolt and nut through the thin air layer existing around the bolt hole. This subject is discussed in section 2.7.

1.5.3 Contact pressure distribution at plate interface and shape of pressure cone

Fig. 1.12 shows the relationship between the contact pressure distribution at the plate interface and the outer diameter of cylinder D_o, for the case of two identical hollow cylinders being clamped using a pair of bolt and nut. The case of Fig. 1.12A, in which D_o is small and nearly equal to mean diameter of nut or bolt head B, is called thin cylinder. The contact pressure is distributed almost uniformly.

The case of Fig. 1.12C with large D_o is called plate, in which the contact pressure does not reach the outer edge of the interface. The contact area subjected to compression force is denoted by pressure cone angle θ_{cn} in the figure. The amount of θ_{cn} is found to be about 40 degrees [20] by finite element (FE) analysis. In this case, contact pressure shows the peak around the bolt hole and decreases radially outwards in a nearly triangular form. The case of Fig. 1.12B is the intermediate state between thin cylinder and plate, called thick cylinder.

In the case of body-side engaged threads, female threads can be regarded as a nut with infinite diameter, where the engaged thread portion and the interface consisting of the clamped plate and the main body are in close proximity. Therefore, it generates a characteristic distribution pattern of contact pressure, as shown in Fig. 1.13A. The amount of θ_{cn} is nearly equal to the case of a bolt-nut connection. The contact pressure becomes very high around the bolt hole and decreases radially outwards rather sharply in an exponential fashion. Fig. 1.13B shows the contact pressure distribution obtained by FE analysis, where θ_{cn} is about 42 degrees [20]. Numerical analyses were conducted using helical thread models, as explained in section 2.5. The parameters that

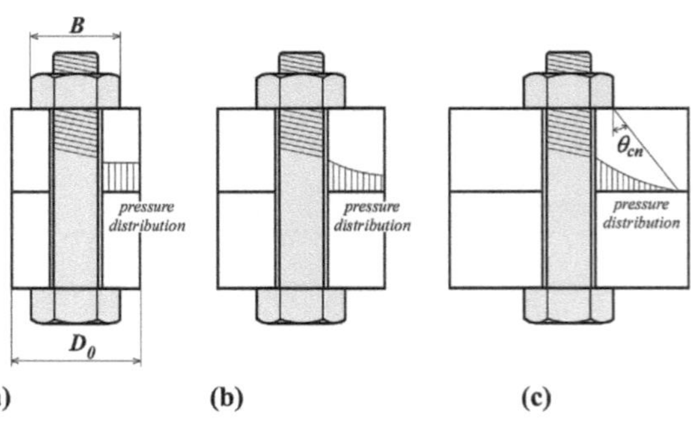

Fig. 1.12 Contact pressure distribution patterns at plate interface. (A) Thin cylinder, (B) thick cylinder, and (C) plate.

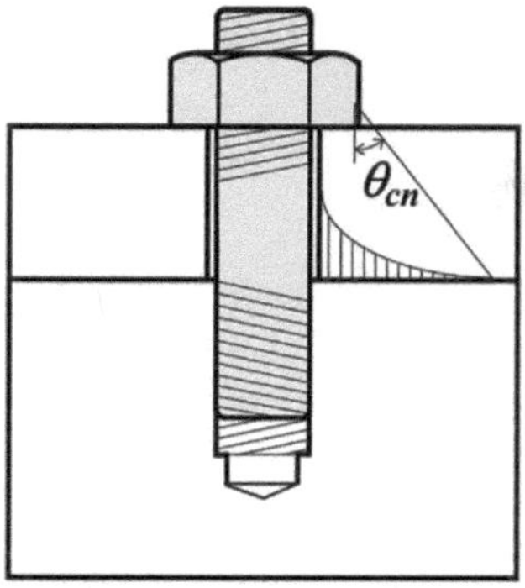

(a)

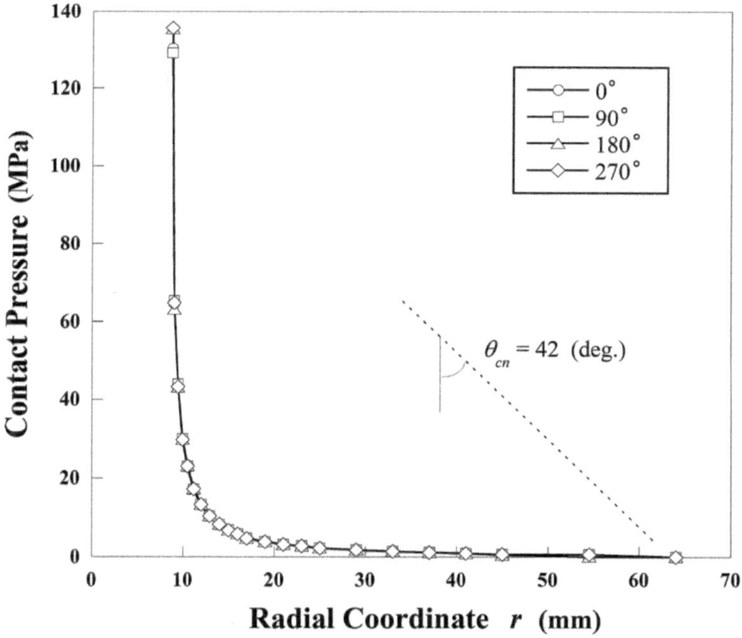

(b)

Fig. 1.13 Contact pressure distribution at plate interface adjacent to body-side engaged threads. (A) Body-side engaged threads and (B) contact pressure distribution by FEM.

appear in the figure—0, 90, 180, and 270 degrees—represent the positions at the interface, which are 90 degrees apart from each other. No significant difference is observed in the circumferential direction. It is predicted that the difference in the pressure distribution patterns, shown in Figs. 1.12 and 1.13, may be one of the major causes due to which bolt-nut connections and body-side engaged threads exhibit different mechanical behaviors under external force.

1.6 Strength, thermal and mechanical properties of threaded fastener materials

1.6.1 Materials used for threaded fasteners

A wide variety of materials are used for threaded fasteners, such as carbon steel, alloy steel, stainless steel, aluminum alloy, brass, titanium, nickel-based heat-resistant alloy, plastics, etc. When selecting materials suitable for the objective threaded fasteners, the judgement is usually made based on mechanical and thermal properties, anticorrosion performance, etc.

Among the various materials, carbon steel and alloy steel are most commonly used. Focusing the target material on carbon steel and alloy steel, Young's modulus E, which represents the material stiffness, is about 200 GPa and Poisson's ratio, ν, is around 0.3, regardless of the content rates of carbon, nickel, chromium, and molybdenum. On the contrary, tensile strength σ_B and yield stress σ_Y show a wide variation; σ_B and σ_Y are the indices for material strength, and they are represented by a single value in threaded fasteners. Tensile strength σ_B widely varies from about 400 MPa for mild steel to 1200 MPa for high-strength steel. For example, in the indication "4.6" the first figure "4" means $\sigma_B = 400$ MPa and σ_Y is calculated to be 240 MPa by multiplying 400 MPa by another figure of "0.6." Therefore, in the case of 4.8, σ_B is kept unchanged at 400 MPa, and σ_Y becomes 320 MPa. In addition to 4.6 and 4.8, bolts with strength grades of 8.8 and 10.9 are frequently used in industrial applications. For a pair of bolt and nut, the recommendation in JIS B 1052 is the selection of the same or a similar grade of bolt and nut.

Stainless steel is roughly classified into austenitic, martensitic, and precipitation hardening type stainless steels, etc. Austenitic stainless steels typified by SUS304, which is equivalent to ASTM S30400, are most commonly used in both elevated and low temperature conditions, and they are also employed in corrosive environments due to their high anticorrosion performance. There exist no significant differences in Young's modulus and Poisson's ratio between carbon steels and austenitic stainless steels. On the other hand, thermal conductivity λ, which governs the behavior under thermal load, is as little as about one-third of that of carbon steel. In contrast, the coefficient of linear expansion α_{ex} is fairly large, about 1.5 times that of carbon steel. The specific thermal properties stated above imply that stainless steels have some essential problems under thermal load, i.e., high probability of troublesome phenomena such as loosening of threaded fasteners, excessive bolt force, increase of contact pressure at the nut-bearing surface, and seizure of thread surface and nut-bearing surface.

Aluminum alloys are extensively used in industrial applications in order to compensate the lack of strength of pure aluminum. Their density ρ_{dn} is slightly larger than one-third of that of carbon steel; thus, they are categorized as light alloys. Young's modulus E is fairly small—about 70 GPa—which is roughly one-third that of carbon steel. Thermal conductivity is about 2.5 times and the coefficient of linear expansion is about twice that of carbon steel, which means that they conduct heat and expand easily. Therefore, aluminum alloys behave in a specific manner under thermal load.

A2024 and A7075 are representative aluminum alloys called duralumin. The former is equivalent to ASTM 2024. They have the same level of tensile strength as carbon steel at room temperature. When used at high temperature, however, care must be taken, because the tensile strength significantly decreases.

Titanium and titanium alloy are classified as light metals because their density is about 60% of that of carbon steel. Due to high anticorrosion performance and low sensitivity to biological reactions, they are extensively used for eyeglass frames and various medical applications. On the other hand, since the raw materials are expensive and difficult to process, the application purposes as threaded fasteners are limited. Young's moduli of titanium and titanium alloy are a little above 100 GPa, which is about half that of carbon steel. Accordingly, when used as bolt materials, the bolt elongation is approximately doubled compared to a bolt made of carbon steel having the same configuration. This represents a specific mechanical property of titanium and titanium alloy bolts—that is, smaller stress amplitudes would be produced under the same repeated loading. In regard to thermal properties, coefficients of linear expansion of titanium and titanium alloy at room temperature are very small, ranging from 0.7 to 0.8 times that of carbon steel. Thermal conductivity of pure titanium at room temperature is slightly greater than that of austenitic stainless steel. A widely used titanium alloy, Ti-6Al-4V, has high strength and low thermal conductivity at room temperature. The value of the latter is about one-third that of pure titanium. However, its thermal conductivity has high temperature dependency and approaches the value of austenitic stainless steel at room temperature as the service temperature is increased. From the abovementioned properties, it is recognized that bolts made of titanium or titanium alloy are further likely to seize than those made of austenitic stainless steel. Nonetheless, if the specific properties of low Young's modulus and small coefficient of linear expansion are utilized in an ingenious manner, the application range as threaded fasteners can be expanded.

An important point to remember when using threaded fasteners made of nonferrous metals is that the strength of raw material is not necessarily consistent with the strength when the same material is processed into threaded fasteners. It is reported in Ref. [21] that when employing magnesium alloy or pure titanium as bolt materials, the strength of a bolt made of magnesium alloy in the clamped state exceeds the one made of pure titanium, although the tensile strength of pure titanium is about one and half times higher than that of magnesium alloy.

In Table 1.4, the Young's modulus E and Poisson's ratio ν of the representative materials for threaded fasteners are shown, i.e., carbon steel, austenitic stainless steel, titanium, and titanium alloy, involving their temperature dependency. In Table 1.5, coefficient of linear expansion α_{ex}, thermal conductivity λ, density ρ_{dn}, and specific heat c_{sp}, are tabulated in the same manner [22–25]. For reference, thermal properties of aluminum alloy at room temperature are also shown. It will be explained in *Learning by numerical examples 5.2 and 6.5* how the difference in coefficient of linear expansion between bolt-nut and fastened plates affects the bolt force variation under thermal load. As reference data for threaded fasteners made of carbon steel and alloy steel used at high temperature, Appendix A of JIS B 1051 provides the values of 0.2% proof stress over the range from room temperature to 300 °C, in which the strength

Table 1.4 Mechanical properties of threaded fastener materials.

Carbon steel						
Temperature (K)	293	373	473	573	673	773
E (MPa)	206,584	201,576	195,500	188,209	180,153	170,284
V	0.279	0.28	0.29	0.292	0.295	0.298
Austenitic stainless steel						
Temperature (K)	297	422	533	583	755	866
E (MPa)	195,020	188,160	182,280	176,400	156,800	147,980
V	0.25	0.27	0.32	0.32	0.27	0.29
Titanium						
Temperature (K)	300	473	673	873		
E (MPa)	110,000	98,067	86,299	71,589		
V	0.323	0.335	0.345	0.359		
Titanium alloy						
Temperature (K)	300	473	673	873		
E (MPa)	112,210	104,931	93,163	78,453		
V	0.301	0.305	0.31	0.322		
Aluminum alloy						
Temperature (K)	300					
E (MPa)	73,108					
V	0.331					

grade of bolts is covered from 5.6 to 12.9. For pipe flanges used at high temperature, bolts made of chrome-molybdenum-based alloy, SNB7, are widely used.

1.6.2 Factors to consider in selecting materials

When selecting materials of which threaded fasteners are made, the power of resistance to the service environment, such as anticorrosion performance, is important in addition to strength, stiffness, and thermal properties. Furthermore, workability sometimes becomes an issue for threaded fasteners used in a normal environment. For instance, when processing female threads into the main body, the dimensional accuracy of thread form becomes lower than that of a bolt-nut connection, due to the difficulty in processing. In particular, the case of the main body being made of high strength materials may generate large geometric errors in female threads. Accordingly, the overall joint strength cannot be improved, regardless of employing high strength materials. Threaded fasteners such as bolts and nuts are usually manufactured by rolling. Meanwhile, a cutting operation is often applied for manufacturing bolts with large nominal diameter and bolts made of processing-resistant materials. The strength of threaded fasteners formed by cutting is weaker than those formed by rolling.

When selecting materials for threaded fasteners, the amount of strain energy, which is to be stored until the rupture occurs, must be taken into account from the mechanical

Table 1.5 Thermal properties of threaded fastener materials.

Carbon steel					
Temperature (K)	300	500	800		
α_{ex} (1/K) $\times 10^{-6}$	11.8	12.3	14.2		
λ (W/mK)	51.5	47.2	36.8		
ρ_{dn} (kg/m^3)	7850	7800	7700		
c_{sp} (J/kgK)	473	520	665		
Austenitic stainless steel					
Temperature (K)	300	400	600	800	1000
α_{ex} (1/K) $\times 10^{-6}$	17.3	17.4	17.8	18.4	19.1
λ (W/mK)	16.0	16.5	19.0	22.5	25.7
ρ_{dn} (kg/m^3)	7920	7890	7810	7730	7640
c_{sp} (J/kgK)	499	511	556	620	644
Titanium					
Temperature (K)	300	600	800	1000	1200
α_{ex} (1/K) $\times 10^{-6}$	8.7	10.4	11.1	11.5	11.3
λ (W/mK)	21.9	19.4	19.7	20.7	22
ρ_{dn} (kg/m^3)	4506	4467	4439	4409	4385
c_{sp} (J/kgK)	522	610	674	732	700
Titanium alloy					
Temperature (K)	300	500	800		
α_{ex} (1/K) $\times 10^{-6}$	8.95	10.3	11.4		
λ (W/mK)	7.6	11.1	17.2		
ρ_{dn} (kg/m^3)	4420	4340	4350		
c_{sp} (J/kgK)	537	557	655		
Aluminum alloy					
Temperature (K)	300				
α_{ex} (1/K) $\times 10^{-6}$	23.2				
λ (W/mK)	120				
ρ_{dn} (kg/m^3)	2770				
c_{sp} (J/kgK)	880				

strength point of view, besides tensile strength and yield strength. The value is evaluated as the area in the stress-strain diagram integrated from the origin to the rupture point. Threaded fasteners made of ductile materials can absorb a large amount of strain energy, even if their yield stress and tensile strength are not so high, because the fastener generally elongates until the rupture occurs. On the contrary, in the case of threaded fasteners made of high strength materials, the strain energy that can be stored may become smaller because the fasteners tend to break under fairly small strains. From the above considerations, when subjected to unexpected large loads, threaded fasteners made of ductile materials may stay within the joint, although producing large plastic deformations. In contrast, threaded fasteners made of high strength materials might be completely broken, resulting in the loss of joint function.

Fundamentals of threaded fasteners

2

2.1 Strength of threaded fasteners

2.1.1 *Occurrence locations of rupture and failure of threaded fasteners*

Fig. 2.1 illustrates the locations where the rupture and failure of threaded fasteners are likely to occur. In the case of bolt-nut connections, bolt ruptures occur in most cases at the first bolt thread root adjacent to the nut-bearing surface and around thread runout and root radius under the bolt head. In contrast, in the case of body-side engaged threads, it is sometimes reported that the rupture initiates from the far end thread root of female threads due to crack propagation. In any event, those accidents can be traced to metal fatigue occurring under repeated external loads in service conditions.

Meanwhile, bolt failures occur during the tightening operation, especially by the torque control method, due to over tightening by applying excessive torque. This produces large plastic deformations in the threaded portion, which may disable the target joint or cause the joint to rupture. The plastic deformations usually initiate from around the first bolt thread root, but the rupture often occurs in the central part of the unengaged threads. This phenomenon can be explained by the expansion of plastically deformed region, as discussed in section 4.2.5. Stress concentration and fatigue failure of threaded fasteners are explained comprehensively in Chapter 4.

To attain the downsizing and lightweighting of jointed components, high strength materials are increasingly used for threaded fasteners. In these cases, attention should be paid not only to stress values in threaded fasteners but also to contact pressure at the nut-bearing surface. Too high a contact pressure produces large plastic deformations around the plate surface in contact with the nut. It causes a permanent set, here termed depression, where the nut-bearing surface bites into the plate surface. Depression could become a serious problem when the plate material is weaker than the threaded fastener material. For jointed components made of carbon steel or chrome-molybdenum-based alloy, the critical contact pressure allowable at the nut-bearing surface is lower than the tensile strength. In contrast, the allowable contact pressure is higher than the tensile strength in the case of cast iron. Concrete values of critical contact pressure are given in [26] for various materials.

Multiplying the mean contact pressure at the nut-bearing surface—determined by its critical contact pressure—by the ratio of bolt cylindrical area to nut-bearing surface area, the amount of allowable bolt stress can be calculated based on the critical contact pressure. The abovementioned area ratio considerably changes for varying nominal diameters and bolt hole grades. It is given in graphical form in section 3.4.3, which deals with the tightening process of direct tension method using hydraulic tensioners.

The Mechanics of Threaded Fasteners and Bolted Joints for Engineering and Design.
https://doi.org/10.1016/B978-0-323-95357-3.00003-4

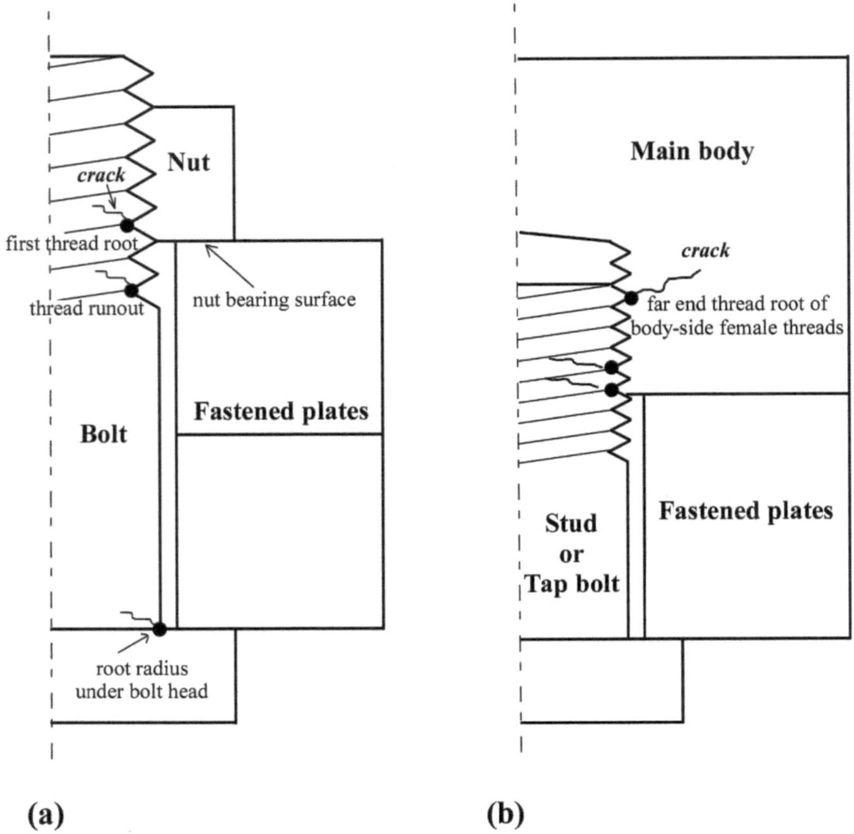

Fig. 2.1 Occurrence locations of rupture and failure in threaded fasteners. (A) Bolt-nut connection and (B) body-side engaged threads.

2.1.2 Strength in tightening process

The target value in the tightening process of threaded fasteners is axial bolt force or axial bolt stress. When tightening is conducted using a pair of bolt and nut, the bolt is elongated by the tensile force and the fastened plates are contracted by the compression force, which is a reaction force. Threaded fasteners are tightened by a variety of methods. For example, when using hydraulic tensioners, the tensile force applied in the tightening process is somewhat larger than the target bolt force. The torque control method is the most widely used because of its easy operation. In the tightening process, however, a twisting moment unnecessary to generate bolt force is also exerted on the bolt body. The twisting moment produces shear stress, which is harmful to bolt strength. Accordingly, plastic deformations are more likely to be produced at the bolt thread root where high stress concentrations occur, even if the bolt stress is relatively lower than the yield stress of the bolt material. That is because the von Mises stress is increased due to the effect of the twisting moment. In this case, if the plastically deformed area is limited around the thread root, the effect of plastic deformation on the mechanical behavior of bolted joints can be negligible.

When applying tensile force directly to the target bolt, commonly employed methods are direct tension, using hydraulic tensioners, or thermal expansion, using bolt heaters. In the former method, the initial tension applied in the tightening process is larger than the target bolt force. Specifically, when tightening a joint with small grip length, a fairly large tensile force is required to achieve the objective bolt force. Therefore, attention must be paid to not causing too large a plastic deformation. In the thermal expansion method, target bolts are hollow and they are heated using bolt heaters until a prescribed amount of elongation is obtained in order to attain the objective bolt force. The magnitude of the attained bolt force increases as heating temperature is increased. Consequently, the upper limit of heating temperature needs to be determined from the mechanical point of view, since too high a heating temperature can significantly alter the material properties of threaded fasteners. Mechanics and problems of the tightening process of bolted joints are discussed in depth in Chapter 3.

2.1.3 Strength in service condition

Problems and accidents concerning threaded fasteners generally occur in service conditions. They arise from a variety of causes, and in many cases multiple causes combined result in fastener troubles. Important considerations to assess the strength of threaded fasteners are described below.

(1) Fatigue strength against repeated loading

Large stress amplitude due to repeated loading sometimes causes fatigue failure of threaded fasteners. Fatigue failure of threaded fasteners is explained in detail in section 4.4 and subsequent sections.

(2) Delayed fracture

Delayed fracture is a kind of brittle fracture, which occurs suddenly after some time has passed from the completion of the tightening process. It becomes a problem in high strength materials with tensile strength of 1200 MPa or higher. The primary cause is the existence of hydrogen, which invades from the surrounding environment due to electroplating and other factors. Accordingly, the upper limit of the material strength of threaded fasteners is commonly grade 12.9. Delayed fracture can be mitigated by a dehydrogenation treatment called a baking process. Detailed discussions on delayed fracture are found in [27].

(3) Loosening of threaded fasteners

Reduction of axial bolt force, due to vibration force and other factors, sometimes induces falling-off of nuts and fatigue failure of bolts, the latter of which is caused by the increase of stress amplitude exerted on threaded fasteners. The loosening phenomenon of threaded fasteners is explained in Chapter 6, where it is categorized into two cases of loosening, with and without rotation of threaded fasteners.

(4) Bolt force variation due to thermal load

When bolted joints are subjected to thermal load, axial bolt force varies due to the difference in thermal expansion among clamped parts and the variation of material properties such as Young's modulus. The amount of thermal expansion is calculated as the product of the coefficient of linear expansion, temperature change, and length. A simple equation derived from the foregoing relationship suggests that in bolted joints subjected to thermal load, attention should be paid to the difference in the coefficient of linear expansion between threaded fasteners and clamped plates. It also should be noted that the bolt force significantly varies, particularly in bolted joints with large grip length under large temperature changes. The difference in thermal expansion around the bolted joint is primarily caused by the differences in temperature and coefficient of linear expansion between bolt-nut connections and clamped plates. The temperature difference reaches its peak in many cases before the steady state has been established after the start of heating.

When combining different materials between bolt-nut connections and clamped plates, particular attention must be paid. This is because the bolt force variation is directly proportional to the difference in coefficient of linear expansion between the two materials, according to the calculation examples by a simple equation given in section 5.1.3. Meanwhile, the amount of coefficient of linear expansion differs to some extent among the materials that belong to the carbon steel group. In this case, the bolt force changes under large temperature variation, even if the difference is small.

Regarding thermal load, most bolted joints are subjected to high temperature thermal load. In contrast, when subjected to low temperature thermal load, the bolt force variation can be predicted with the same equation just by reversing the sign of temperature variation. Thermal and mechanical behaviors of bolted joints under thermal load are discussed in great detail in Chapter 5, including the effect of thermal contact resistance, explained in Section 2.7.

(5) Bolt force variation due to creep

Creep is a phenomenon wherein strain is increased over time under constant load. It is particularly observed at high temperature, while it also becomes an issue at relatively low temperature for low melting point materials. It is not easy to determine the threshold temperature of the carbon steel commonly used for threaded fasteners beyond which the effect of creep must be considered. In [28], it is pointed out that the effect of creep particularly appears in carbon steel materials in the range beyond 350 or 400 °C, after which deformations may progress in the stress condition even below the yield stress. There are some proposals regarding the threshold temperature, e.g., the effect of creep must be considered beyond 500 °C and could be negligible below 300 °C. There is also a proposal that in the range between 300 and 500 °C, the judgment should be made based on the method of applying thermal load and the material.

VDI 2230 (2003) proposes a relationship between creep and embedment [29]. Embedment is a phenomenon wherein bolt force reduces with the progress of plastic deformation of surface asperities. It is explained in detail how to formulate and analyze creep phenomena by FEM in [30]. In [31], the creep phenomena in bolted joints are studied by conducting creep experiments at 600 °C, in which the bolt force variation with time is compared to the one obtained by the evaluation formula for creep rate.

(6) Corrosion and electrolytic corrosion induced by the surrounding environment

Strength and life of threaded fasteners are often significantly decreased due to the surrounding environment, such as seawater and mist. In some cases, electrolytic corrosion becomes a serious issue, which is caused by the difference in ionization tendency among the joint materials. This problem can be prevented by introducing coating treatment or considering the combination of joint materials. In addition, corrosion fatigue sometimes becomes a critical problem for threaded fasteners used in corrosive environments. The generation mechanisms and the prevention methods are explained in great detail in [32], in which the theory of fracture mechanics is successfully applied for the purpose of estimating the fatigue strength against corrosion fatigue.

2.1.4 Friction coefficients affecting mechanical behavior of bolted joints

Bolted joints are composed of threaded fasteners, such as bolts and nuts, and fastened plates. They are regarded as contact structures from the mechanical point of view. Accordingly, it is necessary to analyze contact problems to clarify their mechanical behavior. Contact problems are generally recognized as one of the most difficult mechanical problems, even if the latest finite element theory is introduced [33].

When analyzing contact problems, numerical calculations are usually conducted in an iterative manner to determine the contact conditions, considering the balance of forces among contacting bodies. The contact conditions are typically categorized into three states, i.e., stick, slip, and separation. In stick state, two contacting bodies are regarded as one body. In slip state, relative motion exists at the interface between contacting bodies in the tangential direction. In separation state, some clearance exists between contacting bodies. When dealing with contact structures with complex geometry, nonlinear problems need to be analyzed in most cases, because contact areas and conditions are likely to change according to the increase or decrease of external force.

In three-dimensional structures, slip state is categorized into three patterns depending on the definition of coordinates. They are the state of slip in x-direction and stick in y-direction, the state of stick in x-direction and slip in y-direction, and the state of slip in both x- and y-directions. Assuming that finite element method is used and contact conditions are judged between n pairs of nodes placed on the mating contact surfaces, there could be 5^n combinations of contact conditions in all, corresponding to five possible contact conditions such as stick, separation, and three types of slip state, as explained above. It follows that numerical analyses are required to find only one definite combination out of so many selections. In addition, contact conditions are greatly influenced by the value of the coefficient of friction. Accordingly, since bolted joints are contact structures including all the problems mentioned above, it is tough and intractable, from the viewpoint of numerical analysis, to analyze their mechanical behavior with high accuracy.

Fig. 2.2 shows coefficients of friction on the thread surface, nut-bearing surface, bolt head-bearing surface, and plate interface, which affect the mechanical behavior of bolted joints. μ_r and μ_θ in the figure represent friction coefficients in the radial and circumferential directions, respectively. In any event, the amounts of μ_r and μ_θ are different, slightly or considerably, from each other because friction characteristics vary depending on the direction of the relative motion of contacting bodies. This means that two kinds of friction coefficient, μ_r and μ_θ, must be considered on each contact surface to rigorously evaluate the mechanical behavior of bolted joints. For instance, when tightening a pair of bolt and nut by the torque control method, the relationship between applied torque

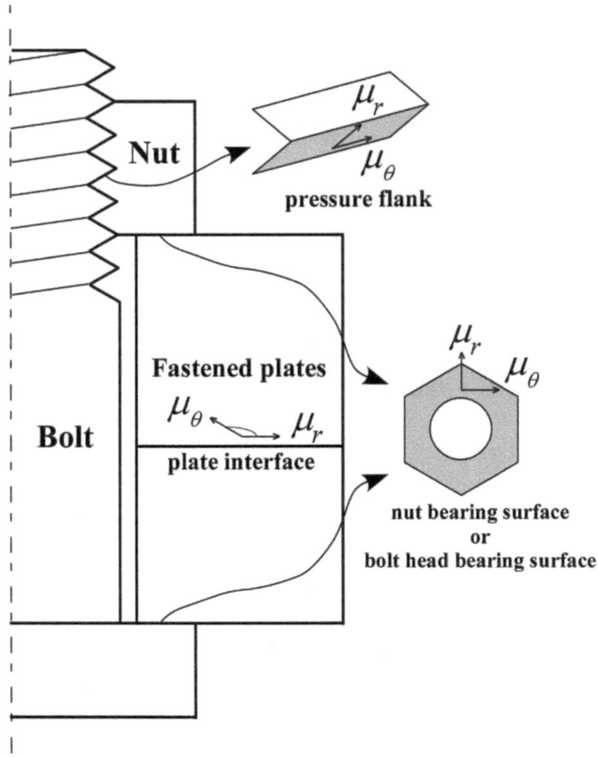

Fig. 2.2 Coefficients of friction to be considered in bolted joints.

and bolt force to be generated is determined by the amounts of μ_θ on thread surface and nut-bearing surface. In the case of the tightening process of tap bolts, shown in Fig.1.9, the relationship can be evaluated just by replacing the coefficient of friction on the nut-bearing surface with that on the bolt head-bearing surface.

When machines, structures, and instruments, clamped by bolts and nuts, are put into service conditions, coefficients of friction on thread surface, nut-bearing surface, bolt head-bearing surface, and plate interface become an issue, because they affect the slip phenomenon on each contact surface. Furthermore, loosening phenomena are greatly influenced by μ_r when subjected to repeated loading perpendicular to the bolt axis and also by μ_θ when subjected to twisting moment around the bolt axis, respectively.

2.2 Stiffness of threaded fasteners

2.2.1 Relationship between stiffness and mechanical behavior in bolted joints

The stiffness of the engaged thread portion is closely related to the fatigue strength. Some past disasters had been caused by improper designs ignoring the importance of the stiffness. In this section, the stiffness of bolted joints is expressed by a simple one-dimensional spring model, by use of which it is explained how to evaluate various mechanical properties of bolted joints, such as tightening characteristics of a variety of methods, fatigue strength, propagation of applied force under impact loading, etc.

Stiffness of jointed components is an important factor to be considered in the design step as well as strength. However, high stiffness is not always desirable, unlike high strength. For example, the stiffness of bolt-nut connection is preferably low, to enhance fatigue strength and mitigate the bolt force reduction due to the embedment at the interface.

It is relatively easy to evaluate the stiffness of a bolt-nut connection under tensile load in both the initial tightening process and the running condition. In the latter case, external loads are applied. On the other hand, only compression force is exerted on the fastened plates in the tightening process. Therefore, it is very difficult to accurately evaluate the stiffness of fastened plates when they are subjected to tensile force due to repeated loading in the running condition.

In general, fatigue strength of the jointed components becomes a problem when subjected to tensile load. In that regard, it is a point of vital importance that the stiffness of fastened plates can be significantly different between the tightening process and in the running condition of the joint under tensile external load. This phenomenon is the primary cause of the low evaluation accuracy of the fatigue strength of bolted joints by means of bolted joint diagram. Detailed discussions of this are found in section 4.5.

Joint stiffness changes with changes in the types of applied loads, such as axial external force, bending, shear, and twisting, even if an identical pair of bolt and nut is used in the joints of a variety of machines, structures, instruments, etc. Among various stiffnesses, the stiffness in the axial direction, i.e., along the bolt axis, is of primary importance. The axial stiffness is closely related to bolt force generation in the tightening process and bolt force variation under repeated loading.

Supposing a straight bar of length L and area A is subjected to axial force F, as shown in Fig. 2.3, the relationship between axial force F and axial displacement u is given in the following equation, considering the definition of stress and Hooke's law:

$$\frac{F}{A} = E\frac{u}{L} \tag{2.1}$$

where E represents Young's modulus.

Transforming Eq. (2.1), proportional factor AE/L between axial force F and displacement u is set to be k:

$$F = \frac{AE}{L}u = ku \tag{2.2}$$

$$k = \frac{AE}{L} \tag{2.3}$$

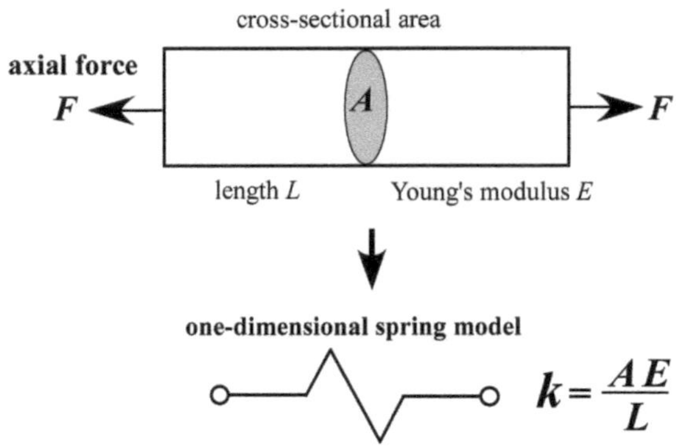

Fig. 2.3 Straight bar subjected to axial force.

where k represents the spring constant of the straight bar under axial force.

Fig. 2.4 shows a pipe flange connection clamped by multiple bolts, in which a gasket is inserted between two flanges. The primary role of pipe flange connections is the prevention of the leakage of contained fluids—leakage which can be traced to the reduction of bolt force due to such factors as thermal expansion or shrinkage of flanges and gaskets and changes in gasket stiffness, all of which are caused by high or low temperature of contained fluids.

The stiffness of gaskets is generally much lower than that of metal flanges. For instance, in the case of sheet gaskets, gasket compression stress is nonlinearly related to gasket strain. If the relationship between stress and strain is tentatively expressed by linear approximation within the range of normal use conditions, the stiffness evaluated by the gradient of the straight line is usually less than hundredths of the Young's modulus of carbon steel. Hence, a pipe flange connection is simply regarded as a structure consisting of two high stiffness plates and a low stiffness gasket. Expressing each part of the pipe flange connection by one-dimensional springs, the whole structure can be

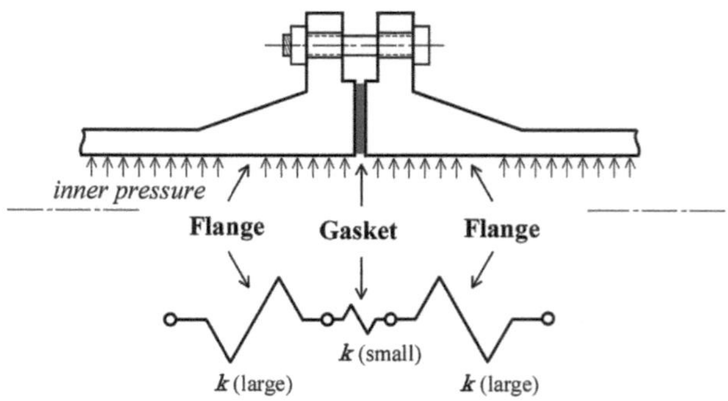

Fig. 2.4 Stiffness of each part of a pipe flange connection.

expressed as an aggregate of elastic bodies, where one low stiffness spring is inserted between two high stiffness springs and connected in series. Accordingly, when subjected to external force or thermal load, large deformations occur intensively in the gasket. It follows that the mechanical behavior of the whole joint is controlled by the gasket stiffness. It is widely recognized that the mechanical behavior of a pipe flange connection with an inserted sheet gasket is difficult to evaluate, regardless of its widespread use. That is because the joint behavior is largely influenced by a low stiffness sheet gasket that exhibits high nonlinearity and hysteresis loop in the stress–strain curve.

2.2.2 Stiffness evaluation using one-dimensional spring elements

When tightening bolted joints, complex deformation patterns are observed around engaged threads, bolt head, nut, and bolt head-bearing surfaces. In that case, the tightening process is primarily affected by the axial deformation of the objective bolted joint. Fatigue strength under external load and bolt force reduction due to nut loosening are also greatly influenced by the axial deformation of each part of the bolted joint. From the above considerations, it follows that the stiffness in the axial direction is of primary importance to evaluate tightening properties, fatigue strength, and loosening phenomena of bolted joints. Axial stiffness of each part of a bolted joint can be expressed by one-dimensional springs, as shown in Eq. (2.3).

A bolted joint can be replaced by five one-dimensional spring models connected in series, i.e., engaged threads, unengaged threads, bolt cylindrical portion, bolt head, and fastened plates, as shown in Fig. 2.5. Five spring constants appearing in the figure—k_{th}, k_s, k_{cyl}, k_{hd}, and k_f—represent the stiffness of each part explained above,

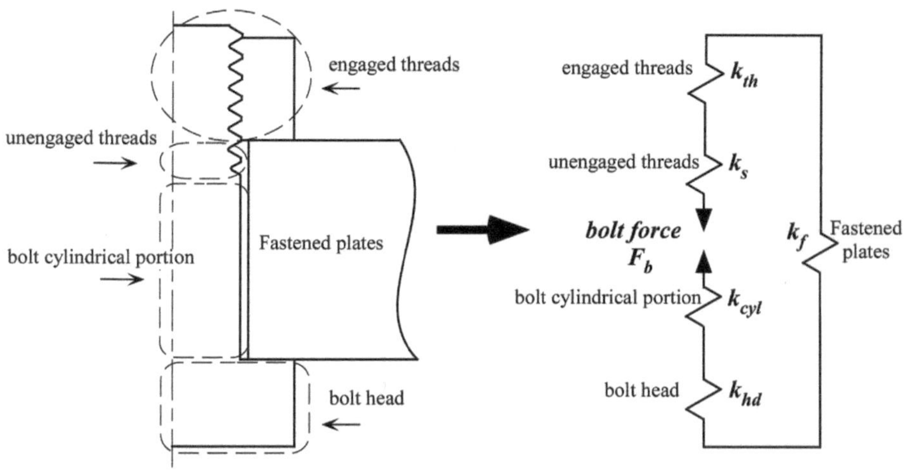

Fig. 2.5 Stiffness of a bolted joint represented by five one-dimensional spring elements.

respectively. Spring constant k_b, calculated by connecting the four spring constants of k_{th}, k_s, k_{cyl}, and k_{hd} in series, represents the entire stiffness of the bolt-nut connection:

$$\frac{1}{k_b} = \frac{1}{k_{th}} + \frac{1}{k_s} + \frac{1}{k_{cyl}} + \frac{1}{k_{hd}} \tag{2.4}$$

By connecting k_b and k_f in series, spring constant k_{total} is produced, which represents the entire stiffness of a bolted joint:

$$\frac{1}{k_{total}} = \frac{1}{k_b} + \frac{1}{k_f} \tag{2.5}$$

The amount of k_{total} significantly affects the accuracy of bolt force obtained by various tightening methods, as introduced in Chapter 3. Since bolted joints with small grip length have large k_{total}, the tightening accuracy is likely to decrease. The ratio of k_b to k_f greatly affects the fatigue strength of bolted joints. Among the five spring constants introduced above, it is quite difficult to evaluate the amounts of k_{th} and k_f with high accuracy, since they are significantly influenced by the processing and assembly errors of threaded fasteners and the complexity of joint configuration. Particular attention should be paid when clamping thin plates with warping beyond a certain level, because the amounts of k_{th} and k_f are changeable. Four spring constants, except for k_f, can be expressed in the same form as Eq. (2.3):

$$k_{th} = \frac{A_S E_b}{L_{th}}, \quad k_s = \frac{A_S E_b}{L_s}, \quad k_{cyl} = \frac{A E_b}{L_{cyl}}, \quad k_{hd} = \frac{A E_b}{L_{hd}} \tag{2.6}$$

where A_s is stress area and A is bolt cylindrical area calculated by nominal diameter d; L_s and L_{cyl} are the lengths of unengaged threads and bolt cylindrical portion, respectively; E_b represents Young's modulus of bolt and nut materials; and L_{th} and L_{hd} are equivalent lengths of engaged threads and bolt head, which are discussed in the next section.

The relationship between the spring constant of an entire bolted joint k_{total} and nominal diameter d is very important when studying various mechanical properties of bolted joints. In order to examine the relationship between nominal diameter d and spring constant k, which represents the stiffness of a straight bar, Eq. (2.3) is transformed as follows:

$$k = \frac{AE}{L} = \frac{\pi d^2}{4}\frac{E}{L} = \frac{\pi d}{4}\frac{E}{L/d} \tag{2.7}$$

Assuming that L is grip length and L/d is constant, which implies that the configurations of target joints are approximately similar each other, it is found that the stiffness of bolted joints with similar form is almost proportional to nominal diameter d. This result is significantly important when considering the tightening characteristics and the strength of bolted joints.

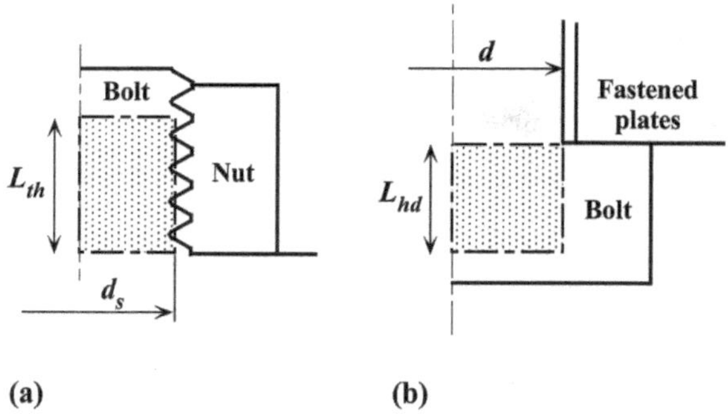

Fig. 2.6 Equivalent lengths of engaged threads and bolt head. (A) Engaged threads and (B) bolt head.

2.2.3 Equivalent lengths of engaged threads and bolt head

Stiffnesses of engaged threads and bolt head are evaluated not by the actual lengths but by the equivalent lengths of cylinders having the same amount of axial stiffness. Fig. 2.6 explains the concept of equivalent length. L_{th} is the equivalent length for engaged threads, which is replaced by a cylinder of diameter d_s, diameter of stress area. L_{hd} is the equivalent length for the bolt head, which is replaced by a cylinder of diameter d, nominal diameter of bolt. Several equations are proposed for L_{th} and L_{hd}. Sawa et al. proposed the following equation based on the theory of elasticity [6]:

$$L_{th} = 0.7d, \quad L_{hd} = 0.5d \tag{2.8}$$

Fig. 2.7A and B shows the amounts of L_{th} and L_{hd} calculated by FEM [34], in which equivalent lengths are normalized with respect to nominal diameter d. Analytical objects are coarse screw threads of M12, M24, and M64. L_{th} is evaluated under the condition of bolt hole diameter d_h of first and second grades, while L_{hd} is evaluated under d_h of second grade. The abscissa represents coefficient of friction μ. Equal coefficients of friction are assumed on the thread surface and nut-bearing surface. As μ is increased, L_{th} and L_{hd} are both decreased, because the relevant deformations are constrained more strongly on the contact surfaces and the stiffness becomes higher. The effect of μ appears more remarkably in L_{th}. That is because the contact on the bolt head-bearing surface occurs along the horizontal plane, whereas contact on the pressure flank of engaged threads occurs along the surface inclined at 30 degrees to the horizontal plane. In the latter case, the amount of relative slip at the interface is liable to change as μ varies. As for the effect of bolt hole diameter d_h on L_{th}, the amount of L_{th} for first grade of d_h is a little smaller than that for second grade, because the deflection of engaged threads is more constrained in the case of first bolt hole, due to the small bolt hole diameter.

The effect of nominal diameter d on L_{th} is insignificant, as found from Fig. 2.7A, although the reduction rate of L_{th} with increasing μ slightly changes for varying nominal diameters. Equivalent length of bolt head L_{hd}, shown in Fig. 2.7B, also decreases with increasing μ, in which the effect of μ is smaller than in L_{th} due to the reason stated above.

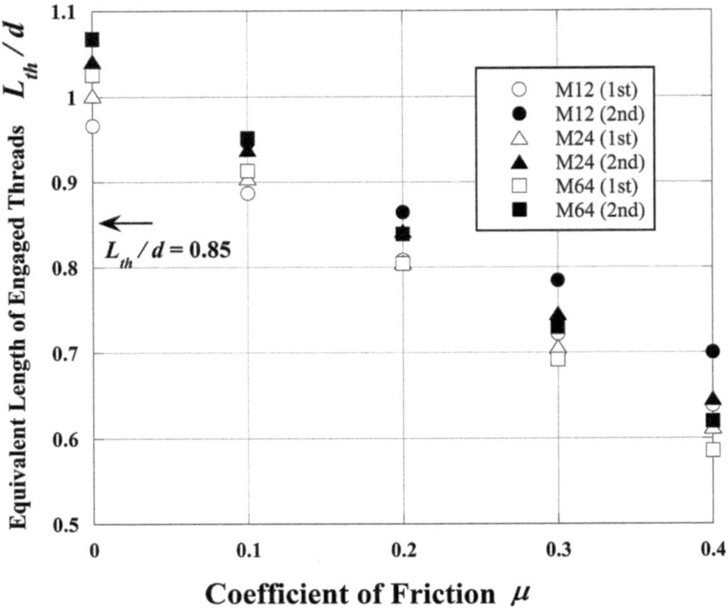

(a)

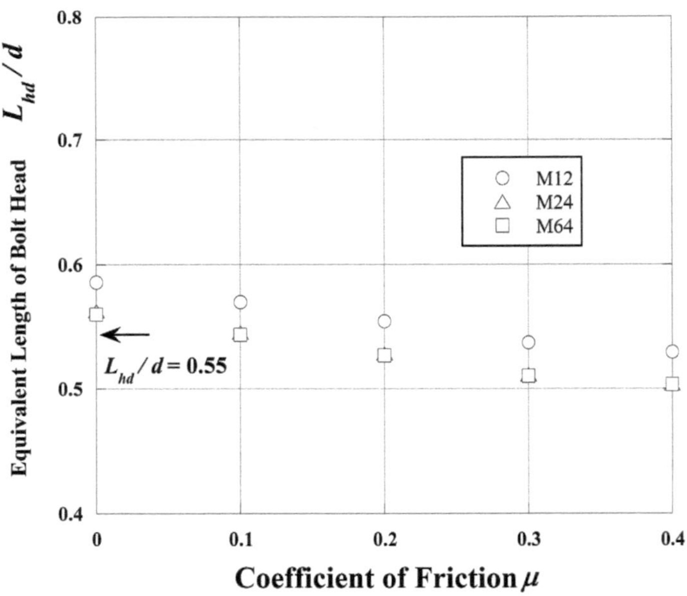

(b)

Fig. 2.7 Effect of friction coefficient on equivalent lengths, L_{th} and L_{hd}.

From the above considerations, when evaluating the mechanical behavior of bolted joints by means of one-dimensional spring models, the following values are employed in this book, unless otherwise stated:

$$L_{th} = 0.85d, \quad L_{hd} = 0.55d \tag{2.9}$$

The equivalent length of engaged threads L_{th} is sensitive to geometric errors existing between mating threads and at the nut-bearing surface, etc. The nut-bearing surface is not always completely flat due to the difficulty in plastic working, as discussed in section 1.5.2. In these cases, a small taper is processed radially inwards, as shown in Fig. 2.8A. Therefore, when tightening a bolt with relatively low torque, the nut-bearing surface might not come into perfect contact with the mating plate surface. This produces an additional axial deflection around the nut bearing surface, leading to a stiffness reduction of the engaged threads. Numerical results by axisymmetric FEM (finite element method) are given in Fig. 2.8B in order to show how the inclination angle existing at the nut-bearing surface θ_{tp} affects L_{th}, where the contact length defined in Fig. 2.8A is set to zero [35]. Positive values of θ_{tp} indicate that the clearance increases toward the bolt axis, which is commonly encountered in actual nuts, while negative values show that the clearance increases toward the outer surface of the nut. Numerical analyses are performed for bolt-nut connections of M24 clamped with bolt stress of 300 MPa. Bolt hole diameter d_h is the second grade. Coefficient of friction μ is changed from 0 to 0.2. The amount of L_{th} shows a minimum of 0.85d when μ is 0.2 and the nut-bearing surface is initially in perfect contact, $\theta_{tp} = 0$. It is also found that the stiffness of engaged threads, represented by the reciprocal of L_{th}, significantly decreases due to the existence of a slight inclination, regardless of its direction. Regarding the effect of μ, L_{th} increases and the stiffness of engaged threads is reduced with decreasing μ for the same reason as in Fig. 2.7A.

Fig. 2.8C shows the numerical results of the effect of bolt stress on L_{th}, where a one-seventh part of the nut-bearing surface is assumed to be in contact from the outer edge of the nut under the condition of zero bolt stress. The bolt stress σ_b is changed from 100 MPa to 200 MPa and 300 MPa. When the nut-bearing surface is partially in contact in the initial state, the effect of bolt stress is rather small and it almost saturates in the region that θ_{tp} exceeds plus or minus 1 degree, as shown in Fig. 2.8C. That is because the contact area between the nut-bearing surface and the plate surface is nearly unchanged beyond a certain value of θ_{tp}, even if bolt stress is increased.

Summarizing the above results, it is concluded that with the existence of a slight taper at the nut-bearing surface, the equivalent length of engaged threads L_{th} becomes significantly larger than 0.85d for the perfect flat case, and L_{th} is sometimes more than twice the forgoing value. Meanwhile, in the case of body-side engaged threads clamped by studs or tap bolts, the equivalent length L_{th} is shorter than 0.85d owing to its high stiffness [36]:

$$L_{th} = 0.75d \tag{2.10}$$

The conclusion that equivalent length of engaged threads L_{th} is changeable, deduced from Figs. 2.7A and 2.8B and C, suggests that when evaluating the mechanical

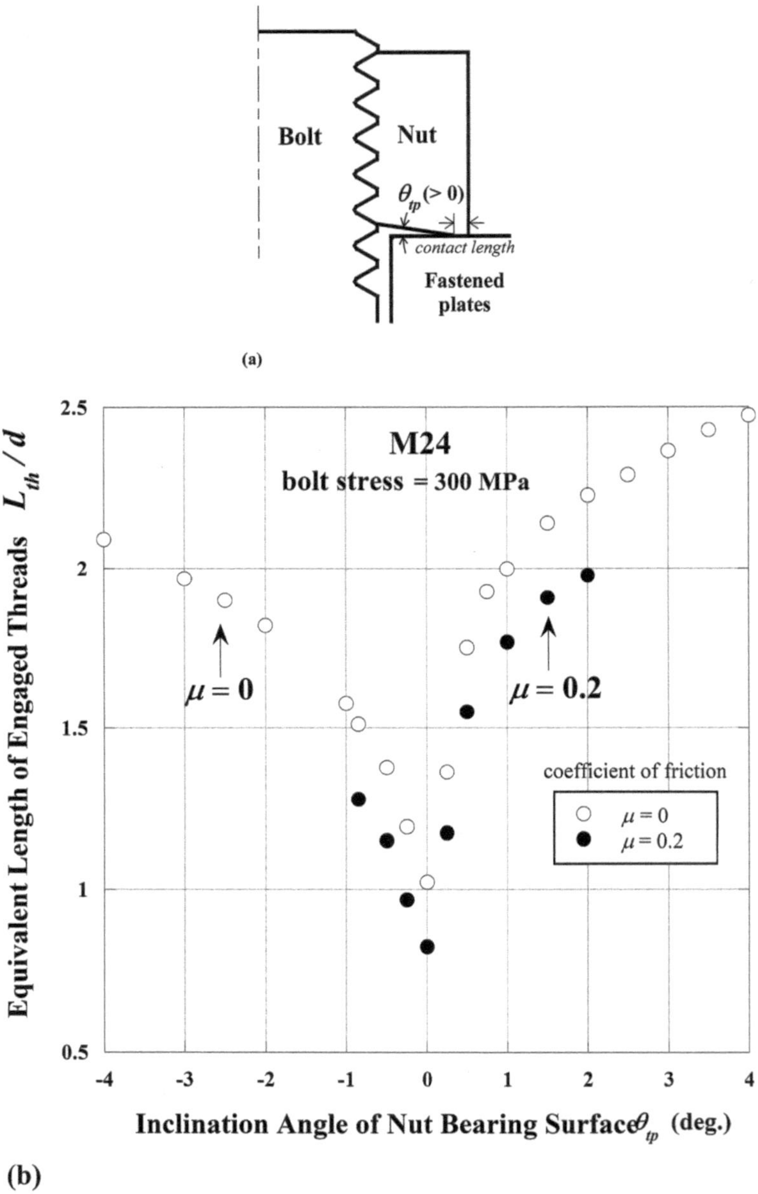

Fig. 2.8 Effect of inclination angle of nut bearing surface on L_{th}. (A) Inclination angle of nut bearing surface, (B) effect of coefficient of friction, and

(Continued)

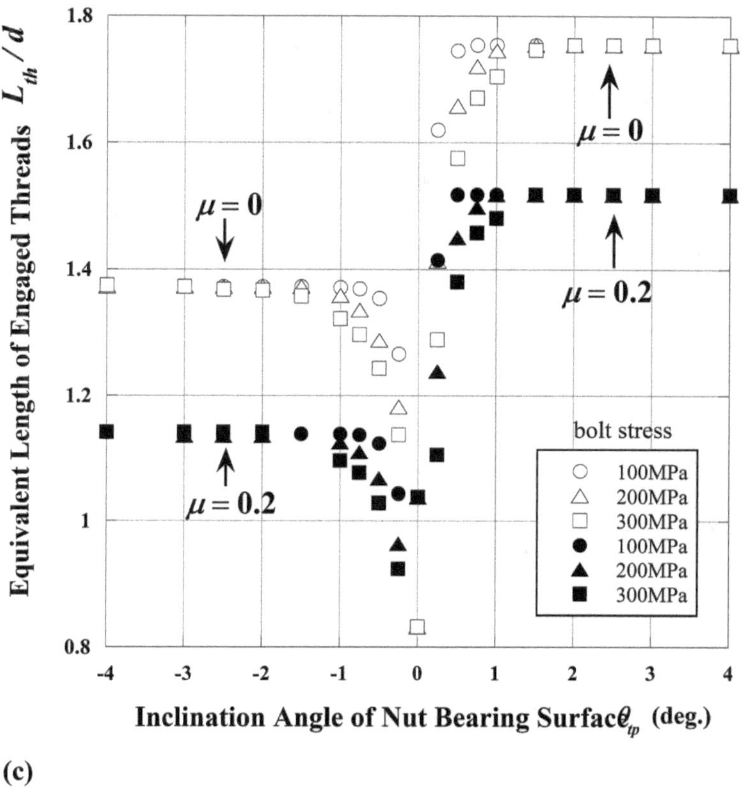

Fig. 2.8, cont'd (C) effect of bolt stress.

behavior of a bolted joint with small grip length, the analytical accuracy is greatly influenced by what value is chosen for L_{th}. There is a concrete example on the tightening accuracy, which is commonly encountered in the direct tension method, thermal expansion method, and elastic angle control method. When clamping relatively thin plates by one of these three methods, the accuracy of the generated bolt force could be remarkably reduced due to error in relation to the selection of L_{th}, compared to the case of clamping thick plates. For the same reason, the decrease in evaluation accuracy occurs when bolt force variation under repeated loading is predicted or bolt force reduction due to embedment is estimated by use of a bolted joint diagram.

2.2.4 Compression stiffness of fastened plates

When tightening bolt-nut connections, tensile force is produced in bolts and compression force is exerted on fastened plates as the reaction force. The stiffness of fastened plates under compression force is of critical importance to evaluate various mechanical behaviors such as tightening characteristics and fatigue strength. It is determined by the shapes, dimensions, and materials of the target fastened plates. Compression stiffness of fastened plates varies according to the contact pressure distribution patterns at the interface, which are classified into thin cylinder, thick cylinder, and plate, as depicted in fig. 1.12. When using bolts and nuts with identical nominal diameter d for a constant grip length L_f, the stiffness of fastened plates increases in the order of thin cylinder, thick cylinder, and plate. In the case of thin cylinder, where the outer diameter of fastened plates D_0 is slightly larger than the mean diameter B of nut or bolt head, the plates are considered to be under uniform compression. Therefore, its spring constant k_f can be calculated by substituting the cross-sectional area of the hollow cylinder and the grip length into A and L, respectively, in Eq. (2.3).

$$k_f = \frac{\pi\left(D_0^2 - d_h^2\right)}{4} \cdot \frac{E_f}{L_f} \tag{2.11}$$

The above equation demonstrates that k_f is proportional to Young's modulus E_f of the plate material and is inversely proportional to the grip length L_f. That is, the stiffness is increased with smaller grip length and is decreased when materials with low Young's modulus, like aluminum alloys, are used. Furthermore, the influences of E_f and L_f on k_f are qualitatively the same for the cases of thick cylinder and plate. Some evaluation formulas for k_f proposed so far are introduced in the following.

For instance, VDI 2230 (2003) proposes an equation for the clamping state of a plate based on the numerical results by FEM, in which the geometry of bolt-nut connection is replaced by a simple form:

$$k_f = \frac{\pi E_f d_h \tan\theta_{cn}}{2\ln\left[\frac{(B + d_h)\left(B + L_f \tan\theta_{cn} - d_h\right)}{(B - d_h)\left(B + L_f \tan\theta_{cn} + d_h\right)}\right]} \tag{2.12}$$

where d_h represents bolt hole diameter.

The pressure cone angle θ_{cn} can be calculated by the following equation:

$$\tan\theta_{cn} = 0.362 + 0.032\ln\left(\frac{L_f}{2B}\right) + 0.153\ln\left(\frac{D_o}{B}\right) \tag{2.13}$$

The detailed calculation procedure of VDI 2230 (2003) is given in [37]. Compression force due to bolt force is not exerted outside the pressure cone at the interface, as explained in section 1.5.3. Hence, no contact pressure exists there, despite mating plates being geometrically in contact. It is pointed out in [38] that the amount of k_f evaluated by the above equation tends to give higher values, because the separation phenomenon outside the pressure cone is not considered in the finite element analysis.

Three-dimensional elastic theory is also applied to derive the equation for k_f by some researchers [39,40]. Replacing the fastened plates by a hollow cylinder and assuming uniform contact pressure distributions on nut- and bolt head-bearing surfaces due to bolt force, Shibahara proposed a simple equation that is applicable to all three contact pressure distribution patterns [39], as shown in fig. 1.12:

$$\frac{k_f}{E_f \cdot d_h} = \frac{\pi\left\{\left(\frac{B}{d_h} + c\frac{L_f}{d_h}\right)^2 - 1\right\}}{4\frac{L_f}{d_h}} \tag{2.14}$$

where

$$c = 0.2\left\{\frac{\frac{B}{d_h} - 0.7}{\frac{L_f}{d_h}} + 0.9\right\}\left(1 - e^{-1.2\left(\frac{D_o}{d_h} - \frac{B}{d_h}\right)}\right) \tag{2.15}$$

The amount of k_f obtained by the above equation also tends to be higher than it actually is. It is presumed that the issue of insufficient accuracy is mainly caused by ignoring the effect of separation at the interface and by assuming uniform contact pressure distributions on the bearing surface.

Although Eq. (2.14) gives somewhat higher values for spring constant of fastened plates k_f than those determined by FEM, it has the advantage of easy calculation and produces a relatively small error for a certain range of joint configuration. Therefore, a practical calculation procedure for k_f is proposed [34], which can be applied to all three types of contact pressure distribution patterns, by combining Shibahara's equations with the elementary theory of solid mechanics. The outline of the method is introduced below, since it provides sufficient accuracy from the practical point of view.

Fig. 2.9 shows three distribution patterns of contact pressure in bolt-nut connections, redrawn referring to fig. 1.12C. Assuming that pressure cone angle θ_{cn}

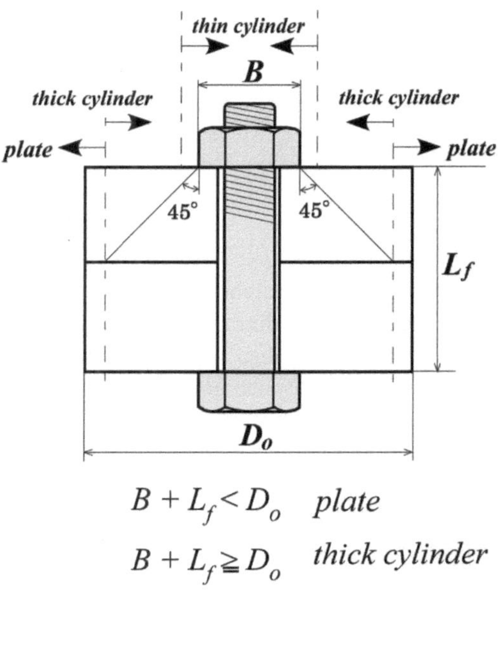

Fig. 2.9 Determination of clamping state by means of dimensions of bolt-nut connection.

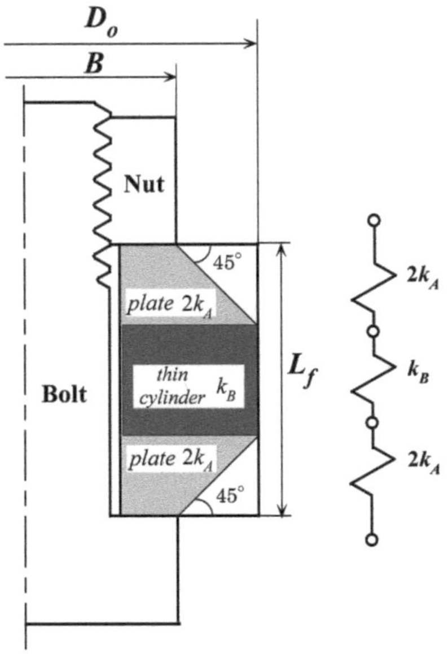

Fig. 2.10 An evaluation method of spring constant for thick cylinder.

is 45 degrees, the determination of clamping state between thick cylinder and plate can be made by the dimensions of mean diameter of nut B, grip length L_f, and outer diameter of fastened plates D_o. The case of the sum of B and L_f being smaller than D_o represents the state of plate and the case of the sum being larger than D_o reduces to the state of thick cylinder. As a result, Shibahara's equation, Eq. (2.14), is simply applied to the case of the plate. In the case of thick cylinder, dividing the fastened plates into three parts, as shown in Fig. 2.10, the aggregation of upper and lower truncated cones is regarded as plate and the central part is assumed to be thin cylinder under uniform compression. Therefore, calculating spring constants for plate k_A and thin cylinder k_B by use of Eqs. (2.14) and (2.11), the spring constant for thick cylinder k_f is evaluated by connecting k_A and k_B in series:

$$\frac{1}{k_f} = \frac{1}{k_A} + \frac{1}{k_B} \tag{2.16}$$

When D_0 is slightly larger than B, the spring constant of fastened plates k_f is simply calculated by Eq. (2.11) as a thin cylinder. The applicable limit of Eq. (2.11) is discussed in the next section by comparing to the numerical results achieved by FEM. The idea of dividing thick cylinder into thin cylinder and plate is basically the same as the method proposed in VDI 2230 (2003), in which fastened plates are divided into two truncated cones and a thin cylinder.

The values of k_{th} and k_{hd} are constant regardless of grip length L_f. Those of k_s and k_{cyl} decrease in inverse proportion to L_f under the condition of a constant ratio of L_s to L_{cyl}. Hence, the entire spring constant of bolt-nut connection, k_b, decreases as L_f is increased, and its reduction rate is slightly smaller than the increasing rate of L_f.

As a result, with increasing L_f, the mechanical behavior of bolted joints comes to be less affected by the evaluation accuracy in equivalent length of engaged threads L_{th}, shown in Fig. 2.8, since the influence of k_{th} on the entire stiffness gets smaller. In contrast, in case of clamping thin plates whose grip length is as much as nominal diameter d, it can be said that the amount of k_b is determined mostly by that of L_{th}, considering that L_{th} varies, possibly from $0.85d$ to $1.8d$, as shown in Fig. 2.8.

– Learning by numerical example 2.1: Spring constants of each part of a bolted joint

Two hollow cylinders of 16 mm thickness are clamped by a pair of bolt and nut with nominal diameter of M16. The outer diameter of hollow cylinder D_o is 28.8 mm, 1.8 times nominal diameter d, which corresponds to the state of the thin cylinder. The inner diameter is 17.5 mm, which is equal to bolt hole diameter d_h of second grade. Assuming equal lengths in L_s and L_{cyl}, the amounts of five spring constants, k_{th}, k_s, k_{cyl}, k_{hd}, and k_f, are calculated. Young's moduli of bolt, nut, and fastened plates are all 200 GPa. Diameter of stress area d_s is cited from Table 1.2. Unless otherwise stated, as for the shape of fastened plates, D_o is selected to be 1.8 times nominal diameter d and bolt hole diameter of second grade is used in the following numerical examples.

From the above condition, grip length L_f is 32 mm. L_{th} and L_{hd} are obtained by Eq. (2.9). Bolt cylindrical area A and stress area A_s are easily calculated by use of d and d_s:

$$
\begin{aligned}
&L_{th} = 0.85d = 13.6(\text{mm}), \quad L_s = 16(\text{mm}), \quad L_{cyl} = 16(\text{mm}),\\
&L_{hd} = 0.55d = 8.8(\text{mm})\\
&A = \frac{\pi}{4}d^2 = 201.06\left(\text{mm}^2\right), \quad A_S = \frac{\pi}{4}{d_S}^2 = 156.67\left(\text{mm}^2\right)
\end{aligned}
$$

Substituting the above values into Eq. (2.6), five spring constants are obtained as follows:

$$
\begin{aligned}
&k_{th} = 2.304 \times 10^6(\text{N/mm}), \quad k_s = 1.958 \times 10^6(\text{N/mm}),\\
&k_{cyl} = 2.513 \times 10^6(\text{N/mm}),\\
&k_{hd} = 4.570 \times 10^6(\text{N/mm}), \quad k_f = 2.568 \times 10^6(\text{N/mm})
\end{aligned}
$$

Then, the entire spring constant of bolt-nut connection k_b is calculated by Eq. (2.4):

$$k_b = 6.405 \times 10^5\ (\text{N/mm})$$

2.2.5 Evaluation of spring constants composing a bolted joint by FEM

Among five spring constants composing a bolted joint, those for unengaged threads k_s and bolt cylindrical portion k_{cyl}, shown in Eq. (2.6), are calculated using the elementary theory of solid mechanics. The other three spring constants, k_{th}, k_{hd}, and k_f, can be evaluated by axisymmetric FEM.

The amounts of L_{th} and L_{hd} obtained by FEM for evaluating k_{th} and k_{hd} are given in Fig. 2.7. In [38], k_f is evaluated by replacing engaged threads and bolt head with cylinders of the same form. In actual bolted joints, however, the stiffness of engaged threads is lower than that of the bolt head, as found from Eq. (2.9), which gives equivalent lengths of those portions. This means that if a finite element analysis is performed under the assumption of equal stiffness, it may produce nonnegligible errors in deformations and contact pressure distributions around the nut-bearing surface. In this section, therefore, a calculation method of spring constants of engaged threads and bolt head, k_{th} and k_{hd}, is presented by means of finite element models considering engaged thread geometry, and how to evaluate spring constant of fastened plates k_f is explained in detail.

(1) Spring constants of engaged threads and bolt head

Fig. 2.11 illustrates how to calculate spring constant of engaged threads. The finite element model is axi-symmetric and composed of a bolt-nut pair and fastened plates. The analytical procedure is as follows.

Step 1: Uniform displacement $\overline{u}$ is applied to the nodes located on the lower end of the bolt model. As the sum of the reaction forces on the nodes, the applied bolt force F_b is calculated.

Step 2: Dividing the bolt force F_b by uniform displacement $\overline{u}$, the spring constant k that connects k_{th}, k_s, k_{cyl}, and k_f in series is obtained:

$$k = \frac{F_b}{\overline{u}} \tag{2.17}$$

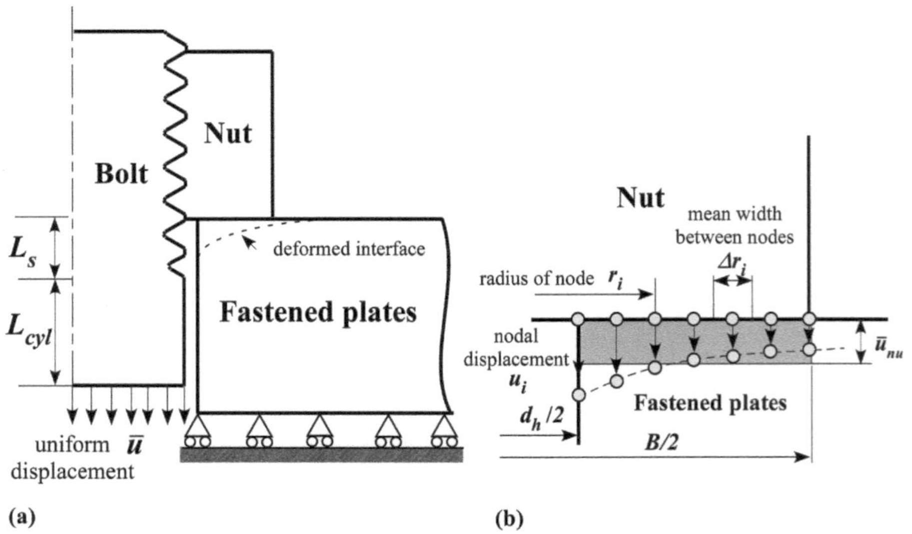

Fig. 2.11 Calculation method of spring constant of engaged threads. (A) Application of tensile force to bolt cylinder by uniform displacement and (B) mean displacement at nut bearing surface $\overline{u}_{nu}$.

$$\frac{1}{k} = \frac{1}{k_{th}} + \frac{1}{k_s} + \frac{1}{k_{cyl}} + \frac{1}{k_f} \tag{2.18}$$

Step 3: In Eq. (2.18), the amounts of k_s and k_{cyl} are calculated by the second and third equations in Eq. (2.6) using unengaged length L_s and length of bolt cylindrical portion L_{cyl} of the analytical model.

Step 4: The deformed volume of fastened plates under nut-bearing surface V, by compression force due to bolt force, is divided by its area A_n. The calculated value is designated as mean displacement under nut-bearing surface $\overline{u}_{nu}$. A_n represents the area of toroidal shape with outer diameter of B, nut mean diameter, and inner diameter of d_h, bolt hole diameter. V corresponds to the volume of the rectangular gray area under the nut-bearing surface shown in Fig. 2.11B. Other symbols are explained in the figure.

$$V = \sum 2\pi r_i \Delta r_i u_i \tag{2.19}$$

$$\overline{u}_{nu} = \frac{V}{A_n} = \frac{V}{\frac{\pi}{4}\left(B^2 - {d_h}^2\right)} \tag{2.20}$$

Step 5: Spring constant of fastened plates k_f is calculated by dividing bolt force F_b by $\overline{u}_{nu}$:

$$k_f = \frac{F_b}{\overline{u}_{nu}} \tag{2.21}$$

It should be noted that the value of k_f obtained by Eq. (2.21) corresponds to the spring constant of fastened plates whose grip length is twice as large as the analytical model, since the deformations of the fastened plates are completely constrained at the lower end.

Step 6: Substituting the values of k, k_s, k_{cyl}, and k_f into Eq. (2.18), the spring constant of engaged threads k_{th} is obtained.

Spring constant of bolt head k_{hd} is obtained in the same manner just by replacing engaged and unengaged thread portions of the analytical model, as depicted in Fig. 2.11, with cylinder-shaped bolt head and the adjacent bolt cylindrical portion.

(2) Spring constant of fastened plates

Spring constant of fastened plates k_f is obtained in Step 5 together with k_{th}, as explained above. However, when the shape of fastened plates is not in vertical symmetry, unlike the case of Fig. 2.11, the entire bolted joint model is required, as shown in Fig. 2.12. The analytical procedure is as follows:

Step 1: The target bolted joint is wholly modeled, including engaged thread portion and bolt head. A bolt model is divided into two parts at the cross-section sufficiently away from both nut- and bolt head-bearing surfaces, as illustrated in Fig. 2.12.

Step 2: Uniform displacements, $\overline{u}_{up}$ and $\overline{u}_{dw}$, are applied to the upper and lower surfaces exposed by cutting the bolt's cylindrical portion. The values of $\overline{u}_{up}$ and

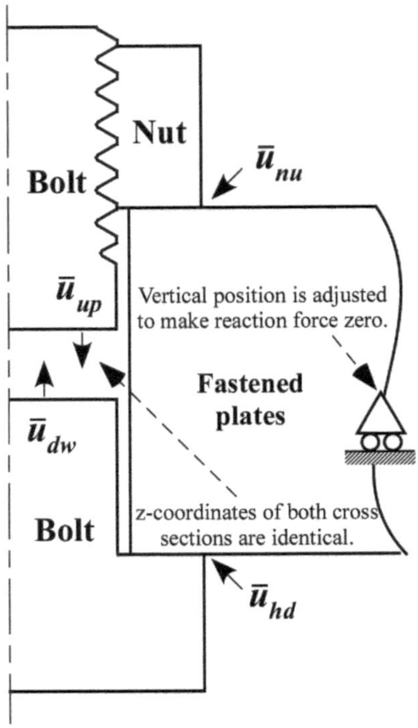

Fig. 2.12 Calculation method of spring constant of fastened plates.

$\bar{u}_{dw}$ are close, but not exactly the same. The two values are selected and adjusted to make the reaction force as close as possible to zero on the nodes placed around the center of the outer surface of fastened plates, at which axial displacements are constrained. This operation is a countermeasure to conform the mechanical property of the cut model introduced here to the noncut real model.

Step 3: Summing up the reaction forces on the nodes, to which $\bar{u}_{up}$ or $\bar{u}_{dw}$ is applied, bolt force F_b is calculated.

Step 4: Mean displacements under nut- and bolt-bearing surfaces, $\bar{u}_{nu}$ and $\bar{u}_{hd}$, are calculated following the same procedure shown in Fig. 2.11.

Step 5: Finally, spring constant of fastened plates k_f is obtained by dividing F_b by the sum of $\bar{u}_{nu}$ and $\bar{u}_{hd}$.

Numerical analyses are performed for M12 and M16 with coarse thread. Second grade of bolt hole diameter d_h is selected in both cases, i.e., 13.5 and 17.5 mm. Mean diameter of nut B is calculated to be 19.015 and 25.375 mm by Eq.(1.14). Grip length L_f and outer diameter of fastened plates D_o are systematically changed based on the value of nominal diameter d.

Fig. 2.13 represents the effect of D_o on k_f by changing the ratio of L_f to d as a parameter. k_f is given as a dimensionless quantity by dividing it by the product of nominal diameter d times Young's modulus of fastened plates E_f. The difference between M12 and M16 is quite small, which means that the spring constant of fastened plates k_f can be evaluated by Fig. 2.13 with sufficient accuracy, regardless of nominal diameter d.

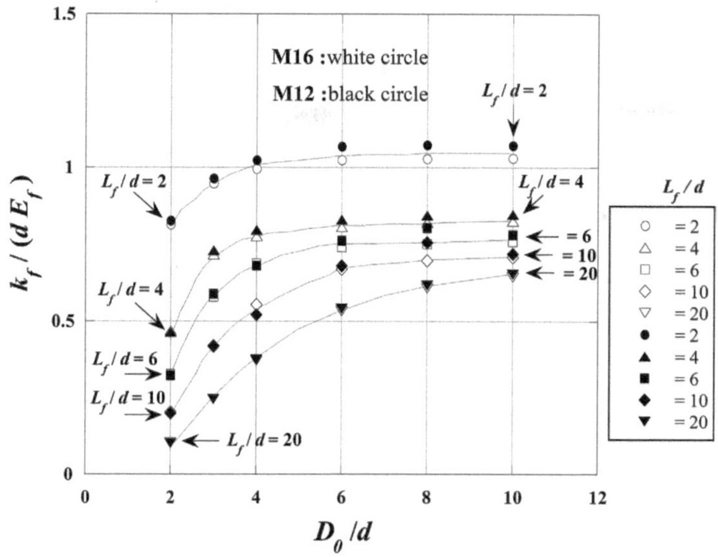

Fig. 2.13 Spring constant of fastened plates evaluated by FE analysis.

This result can be predicted from the statement given in Section 2.2.2 that the stiffness of similar bolted joints is almost proportional to nominal diameter d. k_f increases as D_o is increased and approaches a constant value beyond a certain value of D_o/d, at which the clamping state is regarded as plate. Its threshold value is decreased with decreasing grip length L_f. As for the effect of L_f, the reduction rate of k_f decreases with increasing L_f, as predicted from the formula for thin cylinder, Eq. (2.11), which shows that k_f varies inversely with L_f.

In Fig. 2.14, the amounts of k_f for M16 obtained by FEM are compared to those for thin cylinder calculated by Eq. (2.11). The minimum value of D_o/d used for Eq. (2.11) corresponds to the case that the outer diameter of the nut is almost equal to that of the fastened plates. It is found that the difference between the two results obtained by FEM and Eq. (2.11) shows a rapid increase from around the value of D_o/d exceeding 2. In the case of small grip length, i.e., $L_f/d = 2$, the difference is very large, even for $D_o/d = 2$. The essential reason for the above difference is that fastened plates deform into the shape of a barrel in actual practice, due to the action of friction on contact surfaces, which is found from the numerical results by FEM. In contrast, Eq. (2.11) assumes a completely uniform deformation of fastened plates without considering the effect of friction.

Meanwhile, in actual machines and structures, it is sometimes observed that a number of bolt-nut pairs are attached to the main body with high stiffness in the clamping state of thin cylinder. A typical example is the casing of a steam turbine. In such cases, the spring constant of fastened plates k_f becomes higher than the values given in Fig. 2.14, since the high stiffness of the main body prevents the deformation of the thin cylinder. Accordingly, in order to evaluate the spring constant in such cases with high accuracy, it is necessary to take into consideration how the thin cylinder is attached to the main body.

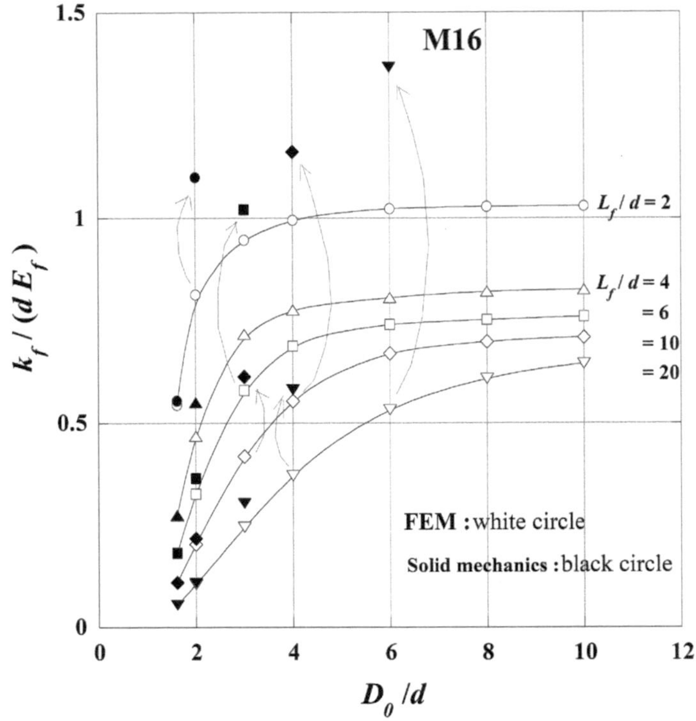

Fig. 2.14 Examination of accuracy of spring constant k_f evaluated by simple equation.

Machines and structures are generally clamped by a lot of bolts. When multiple bolts are uniformly spaced, as shown in Fig. 2.15, the spring constant of fastened plates k_f for a single bolt can be evaluated by constraining the displacements along the symmetric planes between neighboring bolts, which are shown by dashed lines in the figure. Fig. 2.16 shows the numerical results of k_f for the case of symmetry constraints. Comparing to the results without constraints given in Fig. 2.13, the values of k_f are considerably higher in the range of small D_o/d. Then, the difference between the two graphs gets smaller from around the value of D_o/d exceeding six, since the influence of symmetry gradually disappears.

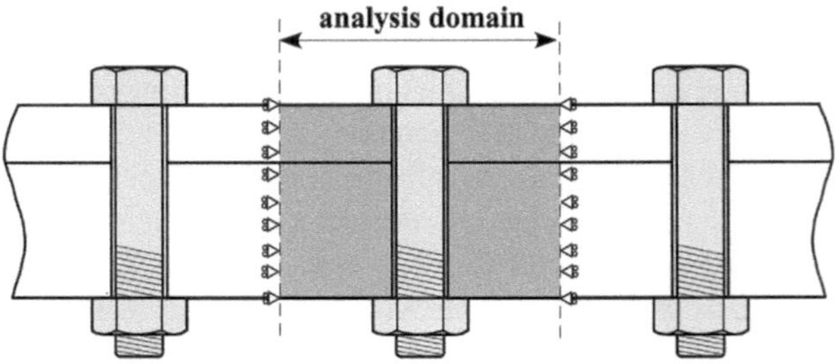

Fig. 2.15 Multiple bolts spaced equidistantly.

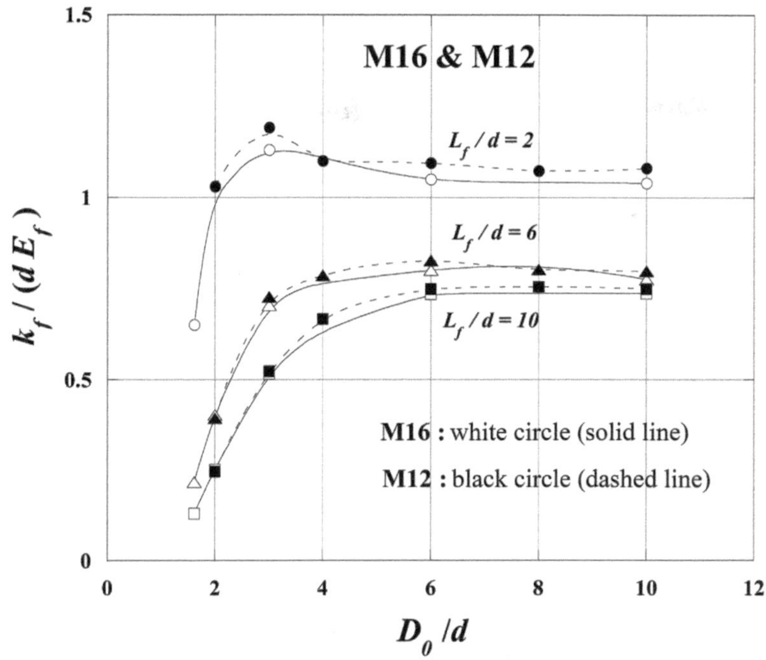

Fig. 2.16 Spring constant of fastened plates clamped by multiple bolts.

2.2.6 Relationship between mechanical behavior and spring constant of each part of a bolted joint

Among various tightening methods (which will be introduced in Chapter 3), the tightening accuracies of the elastic angle control method, direct tension method, and thermal expansion method are greatly influenced by the stiffness of each part of the bolted joint, in comparison with the torque control method. Based on the results discussed so far, the relationships between the spring constants of each part of the bolted joint and its mechanical behavior are summarized in the following.

(1) Among four spring constants composing a bolt-nut connection, the spring constant of engaged threads k_{th} is most changeable due to processing and assembly errors, as shown in Fig. 2.8.

(2) When clamping thin plates, the spring constant of engaged threads k_{th} may have a dominant effect on the mechanical behavior of the target bolted joint. This is because the spring constant of fastened plates k_f is very large and the equivalent length of engaged threads L_{th} is changeable due to the warping or distortion of fastened plates.

(3) It is rather difficult to precisely estimate the tightening characteristics, fatigue strength, and loosening phenomena of thin plates that involve the problem of warping or distortion, because the contact areas and relevant contact conditions vary with increasing bolt force and then k_f is significantly changed.

(4) When parts with low stiffness and hysteresis characteristics, such as gaskets, are involved in the fastened plates, their stiffness may have a dominant effect on the mechanical behavior of the whole joint. The hysteresis phenomena of gaskets are discussed in Chapter 7.

2.3 True profile of cross-section of screw threads

2.3.1 Cross-section of triangular screw threads [41]

Fig. 2.17 depicts the cross-section of triangular screw threads along the axis of the male thread, and the detailed geometry around the thread root is shown in Fig. 2.18A and B. In the figures, ρ is thread root radius and d_r is thread root diameter. Other symbols are defined in section 1.3.2. In Fig. 2.17, the axial position of the thread ridge with one pitch height is indicated by the circumferential angle of θ from $-\pi$ to $+\pi$, in which A–B, B–C, and C–D represent thread root, flank, and crest, respectively. The shape of the thread ridge is symmetric about $\theta = 0$.

The cross-section geometry of the above three parts can be expressed using the distance from the axis r:

$$r = \begin{cases} \dfrac{d}{2} - \dfrac{7}{8}H + 2\rho - \sqrt{\rho^2 - \dfrac{P^2}{4\pi^2}\theta^2} & (0 \le \theta \le \theta_1) \\ \dfrac{H}{\pi}\theta + \dfrac{d}{2} - \dfrac{7}{8}H & (\theta_1 \le \theta \le \theta_2) \\ \dfrac{d}{2} & (\theta_2 \le \theta \le \pi) \end{cases} \tag{2.23}$$

$$\theta_1 = \frac{\sqrt{3}\pi}{P}\rho, \quad \theta_2 = \frac{7}{8}\pi, \quad \rho \le \frac{\sqrt{3}}{12}P, \quad H = \frac{\sqrt{3}}{2}P$$

The geometry of female threads can be expressed in the same manner:

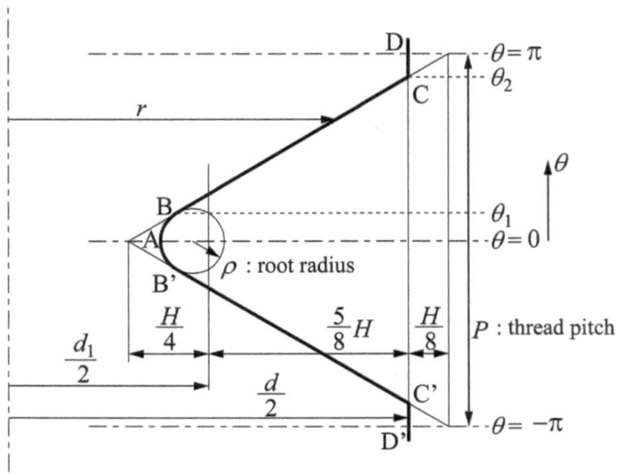

Fig. 2.17 Cross-section of triangular screw threads along the male thread axis.

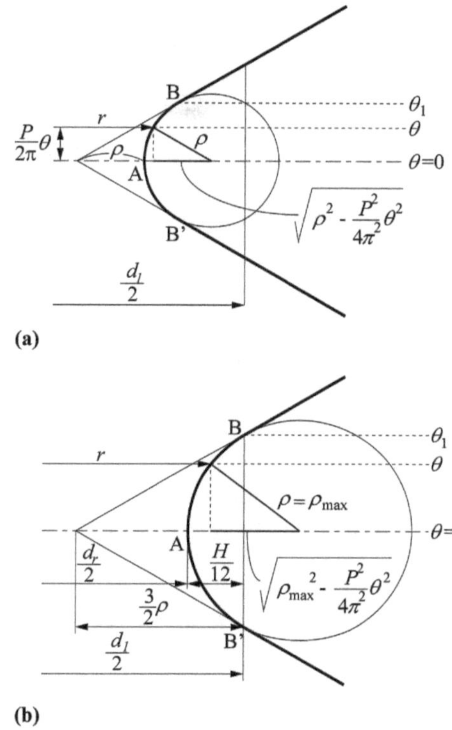

Fig. 2.18 Detailed geometry around the thread root. (A) Details of thread root geometry and (B) in case of $\rho = \rho_{\max}$.

$$r = \begin{cases} \dfrac{d_1}{2} & (0 \le \theta \le \theta_1) \\ \dfrac{H}{\pi}\theta + \dfrac{d}{2} - \dfrac{7}{8}H & (\theta_1 \le \theta \le \theta_2) \\ \dfrac{d}{2} + \dfrac{H}{8} - 2\rho_n + \sqrt{\rho_n{}^2 - \dfrac{P^2}{4\pi^2}(\pi - \theta)^2} & (\theta_2 \le \theta \le \pi) \end{cases} \tag{2.24}$$

$$\theta_1 = \frac{\pi}{4}, \quad \theta_2 = \pi\left(1 - \frac{\sqrt{3}\rho_n}{P}\right)$$

In Eq. (2.24), ρ_n represents the radius of the female thread root. Because of the geometric limitation, to avoid the interference between thread ridge crest and root radius, there exist maximum values for ρ and ρ_n, $\rho_{\max}$, and $\rho_{n|\max}$, shown below:

$$\begin{aligned} \rho_{\max} &= \frac{\sqrt{3}}{12}P \cong 0.1443 \times P \\ \rho_{n|\max} &= \frac{\sqrt{3}}{24}P \cong 0.07217 \times P \end{aligned} \tag{2.25}$$

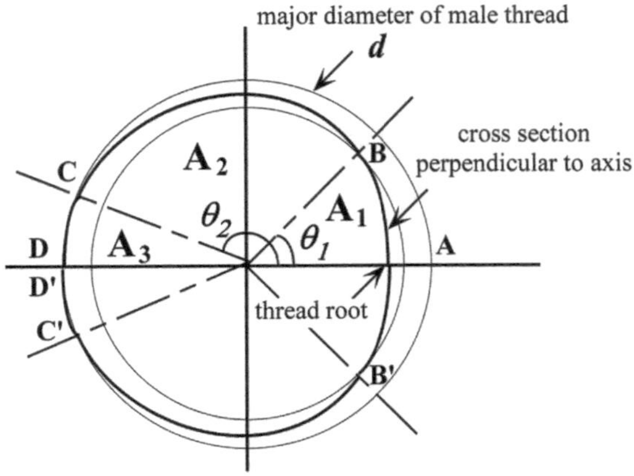

Fig. 2.19 Cross-section of male thread perpendicular to axis.

In Fig. 2.18B, the radius of the male thread root ρ is set to be $\rho_{\max}$. Considering the geometric symmetry of θ between from $-\pi$ to 0 and from 0 to $+\pi$, shown in Fig. 2.17, the contour line of male thread with one pitch height is expanded into a plane by use of Eq. (2.23). Fig. 2.19 shows the cross-section of male thread perpendicular to the axis. Contour lines of A–B, B–C, and C–D in the figure correspond to those in Fig. 2.17. A_1, A_2, and A_3 are areas of each part.

Illustrated in Fig. 2.20A are the cross-sections of male threads, perpendicular to the axis, of M3, M8, M12, M20, M30, M48, and M64 with coarse thread, in which Eq. (2.23) is used for drawing. The contour lines correspond to the upper half of Fig. 2.19. The radius of male thread root ρ is set to be $0.125P$. In the figure, the right end and left end of each contour line represent the thread root and crest of the thread ridge. Solid lines depict the contour of the thread cross-section and dashed lines show a semicircle whose diameter is equal to nominal diameter d. It turns out from the figures that the geometry of a true cross-section of triangular screw threads deviates from the circle whose diameter is equal to nominal diameter d, particularly around the thread root. The tendency is more remarkably observed as nominal diameter is decreased. This is because the ratio of thread pitch P to nominal diameter d, P/d, increases as d is decreased, which is explained in section 1.4.5. Fig. 2.20B illustrates the relationship between the thread pitch size and the contour of cross-section for fine screw threads of M20, M30, M48, and M64. Since the thread ridge size is small, the shape of the true cross-section is relatively close to the circle of diameter d, comparing to the case of coarse screw threads.

– Learning by numerical example 2.2: Maximum diameter of thread root

Maximum values of male thread root diameter d_r are calculated for M16 with coarse and fine threads. Substituting $\theta = 0$, $H = 0.8660P$ and the maximum thread root radius of $\rho = 0.1443P$ into the first equation of Eq. (2.23), the maximum value of d_r is found to be $(d - 1.227P)$. Since thread pitch P of coarse thread is 2 mm and those of fine threads are 1.5 and 1 mm, d_r is calculated to be 13.546 mm for coarse thread, 14.160 mm for $P = 1.5$ mm, and 14.773 mm for $P = 1$ mm, respectively.

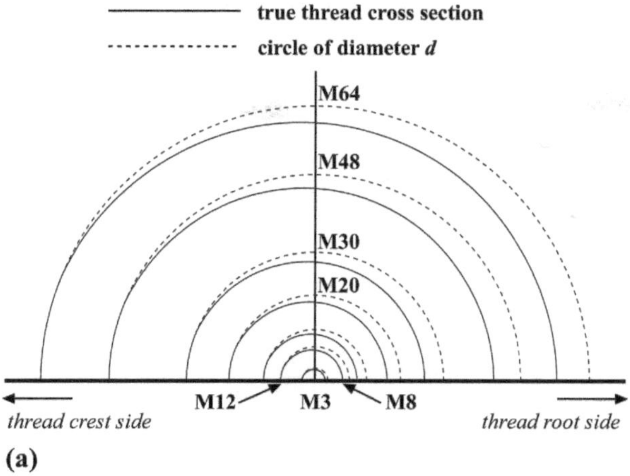

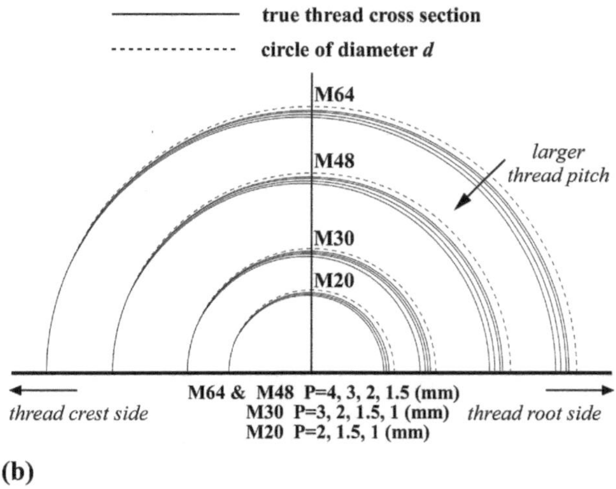

Fig. 2.20 True cross-section geometry of male thread perpendicular to axis. (A) Coarse screw thread and (B) fine screw thread.

2.3.2 Cross-section of screw threads of various shapes [42]

Using the distance from the axis r, cross-section geometry of various types of screw threads can be expressed by mathematical formulas. In the case of single-thread screw, the cross-section is symmetric about θ between from $-\pi$ to 0 and from 0 to $+\pi$. Cross-section of double-thread screw is symmetric between from $-\pi/2$ to 0 and from 0 to $+\pi/2$. The symmetry in triple-thread screw exists between from $-\pi/3$ to 0 and from 0 to $+\pi/3$. Accordingly, the symmetry of cross-section is comprehensively expressed by use of number of threads i. Supposing that the region from 0 to π/i is the basic profile, the same profile with line symmetry appears next to it. The two profiles form a pair, and this is repeated i times. For example, the cross-section geometry of male thread with triple-thread screw is expressed as follows:

$$r = \begin{cases} \dfrac{d}{2} - \dfrac{7}{8}H + 2\rho - \sqrt{\rho^2 - \left(\dfrac{3P}{2\pi}\right)^2 \theta^2} & (0 \le \theta \le \theta_1) \\ \dfrac{d}{2} - \dfrac{7}{8}H + \dfrac{3H}{\pi}\theta & (\theta_1 \le \theta \le \theta_2) \\ \dfrac{d}{2} & \left(\theta_2 \le \theta \le \dfrac{\pi}{3}\right) \\ \dfrac{d}{2} & \left(\dfrac{\pi}{3} \le \theta \le \theta_3\right) \\ \dfrac{d}{2} - \dfrac{7}{8}H + \dfrac{3H}{\pi}\left(\dfrac{2\pi}{3} - \theta\right) & (\theta_3 \le \theta \le \theta_4) \\ \dfrac{d}{2} - \dfrac{7}{8}H + 2\rho - \sqrt{\rho^2 - \left(\dfrac{3P}{2\pi}\right)^2 \left(\dfrac{2\pi}{3} - \theta\right)^2} & \left(\theta_4 \le \theta \le \dfrac{2\pi}{3}\right) \\ \dfrac{d}{2} - \dfrac{7}{8}H + 2\rho - \sqrt{\rho^2 - \left(\dfrac{3P}{2\pi}\right)^2 \left(\theta - \dfrac{2\pi}{3}\right)^2} & \left(\dfrac{2\pi}{3} \le \theta \le \theta_5\right) \\ \dfrac{d}{2} - \dfrac{7}{8}H + \dfrac{3H}{\pi}\left(\theta - \dfrac{2\pi}{3}\right) & (\theta_5 \le \theta \le \theta_6) \\ \dfrac{d}{2} & (\theta_6 \le \theta \le \pi) \end{cases} \tag{2.26}$$

The cross-section geometry of male thread of trapezoidal screw threads is expressed in like manner:

$$r = \begin{cases} \dfrac{d}{2} - \dfrac{P}{2} - \rho & (0 \le \theta \le \theta_1) \\ \dfrac{d}{2} - \dfrac{P}{2} - \sqrt{\rho^2 - \dfrac{P^2}{4\pi^2}(\theta - \theta_1)^2} & (\theta_1 \le \theta \le \theta_2) \\ \dfrac{d}{2} - \left(\dfrac{H}{2} + \dfrac{H_1}{2}\right) + \dfrac{H}{\pi}\theta & (\theta_2 \le \theta \le \theta_3) \\ \dfrac{d}{2} & (\theta_3 \le \theta \le \pi) \end{cases} \tag{2.27}$$

Cross-section geometry of female threads is also expressed in a similar manner. Cross-section geometries of multiple-thread screws, trapezoidal screw threads, and parallel pipe threads are discussed in detail in [42]. The formulas derived for parallel and taper pipe threads, whose thread angle is 55 degrees, are similar to those for triangular screw threads. The formulas for taper pipe threads are slightly more complicated because the threads are processed on the surface along one sixteenth of the taper [43].

Fig. 2.21A and B shows the cross-sections of male threads of triple-thread screw and trapezoidal screw threads in the same form as Fig. 2.20. The parameters in the figures are nominal diameter d. In the case of triple-thread screw, the region from 0 to $\pi/3$ is the basic profile and the region from $\pi/3$ to $2\pi/3$ shows its line symmetric profile. The cross-section of the trapezoidal screw thread looks a bit distorted due to its small thread angle of 30 degrees.

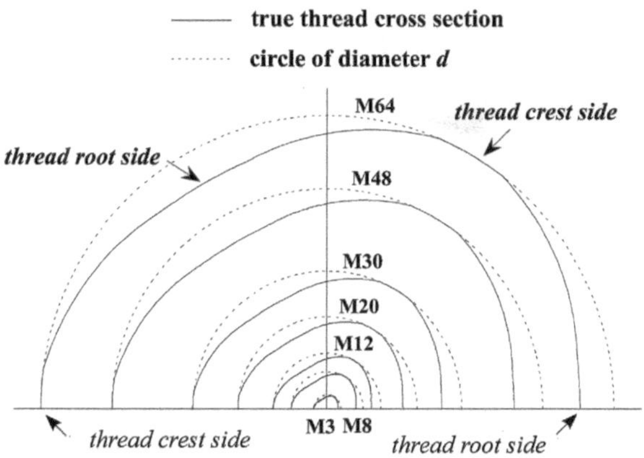

(a)

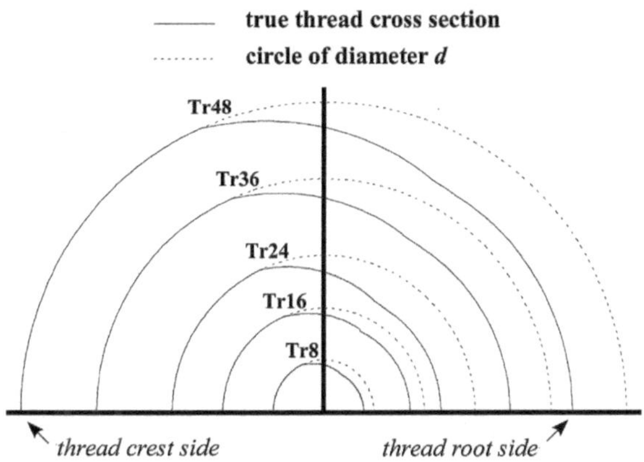

(b)

Fig. 2.21 True cross-section geometry of male threads of triple-thread screw and trapezoidal screw threads. (A) Triple-thread screw and (B) trapezoidal screw thread.

2.4 True cross-sectional area of screw threads

True cross-sectional area of male thread of triangular screw threads, A_e, is obtained as the sum of three areas, as shown in Fig. 2.19 [41]:

$$\begin{aligned} A_e &= 2 \times (A_1 + A_2 + A_3) \\ &= 2 \times \left(\int_0^{\theta_1} \int_0^r r dr d\theta + \int_{\theta_1}^{\theta_2} \int_0^r r dr d\theta + \frac{\pi d^2}{4} \times \frac{\pi - \theta_2}{2\pi} \right) \end{aligned} \tag{2.28}$$

The first and second terms can be integrated analytically. The resulting formulas are shown below:

$$A_e = \left(C_1{}^2 + C_2\right)\theta_1 - \frac{C_3}{3}\theta_1{}^3 - C_1\sqrt{C_3}\left(\theta_1\sqrt{\frac{C_2}{C_3} - \theta_1{}^2} + \frac{C_2}{C_3}\sin^{-1}\sqrt{\frac{C_3}{C_2}}\theta_1\right)$$
$$+ \frac{C_4{}^2}{3}\left(\theta_2{}^3 - \theta_1{}^3\right) + C_4C_5\left(\theta_2{}^2 - \theta_1{}^2\right) + C_5{}^2(\theta_2 - \theta_1) + \frac{d^2}{4}(\pi - \theta_2)$$
$$\text{where } C_1 = \frac{d}{2} - \frac{7}{8}H + 2\rho, \quad C_2 = \rho^2, \quad C_3 = \frac{P^2}{4\pi^2}, \quad C_4 = \frac{H}{\pi}, \quad C_5 = \frac{d}{2} - \frac{7}{8}H \tag{2.29}$$

Calculating true cross-sectional area A_e by Eq. (2.29) and supposing a circle whose area is equal to A_e, true pitch diameter d_e is defined as follows:

$$d_e = \sqrt{\frac{4A_e}{\pi}} \tag{2.30}$$

The true cross-sectional area of female threads can be calculated in a similar manner. First, the area of a hexagon that represents the nut shape is calculated. The true cross-sectional area is obtained by subtracting the area of spatial domain inside the female thread, A_{ins}, from the area of the hexagon. The resulting expression for A_{ins} is given below:

$$A_{ins} = \left(C_{1n}{}^2 + C_{2n}\right)(\pi - \theta_2) - \frac{C_3}{3}(\pi - \theta_2)^3$$
$$+ C_{1n}\sqrt{C_3}\left((\pi - \theta_2)\sqrt{\frac{C_{2n}}{C_3} - (\pi - \theta_2)^2} + \frac{C_{2n}}{C_3}\sin^{-1}\left(\sqrt{\frac{C_3}{C_{2n}}}(\pi - \theta_2)\right)\right)$$
$$+ \frac{C_4{}^2}{3}\left(\theta_2{}^3 - \theta_1{}^3\right) + C_4C_5\left(\theta_2{}^2 - \theta_1{}^2\right) + C_5{}^2(\theta_2 - \theta_1) + \frac{d_1{}^2}{4}\theta_1$$
$$\text{where } C_{1n} = \frac{d}{2} + \frac{H}{8} - 2\rho_n, \quad C_{2n} = \rho_n{}^2 \tag{2.31}$$

Three constants, C_3, C_4, and C_5, appearing in Eq. (2.31) are common with Eq. (2.29). Eq. (2.31) represents A_{ins} for triangular screw threads with single-thread screw. The expressions of A_{ins} for other types of screw threads are given in [42].

In Table 2.1, some calculation examples are given for true cross-sectional area A_e and true pitch diameter d_e of metric coarse threads [41]. For comparison, the diameter of stress area d_s is shown, and the differences between d_s and d_e and between pitch dimeter d_2 and d_e are also evaluated by dividing each difference by d_e. It is found that d_e is about 3%–10% larger than d_s, and the tendency becomes more remarkable as nominal diameter decreases. Regarding pitch diameter d_2, it is naturally closer to d_e comparing to d_s, and d_e is about 0.2%–1.1% larger than d_2. The difference between d_e and d_2 increases as nominal diameter decreases. True cross-sectional area A_e of fine threads, though not shown here, is larger than that of coarse threads because of its small pitch. The difference between d_e and d_s in fine threads is about 0.7%–4.4%.

Table 2.1 True cross-sectional area and true pitch diameter of metric coarse threads.

Coarse thread	M1	M1.2	M1.6	M2	M2.5	M3	M4	M5	M6	M8	M10
A_e (mm^2)	0.564	0.859	1.505	2.412	3.872	5.677	9.980	15.911	22.708	40.939	64.504
d_e (mm)	0.847	1.046	1.384	1.753	2.220	2.689	3.565	4.501	5.377	7.220	9.063
d_S (mm)	0.765	0.965	1.272	1.625	2.078	2.531	3.343	4.249	5.062	6.827	8.593
$(d_s - d_e)/d_e$ (%)	−9.633	−7.667	−8.143	−7.294	−6.424	−5.864	−6.212	−5.589	−5.864	−5.436	−5.184
$(d_2 - d_e)/d_e$ (%)	−1.113	−0.765	−0.844	−0.705	−0.574	−0.495	−0.543	−0.458	−0.495	−0.438	−0.406
Coarse thread	**M12**	**M16**	**M20**	**M24**	**M30**	**M36**	**M42**	**M48**	**M56**	**M64**	
A_e (mm^2)	93.406	170.735	266.773	384.153	606.926	880.441	1204.697	1579.696	2167.137	2847.287	
d_e (mm)	10.905	14.744	18.430	22.116	27.799	33.482	39.165	44.848	52.529	60.210	
d_s (mm)	10.358	14.124	17.655	21.185	26.716	32.247	37.778	43.309	50.840	58.371	
$(d_S - d_e)/d_e$ (%)	−5.018	−4.208	−4.208	−4.208	−3.893	−3.687	−3.540	−3.431	−3.215	−3.055	
$(d_2 - d_e)/d_e$ (%)	−0.386	−0.292	−0.292	−0.292	−0.259	−0.238	−0.223	−0.213	−0.193	−0.178	

In addition, since the difference between d_e and d_2 is very small, they can be regarded as almost equal in fine threads with large nominal diameter. True pitch diameter d_e can be expressed as a function of nominal diameter d and thread pitch P with the help of the least squares method:

$$d_e \approx 0.9985d - 0.6128P \tag{2.32}$$

True cross-sectional area A_e and true pitch diameter d_e of unified threads are calculated in [41]. In [42], theoretical formulas are derived in a similar manner for calculating true cross-sectional areas of triangular screw threads with multiple-thread screw, trapezoidal screw threads, and pipe threads.

2.5 Finite element models with helical shape of screw threads

2.5.1 Previous modeling schemes of helical thread model

A variety of mechanical properties of bolted joints are traced to the helical geometry of screw threads. Therefore, in order to rigorously analyze the mechanical behavior, it is necessary to introduce finite element models of bolted joints that take the effect of helical thread geometry into consideration. For example, it is vital to consider the effect of lead angle for analyzing the tightening process of the torque control method. On the other hand, when evaluating the stress concentration at the thread root, load distributions along engaged threads, and joint stiffness of a bolted joint under tensile force, helical thread geometry can be replaced by axisymmetric abacus-shaped FE models, as shown in Fig. 2.22A. Because of using two-dimensional mesh patterns, axisymmetric analysis by FEM provides high computational efficiency. Axisymmetric models also work well to evaluate the stress amplitude occurring at the thread root when subjected to axisymmetric repeated loading. In contrast, to analyze the tightening process or loosening mechanism of bolted joints, three-dimensional models considering helical thread geometry are required, as shown in Fig. 2.22B.

Meanwhile, when subjected to asymmetric external forces, FE models with axisymmetric thread geometry, as shown in Fig. 2.22A, are of practical use except for the case that the research objective is loosening of threaded fasteners with rotation.

Use of a helical thread model is essential to analyze the loosening phenomena and the tightening process by the torque control method. Some research papers have been

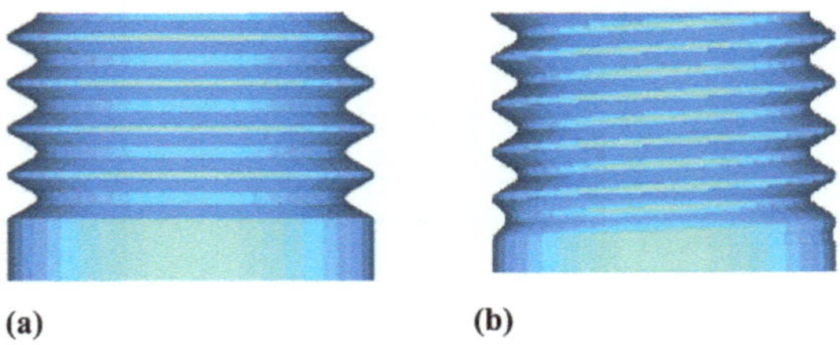

Fig. 2.22 Axisymmetric thread model and helical thread model. (A) Axi-symmetric model and (B) helical thread model.

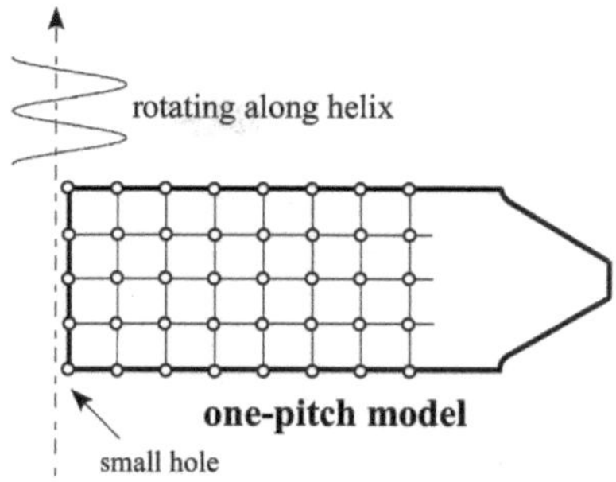

Fig. 2.23 Construction of helical thread model by rotating one-pitch thread cross-section along the helix.

published on the study of loosening mechanisms, such that fastened plates are subjected to repeated shear loading [44–46]. Concerning the tightening process, a notable numerical approach is proposed [47,48], by which the tightening process of the torque control method can be analyzed as elastic or elastic-plastic problems using two-dimensional mesh patterns with each node having three degrees of freedom, instead of using helical thread models. Modeling methods proposed so far that take account of helical thread geometry are classified as follows.

Method-1

Thread cross-section with one pitch height is modeled, which is placed only slightly apart from the axis, as shown in Fig. 2.23. The one-pitch model is rotated along the helix to construct a helical thread model. Although a small hole is produced around the axis, it has a negligible influence on the related mechanical behavior.

Method-2

The center portion of the engaged threads is modeled as a cylinder. The thread portion is separately modeled in the same manner as Method-1, in which the inner diameter of the thread portion is equal to the outer diameter of the above-mentioned cylinder. Then, the two models are combined with the help of a function provided by commercial FEM software. This method is relatively widely used due to its rather simple procedure, although the mesh patterns do not match at the interface between the cylinder and the helical thread model.

Method-3

A solid model considering helical thread geometry is constructed for the entire threaded fastener. The inside of the solid model is properly meshed using a function of automatic mesh generation provided by FEM software.

Method-4

True profiles of cross-sections of screw threads, shown in Figs. 2.20 and 2.21, are meshed using two-dimensional finite elements. The point of this method is to repeat the stacking and rotation of the objective true cross-section, after which the corresponding nodes are connected. Detailed explanation of the procedure is given later.

In Method-1, finite element models to be constructed are inevitably distorted around the far end of the bolt thread and the region that connects thread runout and bolt cylindrical portion. This also becomes a problem in Method-2.

Problems to be tackled when employing Method-1, -2, and -3 are discussed in more detail below. Regarding the meshing of the bolt thread portion, by adding an appropriate number of elements to both sides of the engaged threads to form unengaged threads, the engaged thread portion that governs the joint strength could be meshed into a proper pattern. In contrast, the meshing of the nut thread portion involves a more serious problem. To rigorously model the helical geometry around the nut-bearing surface, where female threads gradually disappear, finite elements with extreme aspect ratio need to be used, which may lead to poor analytical accuracy. Method-3 can accurately reproduce the helical thread geometry using a function of automatic mesh generation. However, it is quite difficult to place small elements, aiming at high analytical accuracy, around the areas under high stress concentration.

For the above reasons, it is considered that helical thread models produced by Method-1, Method-2, and Method-3 cannot necessarily provide sufficient accuracy for the purpose of evaluating the stress concentration at thread root and the distributions of stress and stress amplitude along the thread root on a practical level. In the next section, a modeling scheme of Method-4 is introduced, which can accurately reproduce helical thread geometry by constructing one-pitch thread models successively.

2.5.2 Helical thread modeling using mathematical expressions of cross-section [49]

Utilizing the characteristic that cross-section geometry of screw threads perpendicular to bolt axis is identical at any position, it is possible to construct helical thread models for FE analysis in units of one pitch height, by which fine meshes can be placed around the region under high stress concentration. Using the expressions derived in Section 2.3 that represent true cross-section geometry of screw threads, FE models can be constructed for various types of screw threads, in addition to triangular and trapezoidal screw threads. In the following, it is explained how to model male threads with one pitch height, in which one thread ridge is divided into n_p segments of equal size in the axial direction.

Step 1: Cross-section of male thread with accurate geometry, as shown in Fig. 2.24A, is properly meshed using two-dimensional elements. This is a basic mesh model.

Step 2: The basic model is placed at the reference position, $z = 0$.

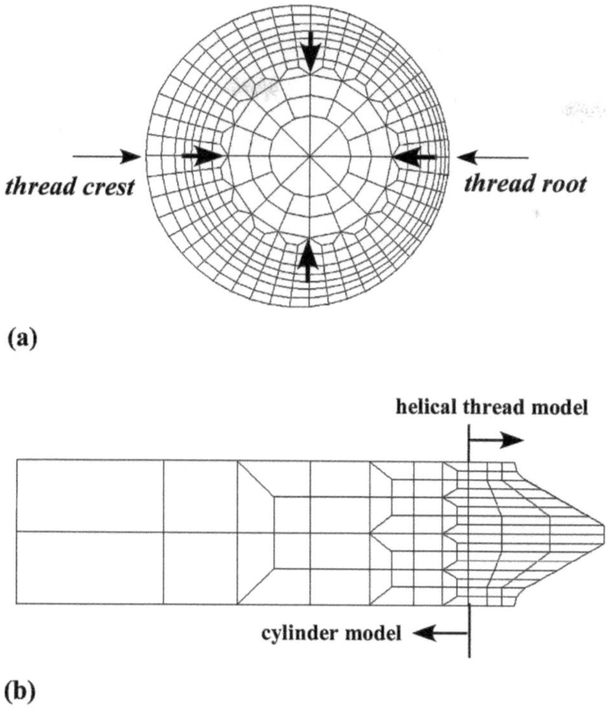

Fig. 2.24 Mesh patterns of cross-sections perpendicular to and along the axis. (A) True cross section perpendicular to axis and (B) one-pitch model of cross section along axis.

Step 3: Rotating the basic model anticlockwise by an amount of $2\pi/n_p$, it is placed at the position of $z = P/n_p$.

Step 4: Connecting the corresponding nodes of the two basic models placed at $z = 0$ and $z = P/n_p$, a three-dimensional model with P/n_p thickness is obtained.

Step 5: The one-pitch helical thread model is completed by repeating Steps 3 and 4 n_p times.

Step 6: An appropriate number of one-pitch models obtained in Step 5 are stacked, according to the number of threads of the objective bolted joint.

If the number of circumferential divisions in the basic mesh model is set to n_p, coarse meshes are inevitably produced around the thread root and crest. As a result, finite elements with extreme aspect ratio, whose side length in the circumferential direction is considerably longer than the other, are to be used, thereby the analytical accuracy being significantly lowered. To cope with the problem, the outside area around the thread portion is modeled by fairly fine meshes and the bolt core portion is simply modeled as a cylinder, from the viewpoints of numerical accuracy and computational efficiency.

The inside area, indicated by an arrow in Fig. 2.24B, corresponds to the bolt core portion, and it is modeled as a cylinder with one pitch height. Then, the one-pitch helical thread model is bonded outside the cylinder. In this case, the mesh patterns are

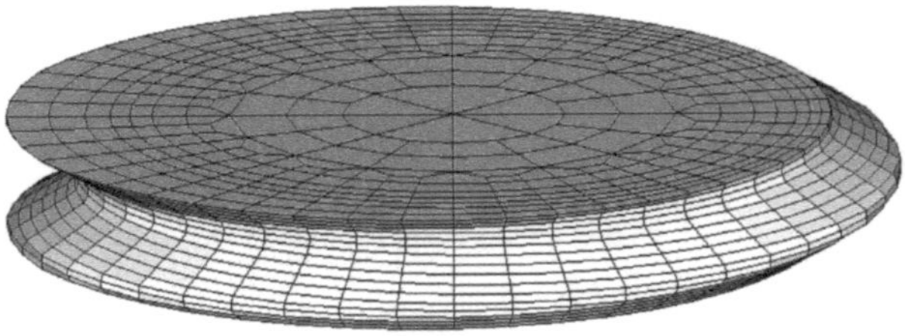

Fig. 2.25 One-pitch helical thread model of male thread.

perfectly matched between the outer surface of the cylinder and the inner surface of the helical thread model. Four arrows in Fig. 2.24A indicate the border between the two models. Fig. 2.25 shows a one-pitch helical thread model of male thread constructed according to the above procedure.

Helical thread models of female threads are constructed in the same manner. Using the method proposed here, well-balanced FE models can be formed, because finite elements with proper aspect ratio can be placed around the nut-bearing surface. For the modeling of thread runout, the depth of thread root of the helical thread model is gradually decreased so as to coincide with the real shape, and then it is connected smoothly to the bolt cylindrical portion.

Fig. 2.26A shows an example of an entire bolted joint model. The nominal diameter is M16 with coarse thread. Fig. 2.26B depicts the mesh pattern of the nut model. It is

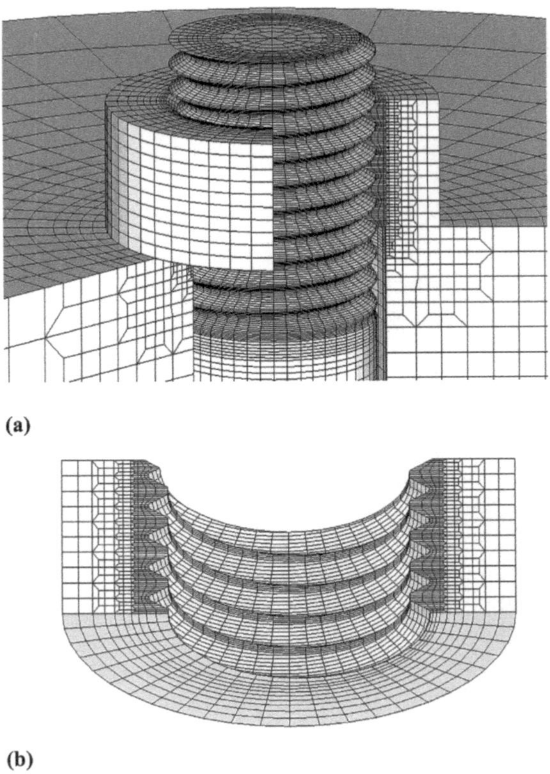

Fig. 2.26 Finite element model of an entire bolted joint and nut model. (A) Entire bolted joint model and (B) mesh pattern of nut model.

(a)

(b)

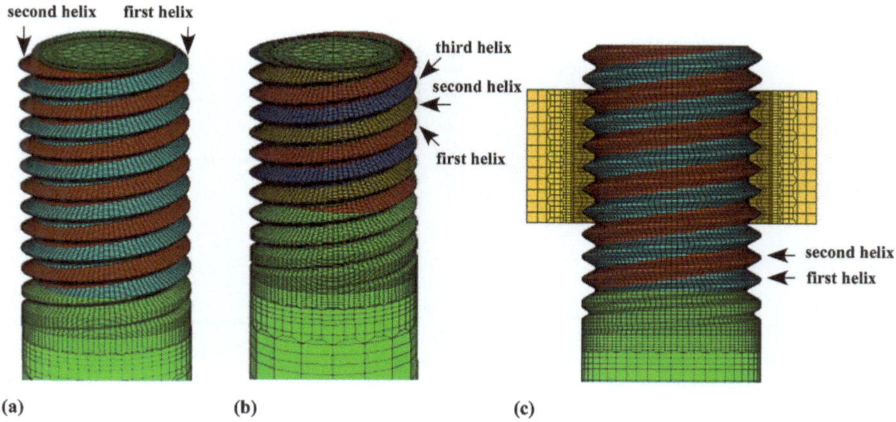

Fig. 2.27 Helical thread models of double- and triple-thread screws. (A) Double-thread screw, (B) triple-thread screw, and (C) cross section of double-thread screw.

found that the heights of the first ridge at two points, 180 degrees apart from one another, are one pitch and a half pitch, which indicates that helical thread geometry is accurately modeled. Although the nut is simply modeled as a round-shaped nut in the figure, it is possible to form a hexagonal nut. In the case of a multiple-thread screw with number of threads i, a helical thread model corresponding to one lead is completed using the height equal to the product of i times thread pitch P. Finite element models of bolts with double-thread and triple-thread screws [42] are shown in Fig. 2.27A and B, respectively. Fig. 2.27C represents the mesh pattern of the cross-section of double-thread screw along the axis. The two positions of thread ridge, 180 degrees apart from one another, of single-thread screw are deviated by a half pitch. In the case of double-thread screw, however, left and right ridges of the first and the second helices are symmetric about the axis. Although a bit difficult to discriminate, it is found from the figures that two helices exist in double-thread screw and three helices exist in triple-thread screw in the same cross-section perpendicular to the axis. Helical thread models of body-side engaged threads [20] and parallel and taper pipe threads [43,50] can be constructed in the same manner, by which their specific mechanical behaviors have been analyzed and reported.

2.6 Interface stiffness in bolted joints

2.6.1 Interface stiffness at mating surfaces

Microscopically, a limitless number of microprojections exist on surfaces even if they appear to be flat. Such surface conditions are commonly evaluated by surface roughness and waviness. In Fig. 2.28A, the relationship between compression force and displacement is shown for the cases of a single cylinder and two cylinders with total height being equal to the former, both of which are under compression [51]. When two cylinders are pressed together, the microprojections existing on the surface, whose stiffness is much

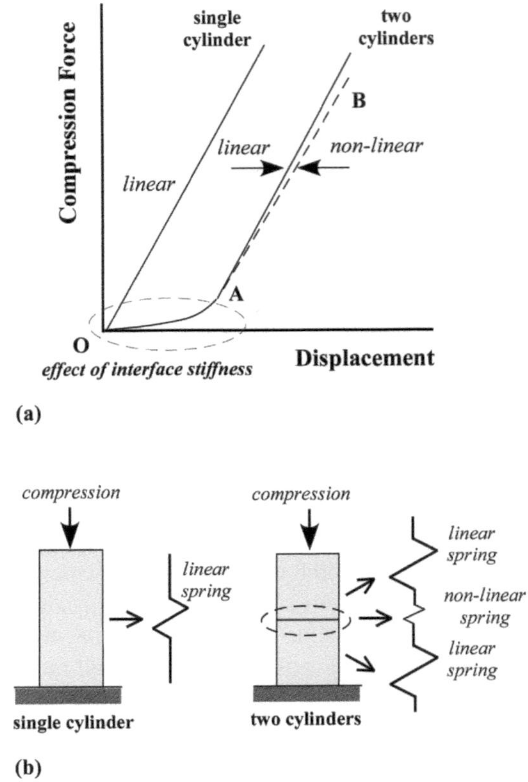

Fig. 2.28 Nonlinear behavior occurring in cylinders under compression due to interface stiffness. (A) Compression characteristics of single cylinder and two cylinders and (B) non-linear behavior due to interface stiffness.

lower comparing to the cylinder body, deform plastically even under low contact pressure, thereby exhibiting the nonlinear behavior shown in the figure. The stiffness due to the plastic deformation of microprojections is called interface stiffness.

Surface roughness greatly affects the interface stiffness. The nonlinearity between compression force and displacement becomes more noticeable as surface roughness increases. In the case of two cylinders being pressed, almost only the deformation of microprojections becomes an issue in the region O–A under low contact pressure. In the region A–B, as the compression force increases, the displacement of the cylinder body gradually governs the total deformation. Since the cylinder body deforms according to Hooke's law, compression force is almost linearly related to displacement in the region A–B. Fig. 2.28B represents a single cylinder and two cylinders pressed together by spring models. In the latter, interface stiffness is represented by a nonlinear spring.

Fig. 2.29 shows an enlarged view around the interface. As the compression force is increased from the initial condition, microprojections deform so as to get the mating surfaces closer. The amount of the deformation is designated the approach of interface herein. Fig. 2.30 shows a typical relationship between approach of interface ζ and contact pressure p_n. Ostrovskii et al. proposed the following equation between ζ (μm) and p_n (MPa) [52]:

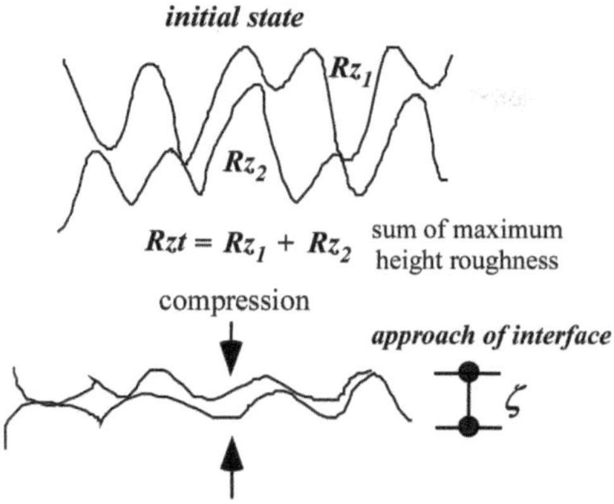

Fig. 2.29 Deformations of microprojections around the interface.

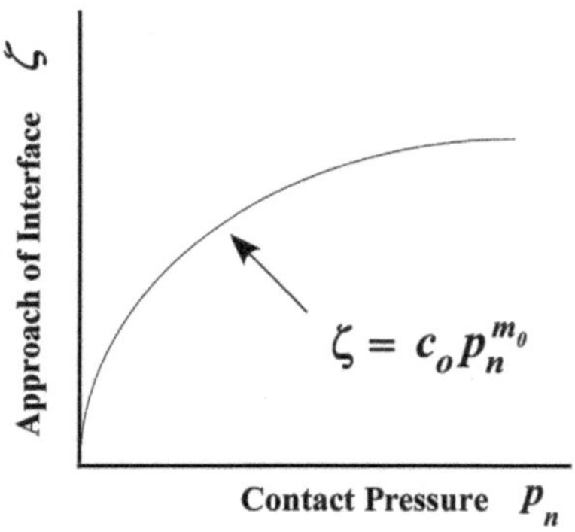

Fig. 2.30 Relationship between approach of interface and contact pressure.

$$\zeta = c_0 {p_n}^{m_0} \tag{2.33}$$

where c_0 and m_0 are constants.

Taniguchi et al. evaluated c_0 and m_0 by systematic experiments, and the values were related to surface roughness [53]. Applying the least squares method to the experimental results, the values of c_0 and m_0 can be related to the sum of maximum height roughness Rz of the mating surfaces, Rzt (μm) [54]:

$$\begin{aligned} c_o &= 0.0674Rzt + 0.413 \\ m_o &= 0.0155Rzt + 0.155 \end{aligned} \tag{2.34}$$

Back et al. [55] proposed the maximum value of m_0 as 0.5. There are four kinds of contact surfaces in a bolted joint, i.e., thread surface, nut-bearing surface, bolt head-bearing surface, and plate interface. Therefore, measuring the surface roughness of each contact surface, c_0 and m_0 can be calculated by Eq. (2.34). Substituting the values of c_0 and m_0 into Eq. (2.33), it becomes possible to evaluate the amount of approach of interface ζ as a function of contact pressure p_n. It is worth noting here that interface stiffness influences the accuracy of various tightening methods, discussed in Chapter 3, and it is also the major cause for the loosening phenomena without rotation of threaded fasteners.

2.6.2 Interface stiffness in normal and tangential directions [56]

Using Eq. (2.33), spring constant in the normal direction k_n, due to interface stiffness, can be expressed by means of normal force F_n acting on the contact surface and contact area A_{cn}:

$$k_n = \frac{dF_n}{d\zeta} = \frac{A_{cn}}{c_o m_o} p_n^{\,1-m_o} \tag{2.35}$$

Spring constant in the tangential direction k_t can be related to k_n by introducing the equation proposed by Kirsanova and Back. Kirsanova suggested that the displacement in the tangential direction ζ_t is expressed as the product of the stress p_t times the compliance $1/k_t$ [57], both of which are in the tangential direction:

$$\zeta_t = \frac{p_t}{k_t} \tag{2.36}$$

Back expressed the relationship between k_t and p_n by using two constants, R and S, determined by the kind of material [55]:

$$\frac{1}{k_t} = \frac{R}{p_n^{\,S}} \tag{2.37}$$

The equation for spring constant in the tangential direction k_t is derived by differentiating p_t with respect to ζ_t and then multiplying it by contact area A_{cn}. Then, following the above procedure, k_t is calculated by use of Eqs. (2.36) and (2.37). Assuming ζ_t is independent of p_n, the equation for k_t is derived by calculating the ratio of k_t to k_n:

$$k_t = \frac{c_o m_o}{R} p_n^{\,m_o+S-1} \times k_n \tag{2.38}$$

Accordingly, it is summarized that the amount of k_t can be calculated by Eq. (2.38) once k_n and p_n and four constants of c_0, m_0, R, and S are given. Back insisted that S can be regarded as 0.5 and R is calculated by the following equation:

$$\frac{R}{c_o m_o} = 2(1 + \nu) \tag{2.39}$$

where ν denotes Poisson's ratio.

As a result, it is concluded that k_n and k_t can be calculated when three factors are given, i.e., the sum of maximum height roughness Rz, contact pressure p_n, and contact area A_{cn}.

2.6.3 Simple formula for evaluating interface stiffness in normal direction

Interface stiffness influences the mechanical behavior of bolted joints in various ways. For instance, in tightening by the direct tension method using hydraulic tensioners, interface stiffness in the normal direction becomes an issue. In those cases, contact pressures at thread surface and nut-bearing surface, which affect the tightening characteristics, are very high; that is, Eq. (2.33) is to be used in the range where the variation of approach of interface ζ is rather small against contact pressure p_n. Considering the above situation, it is possible to estimate the amount of k_n with practical accuracy under the condition of high contact pressure, by using a relatively simple method. Taniguchi et al. obtained the experimental result that ζ is almost linearly related to p_n when exceeding a certain value [53]:

$$\zeta = b_1 p_n + b_2 \tag{2.40}$$

According to their experimental results, constant b_1 is very small and constant b_2 depends largely on the maximum value of approach of interface $\zeta_{\max}$, and the latter constant increases as the sum of maximum height roughness Rzt is increased. Referring to the above experimental result, the relationship between $\zeta_{\max}$ and Rzt is approximated as follows:

$$\zeta_{\max} = 0.25Rzt \tag{2.41}$$

Using Eq. (2.41), interface stiffness at the thread surface, nut-bearing surface, and bolt head-bearing surface can be evaluated by the spring constant K_{cn}:

$$K_{cn} = \frac{F_b}{\zeta_{\max}} \tag{2.42}$$

where F_b is bolt force exerting on each contact surface.

Accordingly, K_{cn} is a nonlinear spring constant that varies depending on F_b. Spring constants of each contact surface are denoted as K_{th}, K_{nu}, and K_{hd}, which are identified by subscripts. At the interface between fastened plates, contact pressure decreases radially outwards from the periphery of bolt hole, as shown in Fig.1.12. The mean approach of interface can be calculated by using the deformed volume around the plate surface subjected to contact pressure. Then, spring constant K_f of plate interface, due to interface stiffness, is calculated [36] just by substituting the mean approach of interface into $\zeta_{\max}$ in Eq. (2.42).

2.7 Thermal contact resistance in bolted joints

When a bolted joint is subjected to thermal load, axial bolt force varies more or less because of the difference in the elongation or shrinkage between threaded fasteners and fastened plates. Factors that significantly influence the bolt force variation are joint geometry, temperature distributions, coefficient of linear expansion, thermal conductivity, elastic moduli of joint materials, etc. Contact surfaces existing in a bolted joint sometimes significantly affect the temperature distributions in the joint, because they play a role in thermal resistance. Accordingly, in order to evaluate the temperature distributions with high accuracy, it is necessary to estimate the thermal resistance as accurately as possible at four kinds of contact surfaces existing in a bolted joint, i.e., thread surface, nut-bearing surface, bolt head-bearing surface, and plate interface. Furthermore, there exist small clearances between the bolt cylindrical portion and bolt hole surface of fastened plates and around the clearance flank of engaged threads, where a certain amount of heat is transmitted through a thin air layer. Thermal conductivity of air is very small and is about one two-thousandth of carbon steel. However, a fair amount of heat flows between the mating surfaces through small gaps by the action of heat conduction, convection, and radiation.

Thermal resistance due to the existence of contact surface is termed thermal contact resistance. Fig. 2.31 schematically depicts the temperature distribution around the interface in steady state, in which the heat flows from left to right and thermal conductivity of the left material is lower than the right one. Around the interface consisting of two contacting bodies, complex temperature distributions are produced owing to surface asperities and waviness. For simplicity, assuming that there exists a temperature difference of ΔT at the interface, thermal contact resistance R_{cn} is expressed by the following equation:

$$R_{cn} = \frac{\Delta T}{q} = \frac{1}{h_c} \tag{2.43}$$

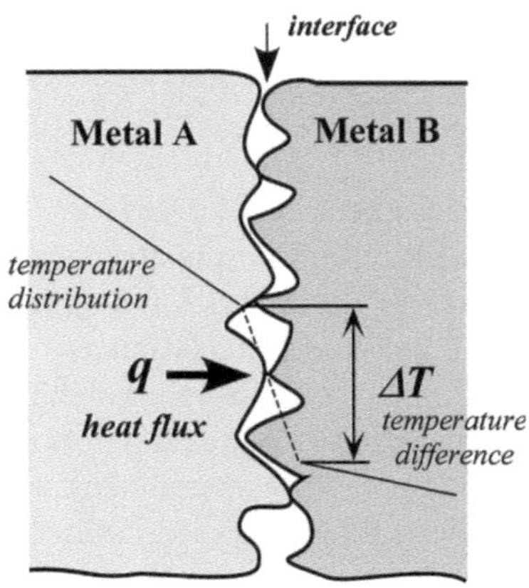

Fig. 2.31 Temperature distribution and thermal contact resistance around the interface.

where q (W/m^2) represents heat flux, and h_c (W/m^{2}K) is the reciprocal of R_{cn}, called the thermal contact coefficient, which can be treated like the coefficient of heat transfer from the viewpoint of heat transfer engineering.

In Fig. 2.31, the temperature gradient in the left body is larger, due to its lower thermal conductivity, than the right one. The thermal contact coefficient is influenced by such factors as contact pressure, surface roughness, thermal conductivity, and hardness of the materials composing the interface.

The mechanism of heat transfer is classified into heat conduction, convection, and radiation. Heat exchange between two surfaces separated by air is usually neglected when there exists a certain amount of gap. However, in the case of very small gaps, e.g., less than a few millimeters, it is empirically known that a certain amount of heat flows through. Heat flow through a small gap can be quantitatively evaluated by apparent thermal contact coefficient h_e, defined below:

$$h_e = \frac{q}{\Delta T_e} \tag{2.44}$$

where ΔT_e represents the temperature difference between the mating surfaces.

Although changing somewhat with nominal diameter, the gap size around the clearance flank of engaged threads is commonly less than 1 mm and the one around the bolt hole is less than a few millimeters.

In JIS, four grades of bolt hole diameter d_h are specified from first to fourth grade. The fourth grade is for cast holes. The gap size increases from first to fourth grade. For bolts with nominal diameters from M6 to M64, the gap size of first grade ranges from about 0.2 to 1 mm, that of second grade from 0.3 to 3 mm, and it ranges from 0.5 to 5 mm in third grade. The heat flowing through small gaps of this degree cannot be ignored from the engineering point of view.

In sections 5.2 and 5.3, it is shown how to measure thermal contact coefficient h_c and apparent thermal contact coefficient h_e. Concrete values and evaluation formulas for h_c and h_e are also presented. In section 5.5, numerical analyses are carried out by incorporating h_c and h_e into finite element formulation. Then, numerical results are presented on the specific thermal and mechanical behaviors of bolted joints when subjected to thermal load in commonly encountered manners.

Mechanics of the tightening process of threaded fasteners

3

3.1 Summary of various tightening methods and comparison of tightening characteristics

When tightening threaded fasteners, various tightening methods are used depending on the shape and dimensions of the fasteners, desired accuracy in bolt force, type of external force exerted on the target joint, etc. In this section, a summary of representative tightening methods is presented, starting with the most commonly used torque control method.

(1) Torque control method

Torque applied by a torque wrench or spanner is converted into axial bolt force utilizing the action of the thread's inclined plane. The torque is usually applied to a nut or bolt head. This is the most widely used tightening method, because the tightening operation is simple and it can be done by human power using inexpensive tools. Hydraulic or pneumatic drive wrenches and impact wrenches utilizing the impact force of air are also tightening tools that follow the principle of the torque control method. It should be pointed out here that a fairly large scatter in bolt force is a major drawback of the torque control method. Even if tightening with the same amount of torque, it is said that the bolt force generated inevitably scatters from 25% to 35% [58,59] due to the scatter in coefficients of friction on the thread surface and nut-bearing surface. The amount of scatter in bolt force is largely influenced by the condition of contact surfaces and the properties of any lubricants used.

(2) Torque gradient method

When tightening threaded fasteners with the torque control method, local plastic deformations occur around the thread root as bolt force is increased. As a result, the relationship between torque and nut rotation angle becomes nonlinear beyond a certain level of bolt force. Hence, the point at which the rate of torque change relative to nut rotation angle starts to decrease distinctly is used as an index for the tightening operation. This tightening procedure is called the torque gradient method and, although higher accuracy is achieved compared to the torque control method, it requires the use of exclusive tools.

(3) Angle control method

The objective bolt force is achieved by controlling the nut rotation angle. The angle control method is classified into elastic angle control method and plastic angle control method. In the former method, the tightening operation is carried out under conditions where the threaded fasteners remain within the elastic region of the stress–strain

The Mechanics of Threaded Fasteners and Bolted Joints for Engineering and Design.
https://doi.org/10.1016/B978-0-323-95357-3.00004-6

curve; whereas, in the latter, the body of the target bolt enters the plastic region and is considerably deformed. The plastic angle control method [60,61] utilizes the characteristics that the rate at which the bolt force changes relative to the nut rotation angle becomes smaller as the bolt force is increased, due to the spread of the plastic region in the bolt body. In this method, bolt force control is relatively easy because the rotation angle used as an index is very large. Therefore, the plastic angle control method is often applied to high strength bolts with small nominal diameter in automobile industries, etc. In the elastic angle control method, by first applying an amount of torque called snug torque, aiming to suppress the negative effects of surface roughness and waviness, the relating contact surfaces, such as thread surface and nut- and bolt head-bearing surfaces, are brought into close contact. Then, the successive tightening operation is carried out by considering that the bolt force is almost in proportion to the nut rotation angle. This method is widely used when tightening critical parts in the marine engineering field [62], etc.

(4) Direct tension method

Bolt force is generated by directly applying tension to the target bolt using hydraulic tensioners. Fig. 3.1 shows a small-sized hydraulic tensioner. Because of it being hardly affected by friction on contact surfaces, it is possible to suppress the bolt force scatter within a few percent; therefore, this method is widely used in industries where high accuracy in bolt force is required. On the other hand, the tightening apparatus is expensive and some space is necessary around the target bolt to install a hydraulic tensioner. Therefore, this method is not applicable to joints where multiple bolts are closely placed. Moreover, in the case of clamping thin plates or fastened plates including low stiffness objects like sheet gaskets, whose compression characteristic is difficult to evaluate, the tightening accuracy may be significantly lowered. When employing the direct tension method, the initially applied tension is higher than the objective bolt force. It follows that care must be taken when the objective bolt stress is set near the yield point of the bolt material.

Fig. 3.1 Hydraulic tensioner.

(5) Thermal expansion method

In this method, bolt force is generated in a similar manner to the direct tension method, in which, instead of a hydraulic tensioner, a bolt heater is used together with a bolt with a concentric hole. The heater is installed into the bolt hole and heated, causing the target bolt to become elongated in the axial direction. Next, the bolt head is brought into contact with the surface of the fastened plates by rotating a nut. Then, the bolt force is generated upon cooling due to the shrinkage of the bolt body. The applicable bolt size is virtually unlimited, and since the tightening apparatus is small-sized and inexpensive, it is possible to tighten multiple bolts located in a small area.

The thermal expansion method has the drawback of requiring fairly long heating time; nevertheless, it is effectively applied in cases where many bolts need to be tightened simultaneously with identical bolt force, for example, the many large bolts used to tighten the casing of large-scale steam turbines in a power plant. In actual applications, to attain high tightening accuracy, it is necessary to preliminarily carry out a precise evaluation of the relationship between heating time and bolt force to be generated. Fig. 3.2 shows an example of a bolt heater. The rod of the bolt heater is inserted into the bolt hole. The rod temperature is raised by Joule heat generated in the coils. Then, the heat is transferred to the bolt body and produces the bolt elongation. Recently, ultrasonic wave bolt heaters, which aim at reducing the heating time, have become commercially available.

There are many common aspects among the angle control, direct tension, and thermal expansion methods. In the direct tension and thermal expansion methods, bolt force is generated by directly applying tension to the bolt, in which the accuracy of bolt force is largely influenced by the stiffness of the joined components. High accuracy is expected in a bolted joint with large grip length and low stiffness. In the angle control method, the tightening operation is carried out by applying torque. However, unlike the torque control method, high tightening accuracy is expected due to its tightening characteristics when the method is applied to bolted joints with large grip length. Meanwhile, the tightening accuracy of the torque control method is generally lower than the direct tension method. However, it has the great advantage that the

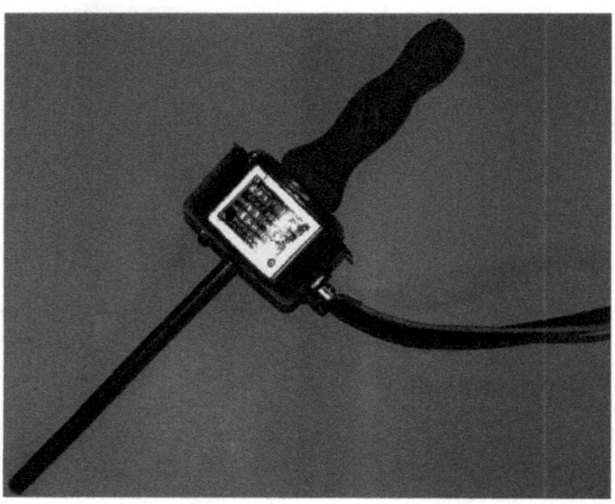

Fig. 3.2 Bolt heater.

relationship between applied torque and bolt force is not influenced by grip length and Young's modulus of joint materials, both of which affect the joint stiffness.

3.2 Torque control method

3.2.1 Relationship between tightening torque and axial bolt force

The torque control method is the most widely used among the various tightening methods because it can produce large bolt force with small applied force, in addition to easy operation. On the other hand, it has been reported that many problems occur due to insufficient or excessive bolt force, because the tightening characteristics are not fully understood. When using the method, the most important point is that even if applying the same torque to the bolts with identical nominal diameter, the produced bolt force significantly varies depending on coefficients of friction on related contact surfaces. It can be expressed in other words that tightening torque is just an index to achieve the objective bolt force, and the relationship between the applied torque and bolt force significantly changes according to coefficients of friction. Tightening torque applied to nut T_t is consumed in the engaged thread portion and nut-bearing surface. Let the two torques be T_1 and T_2; then, the following equation holds:

$$T_t = T_1 + T_2 \tag{3.1}$$

First, the equation for nut-bearing torque T_2 is derived, in which axial bolt force produced by tightening torque T_t is denoted as F_b. T_2 is all consumed as friction work on the nut-bearing surface. Fig .3.3A shows a typical distribution pattern of the contact pressure p_{nu} on the nut-bearing surface. Integrating the compression force, due to the action of p_{nu}, over the contacting area on the nut-bearing surface, the calculated value is equal to axial bolt force F_b:

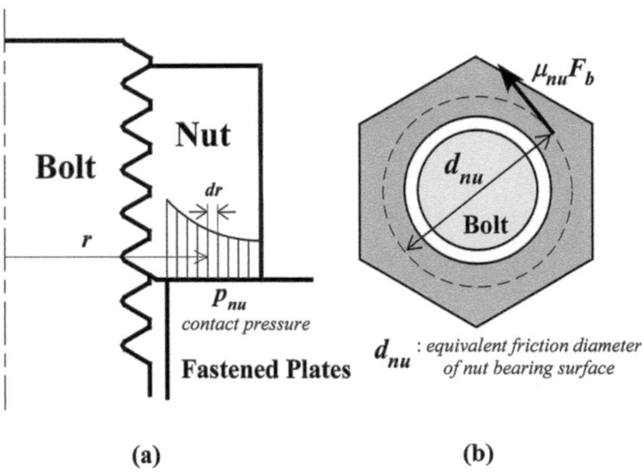

Fig. 3.3 Contact pressure distribution on the nut-bearing surface. (A) Contact pressure distribution pattern and (B) Equivalent friction diameter.

$$F_b = \int_{r_1}^{r_2} 2\pi r \cdot p_{nu} dr \tag{3.2}$$

Multiplying the compression force by the coefficient of friction on the nut-bearing surface μ_{nu} and the distance from bolt axis r and integrating the value, the equation for T_2 is derived as follows:

$$T_2 = \int_{r_1}^{r_2} 2\pi r p_{nu} \mu_{nu} r dr = \frac{1}{2}\mu_{nu} F_b d_{nu} \tag{3.3}$$

where d_{nu} in Eq. (3.3) represents an imaginary diameter, at which the friction force distributing on the nut-bearing surface is replaced by a concentrated force, as shown in Fig. 3.3B. It is an equivalent diameter of the friction circle on the nut-bearing surface, termed equivalent friction diameter of nut-bearing surface in this book. Yamamoto proposed that d_{nu} is about 1.3 times nominal diameter d [63]. Introducing the concept of equivalent friction diameter, T_2 is equal to the product of friction force $\mu_{nu} F_b$ acting on the nut-bearing surface times $1/2 d_{nu}$.

The relationship between thread torque T_1 and bolt force F_b can be derived from the equilibrium of the forces acting on the thread surface. Fig. 3.4A and B show the equilibrium of the forces in square screw threads. β is lead angle and U_f is the circumferential force resulting from T_1. Assuming that U_f acts circumferentially on the circle of pitch diameter d_2, T_1 is calculated as the product of U_f times the arm length of $1/2 d_2$:

$$T_1 = \frac{1}{2} d_2 U_f \tag{3.4}$$

Let coefficient of friction on the thread surface be μ_{th}; the equilibrium equation of the forces along the thread surface is as follows:

$$U_f \cos\beta - F_b \sin\beta = \mu_{th}\left(F_b \cos\beta + U_f \sin\beta\right) \tag{3.5}$$

In the left side of Eq. (3.5), the sign of the forces that act in the direction to go up the inclined plane is set to be positive. The term in the parenthesis of the right side represents the force acting perpendicular to the thread surface. That is, Eq. (3.5) indicates that the force to go up the inclined thread surface is equal to the friction force acting on the thread surface. Denoting friction angle of thread surface as ρ_{th} and rearranging Eq. (3.5), the following equation is obtained:

$$U_f = F_b \tan\left(\rho_{th} + \beta\right) \tag{3.6}$$

where

$$\tan\rho_{th} = \mu_{th} \tag{3.7}$$

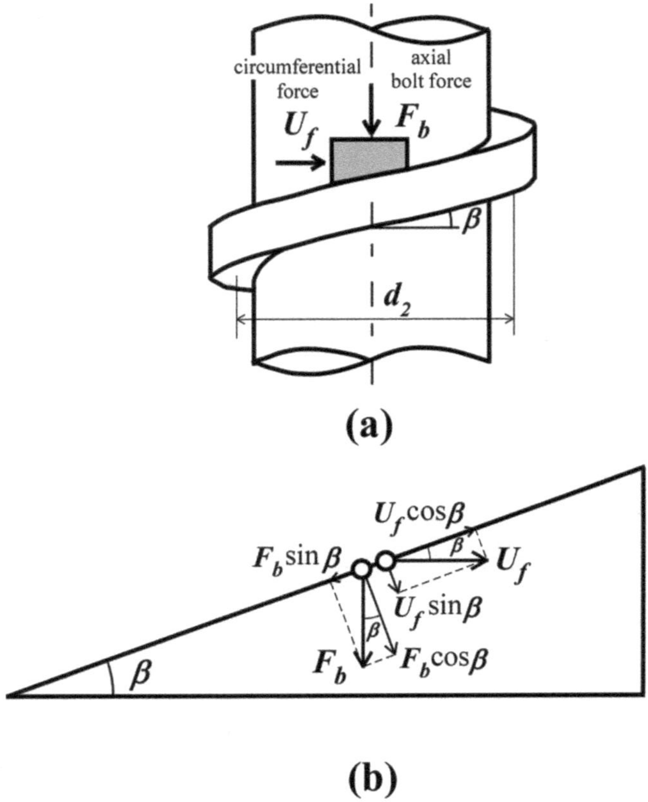

Fig. 3.4 Equilibrium of forces on thread surface in square screw threads. (A) Forces exerted on thread surface and (B) equilibrium of forces along and perpendicular to thread surface.

Substituting Eq. (3.6) into Eq. (3.4), T_1 is related to F_b:

$$T_1 = \frac{1}{2} F_b d_2 \tan(\rho_{th} + \beta) \tag{3.8}$$

Next, the substitution of Eqs. (3.8) and (3.3) for T_2 into Eq. (3.1) yields the equation that relates tightening torque to axial bolt force for square screw threads:

$$T_t = \frac{1}{2} F_b \{ d_2 \tan(\rho_{th} + \beta) + \mu_{nu} d_{nu} \} \tag{3.9}$$

In the cases of triangular and trapezoidal screw threads, the equation for T_1 is different from that for square screw threads, since thread surface is inclined at flank angle α_1 to the horizontal plane. In this case, friction forces must be evaluated in terms of the force components perpendicular to the pressure flank. Fig. 3.5A shows the equilibrium of the forces acting on the thread surface of a triangular screw thread.

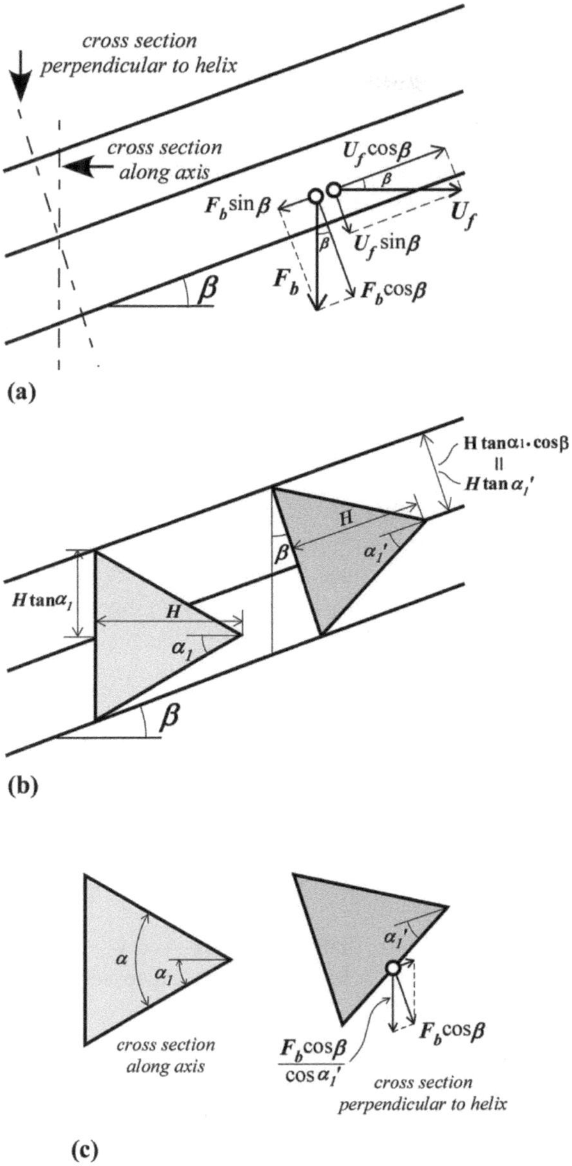

Fig. 3.5 Equilibrium of forces on thread surface in triangular screw threads. (A) Equilibrium of forces on thread surface, (B) relationship between cross sections along axis and perpendicular to helix, and (C) equilibrium of forces on cross section perpendicular to helix.

Flank angle in the plane perpendicular to helix α'_1 is only slightly smaller than α_1 due to the effect of lead angle β. The relationship between α'_1 and α_1 is derived by comparing the side lengths of the triangles in the planes along the axis and perpendicular to the helix, as shown in Fig. 3.5B:

$$\tan\alpha_1' = \tan\alpha_1 \cdot \cos\beta \tag{3.10}$$

For commonly used triangular screw threads, the assumption that $\alpha'_1 \approx \alpha_1$ is allowable with sufficient accuracy. However, when dealing with the thread geometry rigorously, Eq. (3.10) should be considered. In the case of square screw threads, α_1 is zero, and therefore, α'_1 is also zero. Meanwhile, in the case of triangular screw threads, the force acting perpendicular to the thread surface is obtained by dividing $F_b\cos\beta$ by $\cos\alpha'_1$, as shown in Fig. 3.5C. Therefore, the equilibrium of the forces in triangular screw threads along the thread surface is expressed as follows, corresponding to Eq. (3.5):

$$U_f \cos\beta - F_b \sin\beta = \mu_{th}\left(F_b \frac{\cos\beta}{\cos\alpha'_1} + U_f \frac{\sin\beta}{\cos\alpha'_1}\right) \tag{3.11}$$

where

$$\tan\rho'_{th} = \frac{\mu_{th}}{\cos\alpha'_1} \tag{3.12}$$

where ρ'_{th} represents friction angle of thread surface for triangular screw threads. Supposing flank angle α_1 to be 30 degrees, it can be interpreted that the coefficient of friction μ_{th} is virtually increased as much as 15%, because μ_{th} is multiplied by $1/\cos\alpha'_1$. Collecting the terms relevant to U_f and F_b in Eq. (3.11) and rearranging them, an expression of the same form as Eq. (3.6) is derived:

$$U_f = F_b \frac{\tan\rho'_{th} + \tan\beta}{1 - \tan\rho'_{th}\tan\beta} = F_b \tan\left(\rho'_{th} + \beta\right) \tag{3.13}$$

Then, thread torque T_1 can be expressed in the same form as Eq. (3.8):

$$T_1 = \frac{1}{2} F_b d_2 \tan\left(\rho'_{th} + \beta\right) \tag{3.14}$$

Since nut-bearing torque T_2 is identical to that of square screw threads, the torque-bolt force relationship for triangular screw threads is expressed in the following form:

$$T_t = \frac{1}{2} F_b \left\{ d_2 \tan\left(\rho'_{th} + \beta\right) + \mu_{nu} d_{nu} \right\} \tag{3.15}$$

The above equation assumes that the tightening torque is applied to the nut, which is the general procedure when tightening bolt-nut connections. However, the tightening torque is sometimes applied to the bolt head on account of workability and other on-site issues. This problem is solved by simply substituting the coefficient of friction on the bolt head-bearing surface μ_{hd} into μ_{nu}. As for equivalent friction diameter d_{hd}, it can be regarded as 1.3 times nominal diameter d, as in the case of the nut-bearing surface, unless the flatness of the bolt head-bearing surface becomes a problem. The effect of the flatness on equivalent friction diameter is discussed in Section 3.2.3, referring to the numerical results by FEM.

– Learning by numerical example 3.1: Coefficient of friction and friction angle

Eq. (3.10) suggests that the effect of lead angle β on flank angle in the plane perpendicular to helix α'_1 is very small. It means that although β decreases as nominal diameter d increases, α'_1 is hardly influenced by d. Therefore the friction angle of thread surface ρ'_{th}, calculated by Eq. (3.12), increases almost in proportion to the coefficient of friction on thread surface μ_{th}. Calculating friction angle ρ'_{th} for M16 with coarse thread of $P = 2\,\text{mm}$ by changing μ_{th} to 0.05, 0.1, 0.15, and 0.2, it becomes 3.30, 6.59, 9.82, and 13.0 degrees, respectively. In this case, β is 2.48 degrees. Referring to Eq. (3.15), it is found that in the range of small μ_{th}, β and μ_{th} affect the torque-bolt force relationship to the same extent. However, as μ_{th} is increased, the relationship is almost solely governed by μ_{th}.

3.2.2 Simple equation relating tightening torque to axial bolt force through the friction coefficient

The amount of torque to attain the objective bolt force F_b is calculated by Eq. (3.15) for triangular screw threads. However, since the calculation requires some time and effort, a simpler equation is frequently used, in which nut factor K is introduced:

$$T_t = KF_b d \tag{3.16}$$

Once the nominal diameter of bolt d and the objective bolt force F_b are determined, the required tightening torque T_t is calculated by Eq. (3.16). It is a common practice that K is set to be 0.2. This value corresponds to the tightening operation of coarse screw threads with coefficients of friction on thread surface and nut-bearing surface both of 0.15. To clarify the meaning of nut factor, Eq. (3.15) is transformed as follows:

$$T_t = \frac{1}{2}\left\{\frac{d_2}{d}\tan\left(\rho'_{th} + \beta\right) + \mu_{nu}\frac{d_{nu}}{d}\right\}F_b d \tag{3.17}$$

Comparing to Eq. (3.16), it is found that the formula in the brace multiplied by 1/2 corresponds to nut factor K. Supposing that d_{nu} is $1.3d$ [63] and substituting the dimensions of coarse and fine screw threads up to M64, then applying the least squares method to the foregoing values, nut factor K can be expressed as a linear function of coefficients of friction on thread surface μ_{th} and nut-bearing surface μ_{nu} [64]:

$$\begin{aligned} K &= 0.556\mu_{th} + 0.65\mu_{nu} + 0.019(\text{coarse thread}) \\ K &= 0.565\mu_{th} + 0.65\mu_{nu} + 0.011(\text{fine thread}) \end{aligned} \tag{3.18}$$

In the case of coarse screw threads, the error in axial bolt force calculated by use of Eqs. (3.16) and (3.18) is very small, comparing to the results obtained by the rigorous equation of Eq. (3.15). Meanwhile, in the case of fine screw threads, somewhat larger errors arise because multiple pitch sizes are specified for the same nominal diameter. However, the proposed simple equation is still sufficiently practical, considering that the effect of scatter in coefficients of friction is significant. To summarize the above, the amount of coefficient of friction is first estimated by taking account of the type of lubricant to be used and the surface roughness of the jointed components, etc., and then, nut factor K is calculated by the substitution of those values into Eq. (3.18). Consequently, it is possible to control the amount of axial bolt force with higher accuracy compared to the past.

Next, how the tightening torque T_t applied to the nut acts on the bolted joint is examined by using Eq. (3.18). The three terms of the right side of Eq. (3.18) are related to friction on thread surface, friction on nut-bearing surface, and lead angle, respectively. The last term, which is given as a constant, represents the component that is transferred into axial bolt force. Assuming coefficients of friction on thread surface μ_{th} and nut-bearing surface μ_{nu} as being identical, nut factor K is comprehensively calculated [65] in Table 3.1. The percentages of the three terms are shown in the table.

In the case of coarse screw threads with coefficient of friction of 0.15, it is found that about 40% of the applied torque is consumed as friction work on the thread surface, about 50% is consumed as friction work on the nut-bearing surface, and only the

Table 3.1 Nut factor and coefficient of friction.

μ_{th}, μ_{nu}	K (Coarse)	Thread surface (%)	Nut-bearing surface (%)	Lead angle (%)
0.05	0.079	35.1	41.0	24.0
0.075	0.109	38.1	44.5	17.4
0.1	0.140	39.8	46.6	13.6
0.125	0.170	40.9	47.9	11.2
0.15	0.200	41.7	48.8	9.5
0.175	0.230	42.3	49.4	8.3
0.2	0.260	42.7	50.0	7.3
0.25	0.321	43.4	50.7	5.9
0.3	0.381	43.8	51.2	5.0
0.35	0.441	44.1	51.6	4.3
0.4	0.501	44.4	51.9	3.8
μ_{th}, μ_{nu}	**K (fine)**	**Thread surface (%)**	**Nut-bearing surface (%)**	**Lead angle (%)**
0.05	0.072	39.4	45.3	15.3
0.075	0.102	41.5	47.7	10.8
0.1	0.133	42.6	49.1	8.3
0.125	0.163	43.4	49.9	6.8
0.15	0.193	43.9	50.5	5.7
0.175	0.224	44.2	50.9	4.9
0.2	0.254	44.5	51.2	4.3
0.25	0.315	44.9	51.6	3.5
0.3	0.376	45.1	51.9	2.9
0.35	0.436	45.3	52.1	2.5
0.4	0.497	45.5	52.3	2.2

remaining 10% is used for climbing up the lead angle to generate axial bolt force. The percentage of T_t transferred into bolt force decreases as the coefficient of friction increases. Next, the percentage is compared between coarse and fine screw threads. When coefficient of friction is 0.15, nut factors are 0.200 and 0.193, respectively, which means that nut factor is slightly smaller in fine screw threads. The difference in K is due to the difference in lead angle. This indicates that when tightening the two types of screw threads by the same torque under identical coefficients of friction, axial bolt force produced in fine screw threads is about 3% larger than that in coarse screw threads. In actual practice, however, the opposite tendency is sometimes observed. The cause of this is discussed in Section 3.8.

The relationships between coefficient of friction and bolt force to be generated and between the scatters in coefficient of friction and bolt force are of crucial importance from the viewpoint of joint safety. When applying the same torque to bolts with identical nominal diameter, the generated bolt force F_b becomes larger with decreasing nut factor K, as evident from Eq. (3.16). Eq. (3.18) indicates that with decreasing coefficients of friction on thread surface μ_{th} and nut-bearing surface μ_{nu}, K is decreased correspondingly. In the tightening operation of threaded fasteners, it is desirable to produce large bolt force with small torque and to suppress the scatter in bolt force at the same time. Accordingly, it is desirable for lubricants to be used in the tightening operation to provide the coefficients of friction with small values and scatters.

For example, when using molybdenum disulfide as a lubricant, the coefficient of friction is usually less than 0.1. Its value changes to some extent, depending on the elapsed time until the start of tightening operation after applying the lubricants If the tightening operation is carried out just after applying lubricants, coefficients of friction tend to be somewhat larger but the scatter is smaller, because the condition of lubricant is in paste form. On the other hand, when the tightening operation is done after a short interval, under the condition of lubricants in dried form, coefficients of friction tend to be smaller but the scatter is somewhat larger. The phenomenon has been verified by experiments.

From the above results comes the conclusion that it is very difficult to lower both the absolute value and the scatter of coefficients of friction. Measured data of coefficients of friction and the scatter of nut factor, obtained by using various lubricants, are given in [59,66–68].

- Learning by numerical example 3.2: Relationship between axial bolt force and coefficient of friction

The magnitude of axial bolt force produced in M16 with coarse thread is compared for coefficients of friction of 0.15 and 0.30 when applying the same torque. Assuming μ_{th} and μ_{nu} are equal and calculating nut factor K by use of the first equation of Eq. (3.18), K becomes 0.200 and 0.381, respectively. The ratio of the two values is about 52%. This means that axial bolt force is reduced to about half when coefficients of friction are doubled under the same tightening torque. Alternatively, this can be expressed as follows: axial bolt force obtained by the same torque is almost inversely proportional to coefficients of friction.

3.2.3 Advantages of the torque control method and influencing factors on tightening accuracy

The largest advantage of the torque control method is the ease of tightening operation. Another important advantage is that the tightening accuracy is not influenced by the joint stiffness. The equation relating torque to bolt force, Eq. (3.15), has been derived by considering the equilibrium of forces and moments on the thread surface and nut-bearing surface. Therefore, such factors as grip length and Young's modulus of joint materials, which affect the joint stiffness, are not included in the equation. In contrast, the tightening accuracies of the elastic angle control, direct tension, and thermal expansion methods are considerably affected by the joint stiffness, owing to the mechanical properties of the tightening process. Specifically, when tightening bolts with identical nominal diameter to achieve the same bolt force, the tightening accuracy could be significantly reduced in joints with small grip length, due to small bolt elongations. Furthermore, the accuracy is likely to remarkably decrease in joints where a low stiffness object, like sheet gaskets, exhibiting nonlinear and hysteresis behaviors, is inserted between fastened plates.

As stated in the preceding section, the largest cause for the bolt force scatter in the torque control method is the scatter of coefficients of friction on thread surface μ_{th} and nut-bearing surface μ_{nu}. In addition, when the nut-bearing surface or bolt head-bearing surface is not parallel to the mating plate surface, coefficients of friction show a substantial variation due to the nonuniform contact between the mating surfaces. In the following, it is explained how the bolt force scatter occurs due to various factors appearing in torque-bolt force relationship, as given in Eq. (3.15):

$$T_t = \frac{1}{2}F_b\left\{d_2 \tan\left(\rho'_{th} + \beta\right) + \mu_{nu}d_{nu}\right\} \tag{3.19}$$

(1) Lead angle β

As is evident from Eq. (3.19), when tightening with a constant torque T_t, axial bolt force F_b decreases with larger lead angle. Thus, when tightening bolts with coarse and fine threads with an identical torque, axial bolt force produced in fine screw threads is slightly larger, if coefficients of friction are also the same. The amount of lead angle is generally evaluated on the circle of pitch diameter d_2. In other words, even if the corresponding male and female surfaces are not uniformly in contact and the contact position varies between nominal diameter d and minor diameter d_1, the difference in lead angle β at each position is not substantial, as discussed in section 1.4.3.

(2) Flank angle α_1

In Eq. (3.12), which calculates friction angle ρ'_{th}, the flank angle in the plane perpendicular to helix α'_1 is involved. However, the effect of flank angle α_1 is negligible even if α_1 deviates slightly from its standard value of 30 degrees.

(3) Coefficient of friction on thread surface μ_{th}

The amount of coefficient of friction varies depending on the surface condition and the type of lubricant used. The values of μ_{th} are practically changed due to the nonuniform contact on thread surface, which is caused by the difference in flank angles between male and female threads induced by manufacturing errors. Attention should be paid to the above phenomenon when dealing with large bolts or body-side female threads with large nominal diameter, in which processing accuracy is likely to be lowered.

(4) Coefficient of friction on nut-bearing surface μ_{nu}

As in the case of μ_{th}, the amount of μ_{nu} is largely influenced by the surface condition and the type of lubricant. When the nut-bearing surface is not sufficiently parallel to the plate surface to be in contact, μ_{nu} is changed substantially, which causes scatter in axial bolt force. In addition, when clamping thin plates with warping beyond a certain level, a similar problem occurs due to the nonuniform contact.

(5) Equivalent friction diameter of nut-bearing surface d_{nu}

d_{nu} is commonly regarded as 1.3 times nominal diameter d [63]. Aiming at easy processing of nuts, a slight inclination is sometimes added inwards at the nut-bearing surface, resulting in the axial bolt force produced tending to be lowered even if applying the same amount of torque, because the center of contact moves outwards and d_{nu} becomes larger than $1.3d$. When using hexagon nuts with flange, d_{nu} is necessarily larger than $1.3d$ due to larger bearing surface area. Accordingly, the axial bolt force is reduced even if applying identical tightening torque under the same coefficient of friction, μ_{nu}. However, those nuts require larger loosening torque at disassembly, which also provides the advantage of high antiloosening performance. In the following, what factors affect the amount of d_{nu} is discussed in detail, because d_{nu} is an important value that affects the bolt force scatter besides μ_{nu}.

Inclination of nut- or bolt head-bearing surface [69]

Assuming the perfect flatness of bearing surface, the equivalent friction diameter of nut-bearing surface d_{nu} has been evaluated systematically by axisymmetric FEM, where d_{nu} is found to be about 1.3 times nominal diameter d, as already shown in [63]. In actual nuts, however, a slight inclination is sometimes added inwards, as stated above. According to JIS, the inclination angle must be less than one degree or two degrees, depending on the type of nut. The same standard is also specified for bolt head-bearing surface.

Fig. 3.6 shows the measured results for M16, where each curved line represents the shape of nut-bearing surface. Thread pitch P of fine screw threads is 1.5 mm here. The measurements were conducted in the six directions indicated by arrows. The four curved lines correspond to the nuts supplied by four different manufacturers. The shape measurement was performed for six nuts supplied by each manufacturer, and the results averaged. The scatter among the nuts supplied by the same manufacturer was negligible. The maximum inclination angle is found to be about one degree. Accordingly, with the progress of tightening operation, the contact area expands radially inwards, starting from the outer edge of the nut-bearing surface. The existence of such an inclination on the nut-bearing surface, as shown in Fig. 3.6, increases the amount of d_{nu} from the standard value of $1.3d$. Hence, in the actual tightening operation, axial bolt force achieved by the same torque is decreased. To make matters more

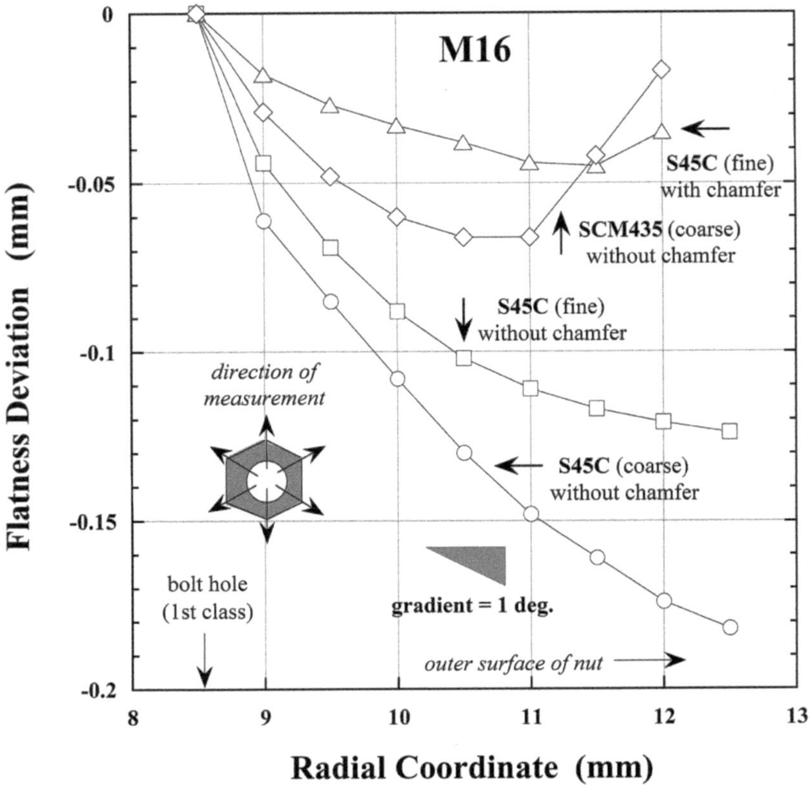

Fig. 3.6 Flatness deviation measured on nut-bearing surface.

complicated, d_{nu} changes with changes in axial bolt force. This can be explained by the phenomenon that the contact area varies nonlinearly with increasing bolt force. Fig. 3.7 shows the calculated results of d_{nu} for M16 by axisymmetric FEM. The axial bolt stress is set to be 100 MPa. The figure indicates that d_{nu} increases rather sharply as the inclination angle θ_{nu} is increased, and it becomes about 1.6d when θ_{nu} reaches 0.6 degrees.

Meanwhile, it is empirically known that the bolt head-bearing surface is sometimes inclined to some extent as well. In this case, the inclination angle usually exists radially outwards as opposed to the nut-bearing surface. Fig. 3.8 shows the measured results of the inclination angle for M16 on bolt head-bearing surface. Two kinds of products, Type-A and Type-B, were selected and the measurements were conducted in the four directions indicated by arrows. Comparing to the nut-bearing surface, the inclination angles are relatively small, in addition to the direction of inclination being opposite. This means that the equivalent friction diameter of the bolt head-bearing surface d_{hd} is decreased with increasing inclination angle.

Fig. 3.9 shows the effect of bolt stress on d_{nu}. As bolt stress is increased, the increase of d_{nu} gets smaller for the same inclination angle θ_{nu}. This is due to the

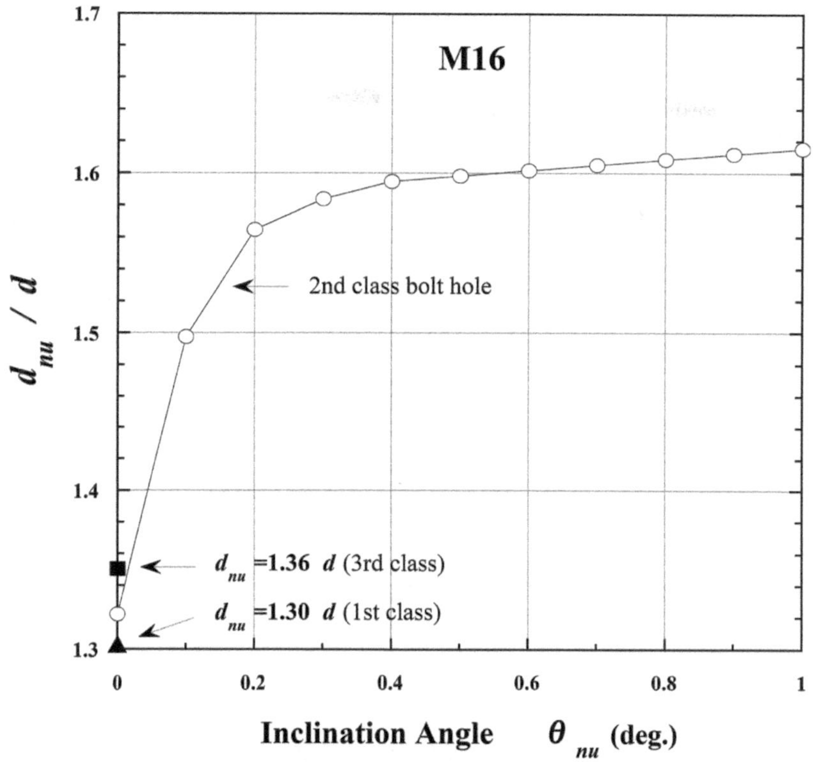

Fig. 3.7 Equivalent friction diameter of nut-bearing surface evaluated by FEM.

increase of contact area when applying larger bolt stress. In Fig. 3.10, it is shown how bolt forces are decreased or increased due to the existence of inclination angle on the bearing surfaces of nut and bolt head. Bolt force F_b is normalized with respect to bolt force F_{b0} for the case without inclination. The parameters in the figure represent bolt stress. Coefficients of friction, μ_{th} and μ_{nu}, are set to be 0.15. It is found that the bolt forces achieved by the same torque under identical contact conditions may be decreased or increased by up to 10% of the objective bolt force. The effect of the inclination of bearing surfaces is summarized as follows.

- **a.** Bearing surfaces of nut and bolt head are sometimes inclined radially inwards and outwards, respectively.
- **b.** The existence of inclination increases the amount of d_{nu} on nut-bearing surface and decreases that of d_{hd} on bolt head-bearing surface.
- **c.** The increase of d_{nu} and decrease of d_{hd} cause the reduction and increase of bolt forces even if applying the same torque under identical contact conditions.
- **d.** The extents of increase in d_{nu} and decrease in d_{hd} significantly change with changes in inclination angle and also to some extent with changes in axial bolt stress.
- **e.** Due to the existence of inclination, the reduction and increase rates of bolt forces reach up to about plus–minus 10% of the target bolt force.

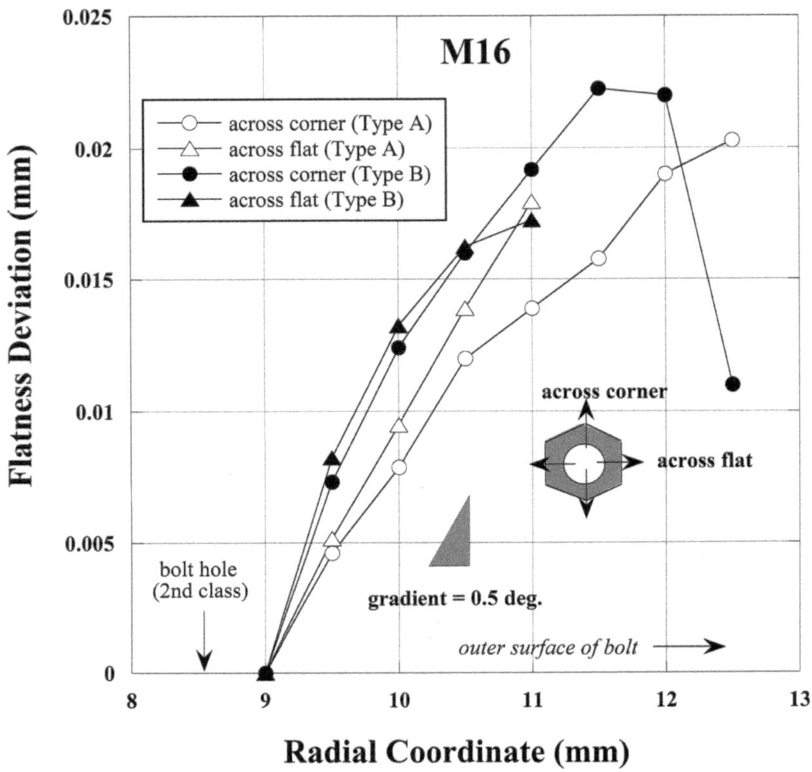

Fig. 3.8 Flatness deviation measured on bolt head-bearing surface.

Warping and flatness on plate surface

Equivalent friction diameters, d_{nu} and d_{hd}, change significantly due to the warping and deviation of the flatness of the plate surface. It is particularly observed when clamping thin plates that are liable to involve processing problems. With the existence of warping, nonuniform contact occurs between nut and plate surface, hence, d_{nu} may change significantly. In such cases, as the bolt force is increased with increasing tightening torque, it causes the changes in contact conditions and equivalent friction diameters, thereby varying the friction coefficients substantially. In other words, the relationship between torque and bolt force is regarded as a highly nonlinear problem, where d_{nu} and coefficients of friction μ_{nu} and μ_{th} vary as a function of contact stress. However, it is practically impossible to evaluate d_{nu} by FEM for each case of clamping thin plates. As a countermeasure, it is recommended to measure the actual values of the coefficient of friction corresponding to the predicted range of axial bolt stress, using the method introduced in Section 3.2.5, and estimate nut factor K in order to employ Eq. (3.16), which relates tightening torque to bolt force by K.

– Learning by numerical example 3.3: Relationship between axial bolt force and equivalent friction diameter

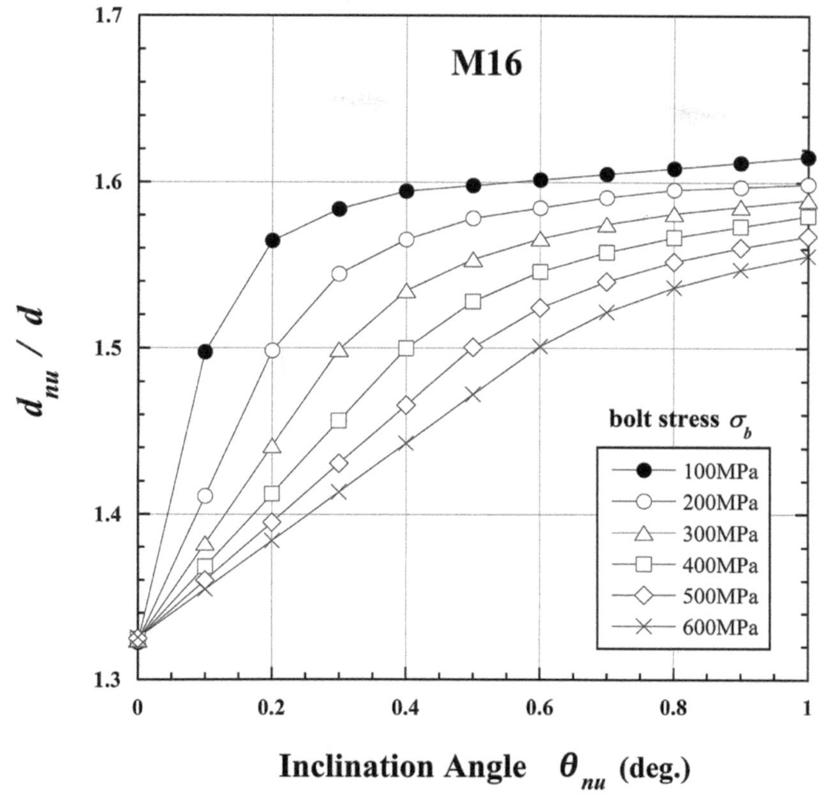

Fig. 3.9 Effect of bolt stress on equivalent friction diameter d_{nu}.

The magnitude of axial bolt force of M16 with coarse thread, achieved by the same torque, is compared when d_{nu} is changed to 1.3*d*, 1.4*d*, 1.5*d*, 1.6*d*, and 1.7*d*. Coefficients of friction, μ_{th} and μ_{nu}, are assumed to be 0.15. Considering that coefficients of the second term in Eq. (3.18) are changed as 0.65, 0.70, 0.75, 0.80, and 0.85 for the equivalent friction diameters selected above, nut factors are calculated to be 0.200, 0.207, 0.215, 0.222, and 0.230, respectively.

For instance, the ratio of nut factors between the cases of d_{nu} being 1.3*d* and 1.6*d* is about 0.9. That is, if d_{nu} is increased to 1.6*d*, axial bolt force would be decreased by as much as 10% even if applying the same torque. The calculation results indicate that the geometric errors existing on bearing surfaces of nut and bolt head significantly affect the tightening accuracy of bolted joints.

3.2.4 Self-locking criteria and efficiency of screw threads

In Section 3.2.1, the equation that relates tightening torque to axial bolt force was derived. When loosening threaded fasteners to disassemble the joint, male threads go down the thread surface as opposed to the tightening operation. Thus, the

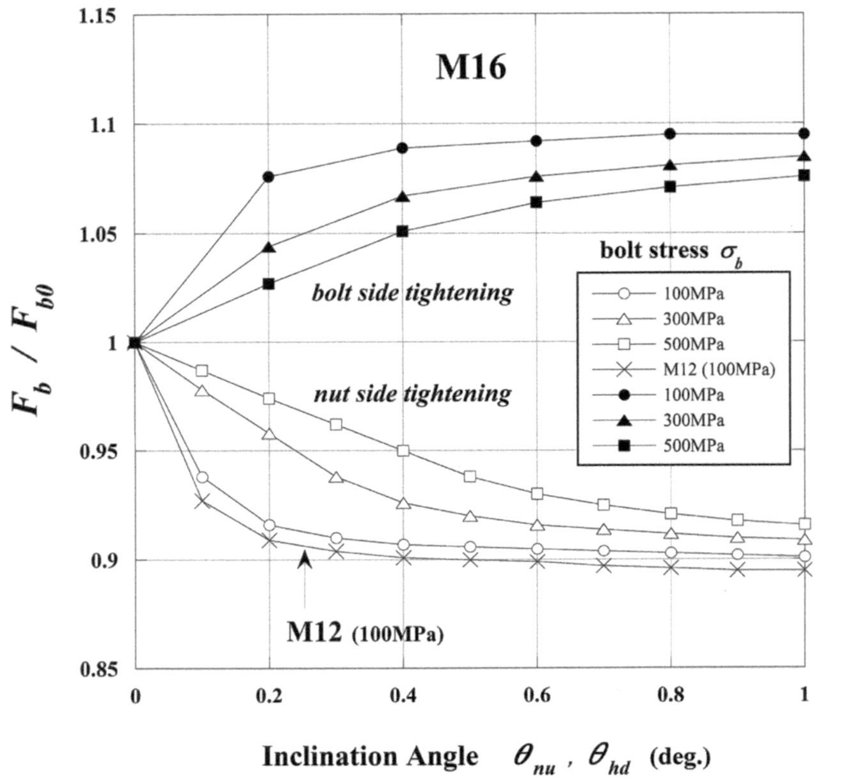

Fig. 3.10 Bolt force variations due to inclination existing on bearing surfaces of nut and bolt head.

equation relating loosening torque T_l to axial bolt force F_b can be obtained by changing the sign of lead angle β in Eq. (3.15).

$$T_l = \frac{1}{2} F_b \left\{ d_2 \tan\left(\rho'_{th} - \beta\right) + \mu_{nu} d_{nu} \right\} \tag{3.20}$$

Assuming that coefficients of friction remain unchanged, loosening torque T_l is smaller than tightening torque T_t, as evident from Eq. (3.20). Paying attention to the term of "tan()" in both equations, the difference is 2β and is larger in coarse screw threads due to larger β. In case of multiple-thread screws having large β, the difference between T_t and T_l becomes even larger. Accordingly, it is usual that the loosening torque T_l is smaller than the tightening torque T_t. If T_l becomes larger than T_t, it is probable that coefficients of friction are increased because of the change of surface conditions or the occurrence of seizure of threaded fasteners.

When coefficient of friction μ_{th} on the thread surface is large, large tightening torque is required. Meanwhile, after the completion of the tightening operation, in order prevent loosening without external force, the first term in a brace of Eq. (3.20) must be positive. That is, to prevent the loosening due to nut rotation under conditions without external force, friction angle ρ'_{th} must be larger than lead angle β.

$$\begin{aligned} &\rho'_{th} > \beta \text{ (triangular thread)} \\ &\rho_{th} > \beta \text{ (square thread)} \end{aligned} \tag{3.21}$$

Eq. (3.21) is called the self-nonloosening condition. When coefficient of friction μ_{th} on the thread surface is identical, ρ'_{th} is about 15% larger than ρ_{th} due to the effect of α'_1 in Eq. (3.12). This means that triangular screw threads are unlikely to loosen comparing to trapezoidal and square screw threads. Incidentally, when rotating a screw thread one turn, it moves one pitch in the axial direction in the case of a single-screw thread. When the tightening operation is carried out by torque control method, efficiency of screw thread is defined as the ratio of the work done by one pitch movement of screw thread to the work required to rotate a screw thread one turn. Efficiencies of triangular and square screw threads, η_3 and η_4, are expressed by the following equations:

$$\begin{aligned} &\eta_3 = \frac{F_b P}{\pi d_2 U_f} = \frac{\tan\beta}{\tan\left(\rho'_{th} + \beta\right)} \\ &\eta_4 = \frac{\tan\beta}{\tan\left(\rho_{th} + \beta\right)} \end{aligned} \tag{3.22}$$

In order to derive the third equation of η_3 from the second one, eq. (1.10), which calculates β, and Eq. (3.13) are used. It is found that η_4 is larger than η_3 under the condition of identical μ_{th}, because ρ_{th} is smaller than ρ'_{th}. To summarize the above:

a. Triangular screw threads are suitable for the purpose of fastening because of high self-nonloosening performance.
b. Trapezoidal and square screw threads are suitable for movement transmission, because the efficiency of the screw thread is high due to small flank angle.
c. Multiple-thread screws provide higher efficiency of screw thread than single-thread screws, due to large lead angle.

– Learning by numerical example 3.4: Effect of coefficient of friction and nominal diameter on efficiency of screw thread

According to Eq. (3.22), efficiency of screw thread η_3 becomes 100% when μ_{th} is zero, and it rapidly decreases as μ_{th} is increased. Paying attention to the effect of β, η_3 increases with smaller nominal diameter due to larger β, and η_3 of coarse screw threads is higher than that of fine screw threads. Calculation examples of η_3 are tabulated in Table 3.2.

– Learning by numerical example 3.5: Relationship between tightening torque T_t and loosening torque T_l

The ratio of T_l to T_t varies with changes in lead angle β and coefficients of friction, μ_{th} and μ_{nu}. The difference between T_l and T_t increases with lager β and smaller μ_{th} and

Table 3.2 Effect of coefficient of friction and nominal diameter on efficiency of screw thread (%).

μ_{th}	0.05	0.07	0.1	0.12	0.15	0.2	0.3
M16	42.8	34.8	27.1	23.7	19.9	15.6	10.9
M12	46.9	38.7	30.6	26.8	22.6	18	12.7
M16 ($P=1$)	26.4	20.4	15.2	13	10.7	8.2	5.6

Table 3.3 Ratio of loosening torque to tightening torque (%).

μ_{th}, μ_{nu}	0.05	0.07	0.1	0.12	0.15	0.2	0.3
M16	49.5	61	70.8	75	79.3	83.7	88.2
M12	43.2	55.7	66.6	71.2	76.1	81.1	86.3
M16 ($P=1$)	71.6	78.8	84.6	86.9	89.3	91.7	94

μ_{nu}. Thus, the difference gets larger as nominal diameter decreases. It is also larger in coarse screw threads than in fine screw threads. Assuming that axial bolt stress is 100 MPa and coefficients of friction remain unchanged between tightening and loosening operations, the ratio T_l/T_t can be calculated using Eqs. (3.19) and (3.20). The calculations are conducted for M16 with coarse and fine threads and M12 with coarse thread. The results, given in Table 3.3, indicate that in the cases of small nominal diameter and low coefficients of friction, the loosening process requires a considerably smaller torque compared to the tightening process.

3.2.5 Measuring method of axial bolt force, tightening torque, and friction coefficient

For the case of tightening bolt-nut connections by the torque control method, a simple testing method is introduced, by which axial bolt force F_b and thread torque T_1 can be relatively easily measured. Furthermore, it is explained how to calculate coefficients of friction on thread surface μ_{th} and on nut-bearing surface μ_{nu}, by using the measured values of F_b and T_1. The magnitude of tightening torque T_t applied to nut is read from the scale of a torque wrench.

(1) Measurement of bolt force F_b and thread torque T_1

Fig. 3.11 shows an example of bolt specimen, to which strain gages and crossed gages are glued on the bolt cylindrical surface to measure F_b and T_1, respectively. Thread torque T_1 is defined as the torque exerted on thread surface in the tightening process, and it is transmitted to the bolt body. Therefore, T_1 is measured on the bolt cylindrical portion to simplify the experiment. Two strain gages are located 180 degrees apart from one another. Bolt force F_b is obtained by averaging the two measured values. The bending moment exerting on the bolt is calculated by use of a simple theory

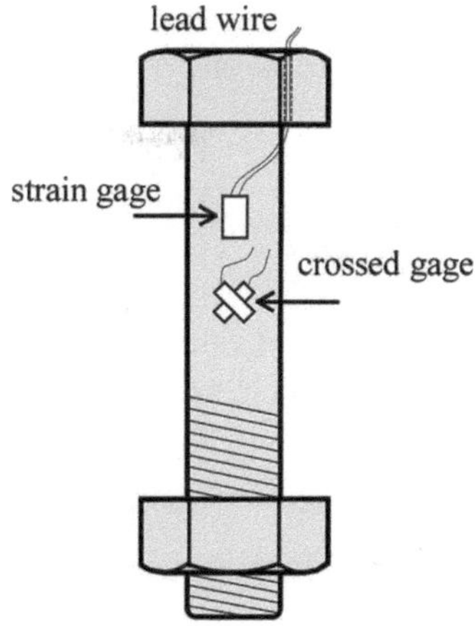

Fig. 3.11 Bolt specimen for measuring bolt force and thread torque.

of solid mechanics, in which the difference of the two values is divided by two. Two strain gages can be replaced by a strain gage for the exclusive use of bolt force measurement. The special gage has a cylindrical shape and is inserted into a hole machined along the bolt axis. Although it is impossible to measure the bending moment, it has a great advantage of easy handling of lead wires. A crossed gage consists of two strain gages arranged at right angles to each other, which enables the measurement of shear strains induced by thread torque T_1. Calculating T_1 using the measured shear strains, nut-bearing torque T_2 is obtained by the equation $T_2 = T_t - T_1$, which is transformed from Eq. (3.1). Thus, four quantities of F_b, T_t, T_1, and T_2 are all obtained.

(2) Measurement of μ_{th} and μ_{nu} separately [70]

Transforming Eq. (3.3) and substituting $d_{nu} = 1.3d$, the amount of μ_{nu} is calculated as follows:

$$\mu_{nu} = \frac{2T_2}{F_b \cdot (1.3d)} \tag{3.23}$$

The formula for calculating μ_{th} is derived from Eqs. (3.12) and (3.14):

$$\mu_{th} = \tan\left\{\tan^{-1}\left(\frac{2T_1}{d_2 F_b}\right) - \beta\right\} \cdot \cos\alpha_1' \tag{3.24}$$

If the measurement of T_2 is possible instead of T_1, μ_{th} and μ_{nu} are calculated in a similar manner, using the relationship $T_1 = T_t - T_2$.

(3) Case of assuming μ_{th} and μ_{nu} as being identical

If the purpose is to find the average value of μ_{th} and μ_{nu}, under the assumption of two values being identical, it is enough to just measure F_b and T_t. In this case, the experiment is quite simple, because the installation of crossed gages is not required. Substituting Eq. (3.12) into Eq. (3.15) and putting $\mu_{th}=\mu_{nu}=\mu$, the following equation is derived.

$$T_t = \frac{1}{2}F_b\left[d_2 \tan\left\{\tan^{-1}\left(\frac{\mu}{\cos\alpha_1'}\right)+\beta\right\}+\mu\times 1.3d\right] \tag{3.25}$$

In Eq. (3.25), since T_t and F_b are measured values, the average coefficient of friction μ is found using a trial-and-error procedure by changing μ systematically. Evaluation of μ corresponds to obtain nut factor K in Eq. (3.16).

(4) Test equipment with simple structure for measuring bolt force and coefficient of friction

In order to measure coefficients of friction with high accuracy by repeating the experiments using a lot of specimens, it is considered that the installation accuracy of strain gages on the bolt cylindrical surface significantly affects the measurement accuracy. In particular, if the direction of strain gages deviates from the bolt axis, it necessarily leads to a decrease in the measurement accuracy of the bolt force. Fig. 3.12 shows test equipment manufactured to solve the problem. Fig. 3.12A illustrates the construction and the measurement principle. Fig. 3.12B illustrates the three parts composing the test equipment. Strain gages and crossed gages are attached not to each test specimen, but to Part-I of the testing equipment. Therefore, bolt force F_b and thread torque T_1 can be simultaneously measured [71] in each test with constant measurement accuracy. The experimental procedure and measurement principle are as follows.

Attaching a bolt, a nut, and a fastened plate to the test equipment, tightening torque T_t is applied to the nut, then the bolt force F_b is generated. The fastened plate is thin and is carefully machined so that the top and bottom surfaces are completely parallel. During the tightening operation, nut-bearing torque T_2 is consumed as friction work and the remaining torque, $T_1=T_t-T_2$, is transmitted to the bolt body as a twisting moment. Inside the top and bottom of Part-II, two grooves are machined so as to insert the bolt head and the top portion of Part-I, by which Part-II can transmit bolt force F_b and thread torque T_1 to Part-I. The amounts of F_b and T_1 are measured by strain gages and crossed gages mounted on Part-I, respectively. This test equipment makes it possible to measure F_b and T_1 without attaching strain gages to the bolt body. The amount of tightening torque T_t is directly read from the scale of a torque wrench. The calculation procedure of μ_{th} and μ_{nu} have already been explained in (2).

3.2.6 Behavior of torque and axial bolt force after releasing tightening torque

In this section, how torque and bolt force behave after the completion of tightening operation in the torque control method is discussed. Completing the tightening work and releasing the hand from a torque wrench, the wrench slightly rotates in the opposite direction, and then the tightening torque T_t becomes zero [72,73]. As shown in Eq. (3.1), the tightening torque applied to nut T_t is the sum of thread torque T_1, which

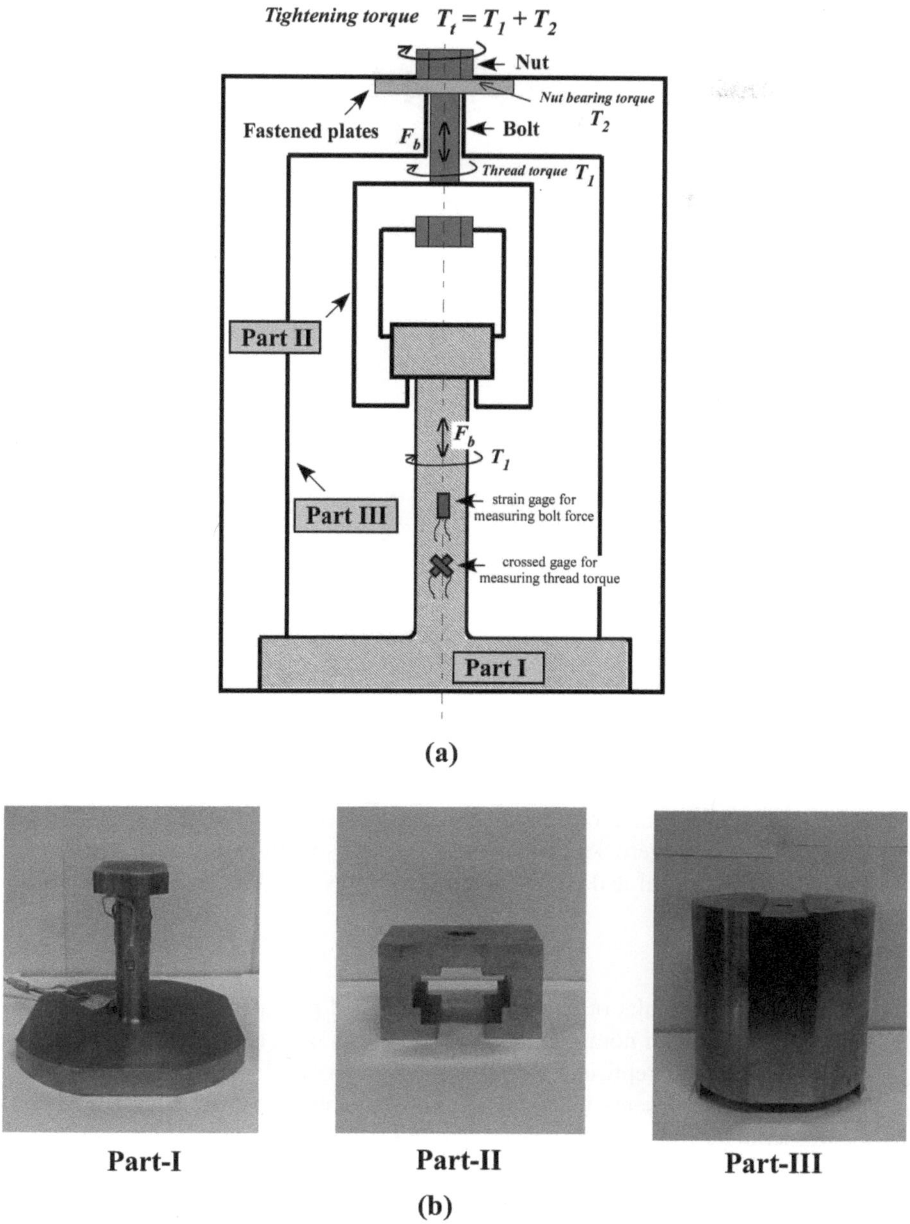

Fig. 3.12 Test equipment with simple structure for measuring bolt force and coefficient of friction. (A) Construction and measurement principle of test equipment and (B) three parts composing test equipment.

is transmitted to the bolt body, and nut-bearing torque T_2 consumed as friction work. In Fig. 3.13, it is schematically shown how the torques exerted on each part vary during a very short period between the achievement of the objective tightening torque T_t and the disappearance of T_t after releasing the grip of the wrench [74].

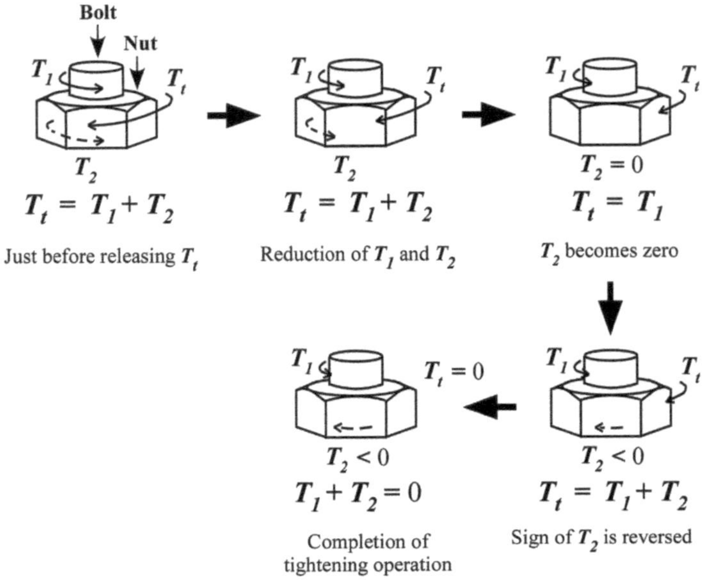

Fig. 3.13 Behavior of torques exerted on each part during the whole tightening process.

As tightening torque T_t starts to reduce, thread torque T_1 and nut-bearing torque T_2 also decrease correspondingly. Eq. (3.1) holds true during this time. Eventually, T_2 becomes negative past zero. The reversal of the sign implies that the direction of friction force is reversed on the nut-bearing surface. After that, T_t continuously decreases and becomes zero. At that moment, T_1 and T_2 are in balance. In other words, the two torques are equal and opposite and satisfy the following equation:

$$T_1 + T_2 = 0 \tag{3.26}$$

Fig. 3.14 shows the results of experiments conducted to verify the phenomenon. A bolt-nut connection with nominal diameter of M16 is used and the grip length L_f is 35 mm. Two ordinates represent bolt force F_b and coefficients of friction, μ_{th} and μ_{nu}. The abscissa represents the elapsed time concerning the tightening operation. As the tightening torque T_t is increased, μ_{th} and μ_{nu} decrease and approach constant values. When the grip of the wrench is released and T_t approaches zero, the sign of μ_{nu} is changed, because the sliding direction on the nut-bearing surface is reversed. Although not shown here, the magnitudes of T_1 and T_2 drop to about 70%–80% of the values at the completion of tightening operation. As for F_b, it increases only slightly by showing a tiny peak just after releasing the hand, and then F_b becomes a constant value. This specific behavior is presumably caused by the release of the twisting deformation of the bolt body. That is, when letting go of the wrench, T_1 decreases and the bolt body rotates only slightly so as to release the twisting deformation, and the rotation occurs in the direction to climb the thread surface.

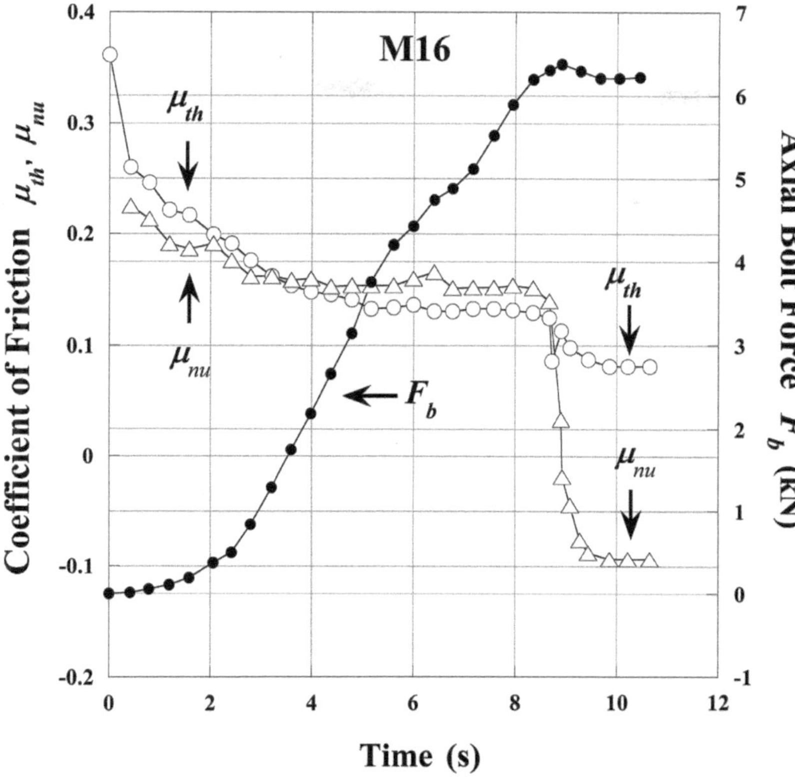

Fig. 3.14 Variations of bolt force and coefficient of friction with time during the tightening operation.

3.2.7 Simple strategy for reducing bolt force scatter [75]

It is of critical importance to reduce the scatter in bolt force when employing the torque control method. Referring to Eq. (3.16), once the objective bolt force F_b and nominal diameter of bolt d to be used are determined, the amount of tightening torque T_t is calculated by assuming nut factor K. This means that high accuracy in bolt force can be achieved by choosing an appropriate value of K. Meanwhile, as found from Eq. (3.18), K can be expressed as a function of coefficients of friction μ_{th} on thread surface and μ_{nu} on nut-bearing surface. Accordingly, the point is to reduce the scatters in μ_{th} and μ_{nu}.

Incidentally, it is a common practice to apply an identical surface treatment to a pair of bolt and nut, which means the thread surface is composed of two surfaces having a similar surface condition. On the other hand, the nut-bearing surface is generally composed of two surfaces subjected to different surface treatments. As a result, it is presumed that the scatter in μ_{nu} is usually larger than that in μ_{th}. Use of flat washers, to which the same surface treatment is applied, is a simple solution to lower the scatter in μ_{nu}. The strategy is that the flat washer is used to make the nut-bearing surface being

Fig. 3.15 Test specimen consisting of bolt, nut, flat washer, and plate.

composed of two surfaces under the same treatment. The validity of the use of such flat washers is examined by tightening experiments using the test equipment shown in Fig. 3.12.

100 pairs of bolt and nut are prepared together with 50 flat washers. Test bolts and nuts are M16 with coarse thread and are made of carbon steel with strength grade of 8.8. Flat washers are also made of the same material. The surfaces of bolt and nut threads and flat washers are all processed using black oxide film and antirust oil. Fig. 3.15 shows a set of test specimens consisting of a bolt, a nut, a flat washer, and a fastened plate. To examine the effectiveness of the flat washer, tightening experiments are conducted 100 times, using 50 sets of test specimens with flat washer and 50 without flat washer. Fig. 3.16A and B represent the measured results of μ_{th} and μ_{nu} for the cases with or without flat washers. Noticeable differences are not observed in μ_{th} between the cases with and without flat washers. However, the scatter in μ_{nu} is significantly reduced by use of flat washers. Next, the scatter in bolt force F_b is evaluated using nut factor K, where the measured values of μ_{th} and μ_{nu} are substituted into Eq. (3.18). Tightening experiments corresponding to Fig. 3.16 are repeated five times. It is shown in Fig. 3.17 how much scatter in bolt force is predicted using the standard deviation σ of nut factor K. Supposing that the range represented by 2σ, which involves about 95% of the total data, is an index for the scatter, the bolt force scatter ranges from about 15% to a little over 20% without flat washers. In contrast, the scatter is wholly kept around 10% with the use of flat washers. Additionally, antiloosening performance of flat washer is discussed in Chapter 6.

3.2.8 Tightening characteristics and strength of bottoming studs

When clamping studs, the thread runout portion is sometimes screwed into body-side female threads to prevent the falling off from the main body in the disassembly process, caused by the co-rotation with a nut. In that case, high stress concentration occurs around the thread runout. Instead of screwing into the female threads, there is an intriguing countermeasure: the far end of the stud is pushed onto the bottom surface

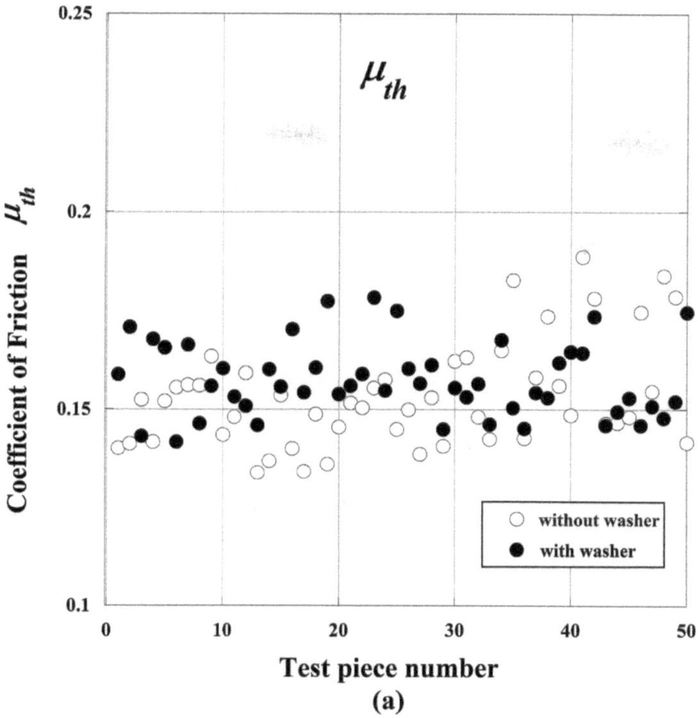

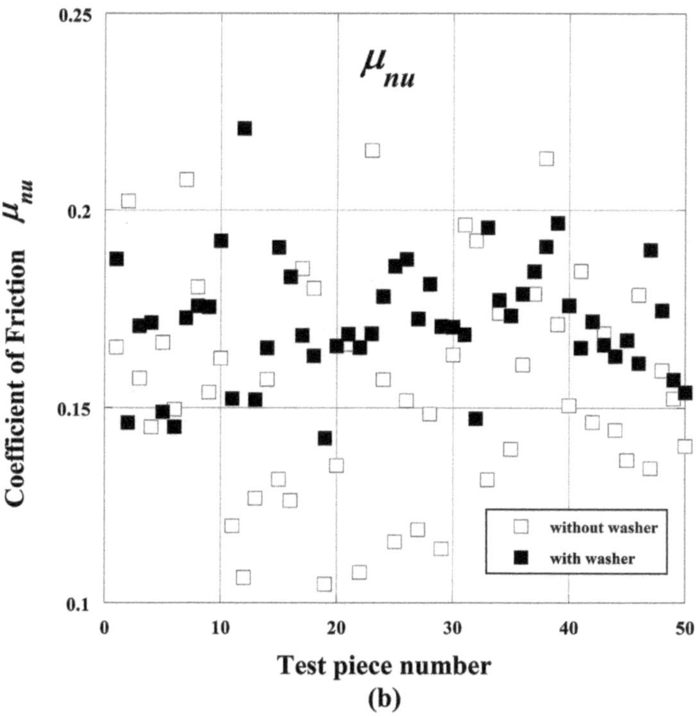

Fig. 3.16 Measured values of μ_{th} and μ_{nu} with or without flat washer. (A) μ_{th} and (B) μ_{nu}.

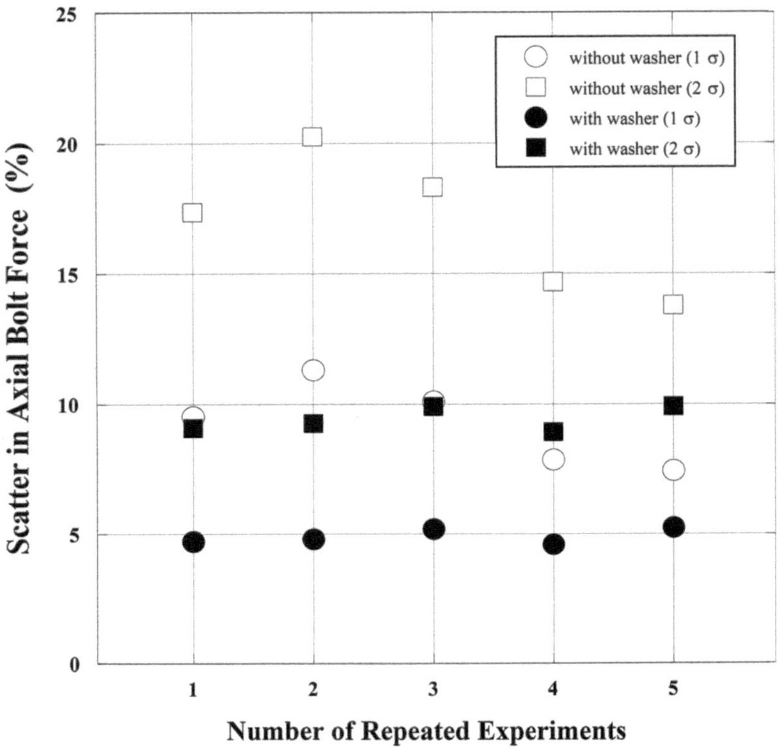

Fig. 3.17 Bolt force scatter predicted by use of standard deviation of nut factor.

of the bolt hole. This type of stud is called a bottoming stud. When a bottoming stud is pushed onto the bottom surface by applying a torque above a certain level, it never falls off in the disassembly process caused by co-rotation. Therefore, it has been said that bottoming studs are superior to ordinary studs in both workability and strength.

In Fig. 3.18, the shape of a bottoming stud is schematically shown. The end of the projection part with a diameter of d_c has a conical shape so as to be smoothly in contact with the bottom surface. The tip of the projection is often machined to be hollow, as shown in the figure, to prevent crack initiation. Fig. 3.19 shows the assembly and disassembly processes of bottoming studs [64]. When applying the bottoming torque T_{bt}, a certain amount of compression force F_{bt} is generated between the lower engaged threads and the bottom surface of bolt hole, thereby subjecting the engaged thread portion to compression stress. Next, a nut is attached to the top of the stud and the tightening torque T_t is applied to produce the objective bolt force F_b. The absolute value of F_b is larger than that of F_{bt}. Accordingly, the application of tightening torque decreases the compression force occurring around the lower engaged threads in the bottoming process and changes the compression force into a positive value past zero. At the same time, it reduces the compression force F_{bt} acting on the bottom surface of the bolt hole.

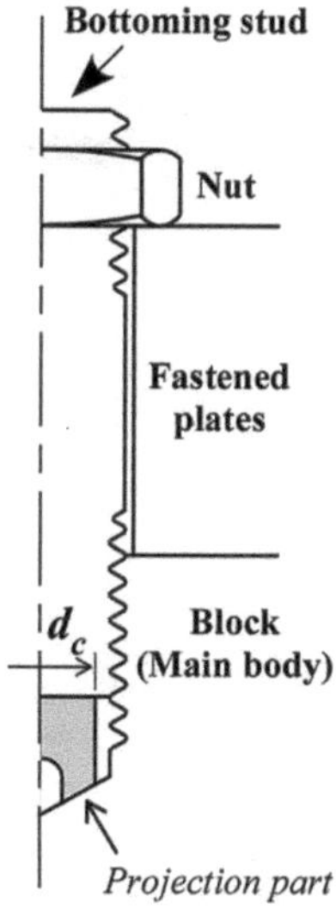

Fig. 3.18 Shape of bottoming stud.

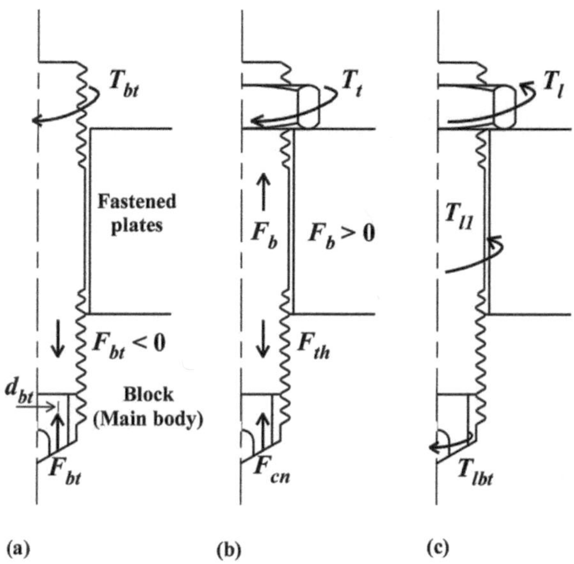

Fig. 3.19 Assembly and disassembly processes of bottoming stud. (A) Bottoming process, (B) tightening process, and (C) disassembly process.

From the above consideration, it is found that at the completion of the total tightening process, the objective bolt force F_b, the axial force generated in the lower engaged threads F_{th}, and the compression force acting on the bottom surface F_{cn} are related by the following equation:

$$F_{th} = F_b + F_{cn} \tag{3.27}$$

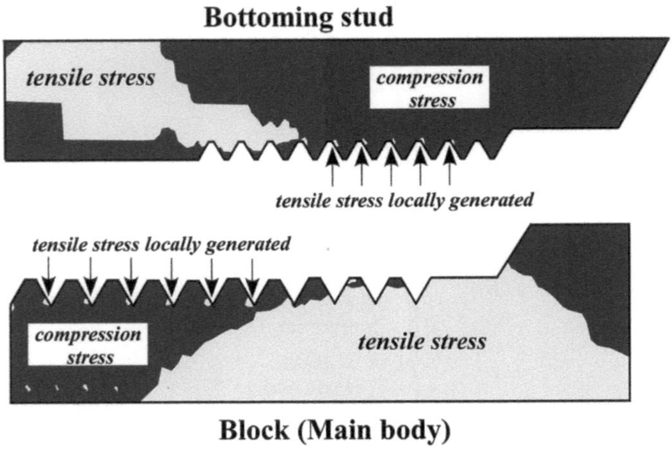

Fig. 3.20 Tensile stresses locally produced around thread root in bottoming process.

Since the bottoming operation produces a compression force in the lower engaged threads, the tensile force occurring there, at the completion of the tightening operation, is necessarily smaller comparing to the case without bottoming. Therefore, it is predicted that the maximum stress is decreased and the strength of the engaged threads increased. However, it has been reported that although its static strength is higher than the studs used without bottoming, the fatigue strength is likely to be lower [76]. These two phenomena are contradictory from the engineering point of view. The apparently conflicting consequences can be explained by examining the stress distributions around the engaged threads in the bottoming process.

In Fig. 3.20, the regions under compression and tension in the bottoming process are indicated by dark gray and light gray, respectively [64]. The engaged thread portion of the bottoming stud is almost entirely under compression. However, tensile stresses are locally produced in the vicinity of the thread root. This means that in the bottoming process, a certain amount of tensile stress is already produced around the thread root and its magnitude is further increased when tightening a nut. From the above considerations, it is presumed that the static and fatigue strengths of bottoming studs are both likely to be lower than those of ordinary studs. It is shown in [64], based on the numerical results by FEM, that the maximum stress occurring at thread root is slightly raised by the bottoming operation.

In order to take advantage of bottoming studs, therefore, it is necessary to obtain the minimum bottoming torque T_{bt} that won't cause the co-rotation and falling-off of the bottoming stud in the disassembly process. The condition needed to prevent the co-rotation is given in a simple form, using the symbols in Fig. 3.19:

$$T_{l1} < T_{lbt} \tag{3.28}$$

In actual application, it is important to predict how much bottoming torque T_{bt} is required against the tightening torque T_t to be applied for achieving the target bolt

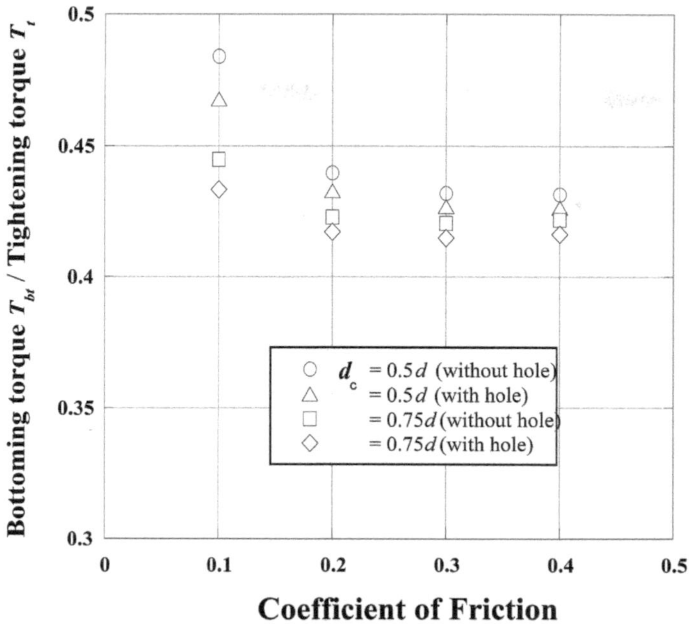

Fig. 3.21 Minimum bottoming torque to avoid co-rotation of studs.

force F_b. Fessler et al. evaluated the mechanical behavior of bottoming studs by means of photo-elastic experiments [77]. They recommend that T_{bt} should be as large as three times thread torque T_1. Such a large bottoming torque probably degrades the strength of engaged threads in the bottoming process. Hence, aiming at providing a guideline for the bottoming operation, the minimum value of T_{bt}/T_t has been evaluated by FEM [64].

In Fig. 3.21, the ordinate represents the minimum ratio of T_{bt} to T_t that won't cause co-rotation and the abscissa represents the coefficient of friction. The relationship between torque and bolt force in the bottoming process is evaluated by FEM, and the equation relating T_t to F_b in the tightening process is also taken into account. The value of T_{bt}/T_t is changed according to the diameter of the projection part d_c and the tip shape, which is solid or hollow. As a result, T_{bt}/T_t is less than 0.5 overall. It is therefore concluded that to prevent the co-rotation of bottoming studs in the disassembly process, the magnitude of bottoming torque T_{bt} should be more than 50% of the tightening torque T_t.

3.2.9 Bolt strength in tightening process

When tightening a bolt-nut connection with the torque control method, the threaded portion of the bolt is subjected to tensile stress σ_{th} and shear stress τ_{th}, due to bolt force F_b and thread torque T_1, respectively. Calculations of σ_{th} and τ_{th} are performed with the use of diameter of stress area d_s.

$$\sigma_{th} = \frac{4F_b}{\pi {d_S}^2} \tag{3.29}$$

$$\tau_{th} = \frac{16T_1}{\pi {d_S}^3} \tag{3.30}$$

In the case of a bolt-nut connection made of ductile materials, by applying shear strain energy theory, the strength of engaged threads can be evaluated by von Mises stress $\overline{\sigma}$ defined below:

$$\overline{\sigma} = \sqrt{{\sigma_{th}}^2 + 3{\tau_{th}}^2} \tag{3.31}$$

When $\overline{\sigma}$ reaches the yield stress σ_Y in the simple tension test, it can be judged as the start of plastic deformations. Yamamoto proposed that Eq. (3.31) agrees well with the experimental results by changing the coefficient of τ_{th} from 3 to 1.8 [78]. The value of 1.8 corresponds to the case that τ_{th} is replaced by $0.78\tau_{th}$ in Eq. (3.31). It is worth noting that the value of 0.78 is nearly equal to the residual rate of thread torque T_1 at the completion of tightening operation, ranging from 70% to 80%, as discussed in Section 3.2.6. von Mises stress, defined by Eq. (3.31), represents the mean stress of engaged threads. To evaluate the strength with higher accuracy, the stress concentration around thread root must be considered, which is explained in section 4.2. In the finite element analyses discussed in later sections, von Mises stress $\overline{\sigma}$ is evaluated by the equation using three components of principal stress, σ_1, σ_2, and σ_3:

$$\overline{\sigma} = \frac{1}{\sqrt{2}}\sqrt{\left\{(\sigma_1 - \sigma_2)^2 + (\sigma_2 - \sigma_3)^2 + (\sigma_3 - \sigma_1)^2\right\}} \tag{3.32}$$

- Learning by numerical example 3.6: Relationship between von Mises stress and coefficient of friction (part 1)

Assuming the objective bolt stress σ_b to be 100 MPa for M16 and M12 with coarse thread, von Mises stresses $\overline{\sigma}$ in the engaged threads are calculated for coefficient of friction μ_{th} on thread surface, ranging from 0.05 to 0.30. First, bolt force F_b for $\sigma_b = 100$ MPa is calculated. Then, thread torque T_1 is obtained by use of Eq. (3.14). Tensile stress σ_{th} and shear stress τ_{th} produced in the engaged threads are calculated by Eqs. (3.29) and (3.30), then von Mises stress $\overline{\sigma}$ is obtained by substituting σ_{th} and τ_{th} into Eq. (3.31). The results are given in Table 3.4. As μ_{th} is increased, shear stress τ_{th} increases due to the increase of thread torque T_1. When μ_{th} exceeds 0.25, in either case, von Mises stress $\overline{\sigma}$ becomes more than twice as large as the objective bolt stress σ_b.

- Learning by numerical example 3.7: Relationship between von Mises stress and coefficient of friction (part 2)

Assuming that the objective bolt stress σ_b is 300 MPa for M16 with coarse thread and μ_{th} and μ_{nu} are both 0.12, the required tightening torque T_t is calculated to be

Table 3.4 Relationship between coefficient of friction and von Mises stress (MPa) in engaged threads.

μ_{th}	0.05	0.07	0.1	0.12	0.15	0.2	0.3
M16	136.6	140.7	148.1	153.7	163.3	181.4	223.6
M12	144.4	149.0	157.2	163.4	173.8	193.4	238.7

Table 3.5 Variations of bolt stress under constant torque for varying μ_{th} and μ_{nu}.

μ_{th}, μ_{nu}	0.05	0.07	0.1	0.12	0.15	0.2	0.3
Bolt stress (MPa)	615.1	473.3	351.5	300.0	245.9	189.0	129.0
von Mises stress (MPa)	840.4	665.8	520.4	461.2	401.5	342.8	288.5

156.4 Nm. In the actual application, the torque value is commonly used as a guideline for the tightening operation. However, the bolt stress to be produced changes according to the variation of coefficients of friction, even if applying a constant torque. Table 3.5 shows the bolt stress σ_b and von Mises stress $\overline{\sigma}$ in the engaged threads calculated for coefficients of friction varying from 0.05 to 0.30. In the calculations, the tightening torque T_t is supposed to be a constant value of 156.4 Nm. Using Eqs. (3.15) and (3.14), bolt force F_b and thread torque T_1 are calculated for each coefficient of friction. The subsequent calculation procedure is the same as numerical example 3.6. It is found from Table 3.5 that if coefficients of friction substantially deviate from the expected values, bolt stress σ_b and von Mises stress $\overline{\sigma}$ significantly deviate from the objective values.

3.3 Elastic angle control method

3.3.1 Tightening principle

In this section, the tightening process of the elastic angle control method is explained from the mechanical point of view. It is commonly used for clamping critical large bolts, such as crank pin bolts of large-scale marine diesel engines. Among the various types of joints clamped by threaded fasteners, some joints are disassembled periodically after a certain period of running time. The necessity of the repetition of assembly and disassembly is determined by the type and function of the target machines and structures.

The elastic angle control method is mainly used for critical joints, in which a failure or rupture of a single threaded fastener can lead to a catastrophic accident of structures or machines. For example, crank pin bolts, which are used to tighten a connecting rod and a crank shaft in large diesel engines, need to be inspected regularly. Since the

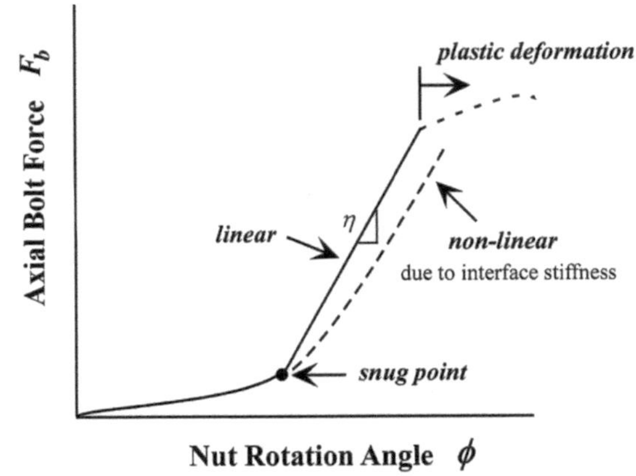

Fig. 3.22 Relationship between bolt force and nut rotation angle in the elastic angle control method.

assembly and disassembly operations are repeated, crank pin bolts are supposed to be reused. In those cases, the target bolts are basically tightened within an elastic range to avoid excessive plastic deformations.

Fig. 3.22 shows the relationship between bolt force F_b and nut rotation angle ϕ in the elastic angle control method. According to JIS B 1083 "General rules for tightening screw threads," the objective screw thread is first tightened by the torque control method to the point termed snug point, where the linear relationship between F_b and ϕ is almost established. It is recommended that the value of snug torque T_{sng} should be selected to generate a bolt force as small as possible, in order not to introduce the disadvantage of the torque control method.

The relationship between F_b and ϕ is expressed by the gradient η in the region that is assumed linear in Fig. 3.22, using spring constants k_{th}, k_s, k_{cyl}, and k_{hd} of a bolt-nut connection and k_f of fastened plates. The formula relating ϕ to F_b is shown below:

$$\begin{aligned}\phi &= \frac{2\pi}{P}F_b\left(\frac{1}{k_{th}}+\frac{1}{k_s}+\frac{1}{k_{cyl}}+\frac{1}{k_{hd}}+\frac{1}{k_f}\right)=\frac{2\pi}{P}F_b\left(\frac{1}{k_b}+\frac{1}{k_f}\right)\\ &= \frac{2\pi}{P}\frac{F_b}{k_{total}}\end{aligned} \tag{3.33}$$

where P is thread pitch, and k_b and k_{total}, respectively, represent the stiffness of bolt-nut connection and the entire stiffness of bolted joint given in eqs. (2.4) and (2.5).

Eq. (3.33) is derived by considering the relationship that when tightening a bolt by rotating a nut, one rotation (2π) produces a one-pitch movement in the axial direction, which is equal to the sum of the axial displacements obtained by dividing bolt force F_b by the spring constants corresponding to each part of the bolted joint. It is found from Eq. (3.33) that the rotation angle ϕ, required for producing the same amount of F_b, increases with smaller thread pitch P and lower stiffness of the jointed portion. Since higher tightening accuracy is expected as the required rotation angle becomes larger, the elastic angle control method is suited for joints with large grip length. In other words, the method is not suitable for clamping thin plates.

According to Eq. (3.33), the bolt force F_b to be produced increases in proportion to nut rotation angle ϕ. In the actual application, however, the relationship between F_b and ϕ is not perfectly linear, owing to the effects of surface roughness, warping, and geometric errors of the jointed portion. Among the foregoing influencing factors, it is considered that the effect of interface stiffness due to surface roughness, discussed in section 2.6, is fairly large. Consequently, the rotation angle ϕ required to achieve the objective bolt force F_b is considerably larger in some cases than the angle calculated by Eq. (3.33). To cope with this problem, it is recommended in JIS that after making a complete contact on the nut-bearing surface by applying snug torque T_{sng}, the target bolt force should be generated by controlling the nut rotation angle ϕ. In that sense, it is interpreted that Eq. (3.33) represents the relationship between nut rotation angle ϕ and bolt force F_b in the tightening operation after applying snug torque T_{sng}. However, no concrete guideline is given in JIS for the amount of snug torque. From the above considerations, it is concluded that the following information is required in order to perform the elastic angle control method with high accuracy in application:

(1) Evaluation method of spring constants composing a bolted joint, which are involved in Eq. (3.33), and its accuracy.
(2) Evaluation method of interface stiffness due to surface roughness.
(3) Guideline for an appropriate amount of snug torque.

In the next section, an equation relating nut rotation angle to bolt force is derived by considering the effect of interface stiffness. Furthermore, in Section 3.3.3, a guideline for the elastic angle control method is proposed, in which a concrete procedure to determine the amount of snug torque is explained.

Next, to examine the effect of nominal diameter d, the nut rotation angle ϕ required to produce the bolt stress of 100 MPa is calculated for the case of the ratio of grip length L_f to d being 6. For bolted joints with similar forms tightened by a bolt-nut pair of M12, M16, M24, and M36, the amounts of ϕ are calculated to be 14.5, 16.5, 16.4, and 18.2 degrees, respectively. This result implies that the rotation angle required to produce the same bolt stress slightly increases as nominal diameter is increased.

The relationship between nut rotation angle ϕ and nominal diameter d can be explained using Eq. (3.33) by focusing on the mutual relationship among nominal diameter d, thread pitch P, and spring constants k_b and k_f. Regarding the effect of joint materials, the required rotation angle increases in inverse proportion to Young's modulus, since the amounts of k_b and k_f are proportional to Young's modulus. For example, when applying the elastic angle control method to the joints made of titanium or aluminum alloy, the required rotation angles become almost twice or three times as large as that of carbon steel joints.

– Learning by numerical example 3.8: Effect of grip length and nominal diameter in the elastic angle control method

For a bolt-nut connection of M16 with coarse thread, nut rotation angles required to produce bolt stress of 100 MPa are calculated using Eq. (3.33), where L_f/d is changed to 2, 4, 6, 8, and 10. Other conditions are supposed to be the same as numerical example 2.1. The required rotation angles ϕ are calculated to be 7.1, 11.8, 16.5, 21.2, and

25.9 degrees, respectively, for each value of L_f/d. These results indicate that the nut rotation angle required to produce the same bolt stress increases with larger grip length, but the rate of increase is smaller compared to the rate of increase in grip length. This nonlinear behavior between ϕ and L_f can be explained in Eq. (3.33) by considering the relationship between spring constants, k_b and k_f, and grip length L_f, which is discussed in numerical example 2.1.

3.3.2 Equation relating axial bolt force to nut rotation angle by taking account of surface roughness

Considering the approach of interface ζ due to surface roughness of each contact surface, Eq. (3.33) is rewritten in the following form [54]:

$$\phi = \frac{2\pi}{P}\left(\frac{F_b}{k_{total}} + \zeta_{th} + \zeta_{nu} + \zeta_{hd} + \zeta_f\right) \tag{3.34}$$

where ζ_{th}, ζ_{nu}, ζ_{hd}, and ζ_f represent the approaches of interface on thread surface, nut-bearing surface, bolt head-bearing surface, and plate interface, respectively.

The relationship between surface roughness and approach of interface is explained in section 2.6, and detailed discussions are given in [54]. In the case of surface roughness of each contact surface being sufficiently small, the nut rotation angle ϕ required for the objective bolt force F_b can be calculated by Eq. (3.33) with practical accuracy, because the effect of interface stiffness is negligible.

Fig. 3.23 shows an experimental setup for measuring the relationship between nut rotation angle ϕ and bolt stress σ_b. Nut rotation angle is measured using a potentiometer. Some measurement results of the relationship between σ_b and ϕ are given in Fig. 3.24. The experiments are conducted using bolt-nut connections of M12. The experimental results show a considerable scatter, despite using test specimens of the same specifications. However, by moving the reference points for measuring ϕ, as shown in the figure, three experimental results are in approximate agreement. It is found from the measurement results that the relationship between σ_b and ϕ shows a noticeable nonlinearity in the region of low σ_b. Accordingly, when tightening operation is performed based on Eq. (3.33) or Eq. (3.34), it is vital that snug torque T_{sng} is first applied in order to make close contact on thread surface, nut bearing-surface, and bolt head-bearing surface. Hence, the target bolt force is achieved with sufficient accuracy:

In Fig. 3.25, the experimental results given in Fig. 3.24 are compared to three types of numerical results, which are obtained by Eq. (3.33), represented by linear spring models and finite element analysis, and Eq. (3.34), which accounts for interface stiffness. The gradients given in the figure represent the bolt stress produced by one degree of rotation angle, which are obtained by connecting the measurement results of $\sigma_b = 100$ MPa and $\sigma_b = 400$ MPa with a straight line. The experimental results suggest that the relationship between σ_b and ϕ is still somewhat nonlinear even under high bolt stress, and the numerical results obtained by Eq. (3.34), in which interface stiffness is taken into account, agree well with the experimental ones.

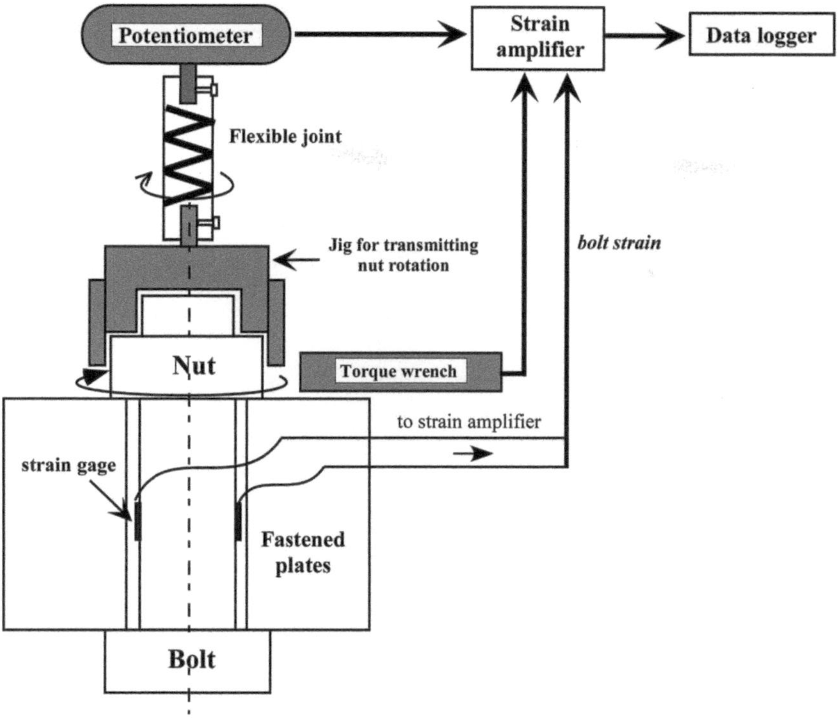

Fig. 3.23 Experimental setup for simultaneously measuring nut rotation angle and bolt stress.

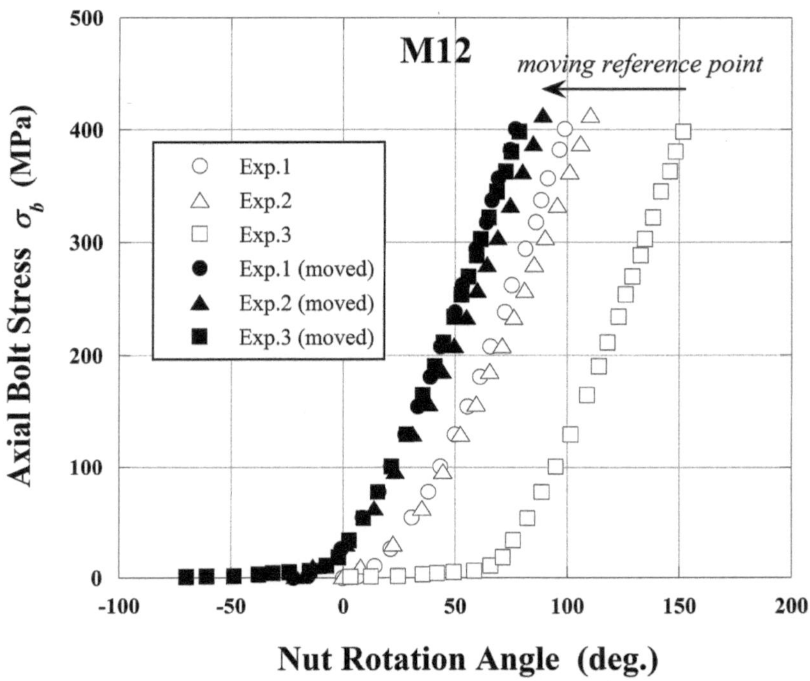

Fig. 3.24 Measurement results of bolt stress and nut rotation angle.

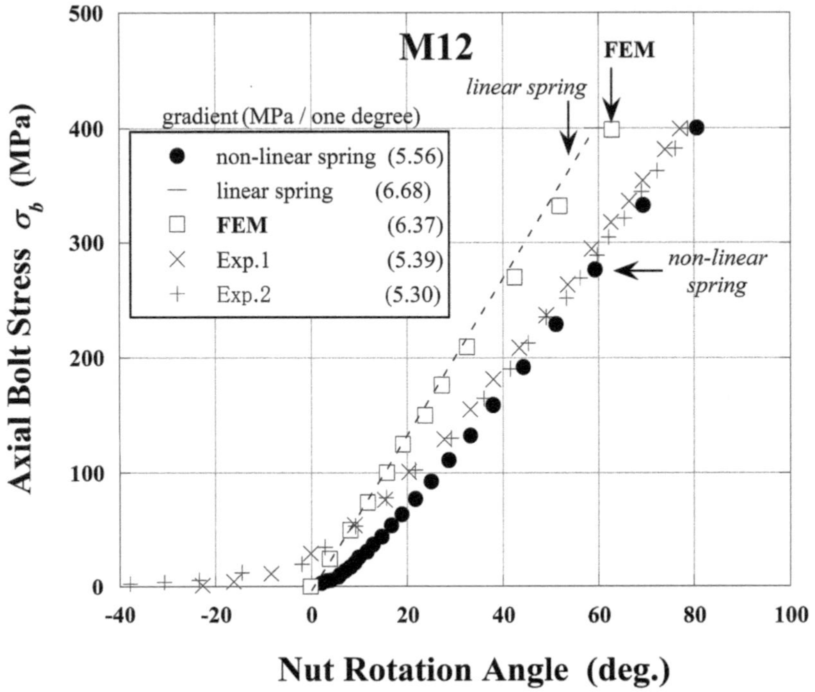

Fig. 3.25 Comparison of experimental results with three types of numerical results.

3.3.3 Application range and tightening guidelines

In Fig. 3.26, the relationship between bolt stress σ_b and nut rotation angle ϕ calculated by Eq. (3.34) are presented, in which the sum of maximum height roughness Rz of the mating surfaces, Rzt, is systematically changed. The grip length L_f is set to be eight times the nominal diameter d. The nut rotation angle ϕ required to produce the same bolt stress σ_b increases with increasing Rzt. Fig. 3.27 represents the effect of grip length L_f on the relationship between σ_b and ϕ. As in the case of Rzt, the rotation angle required for the same bolt stress increases as L_f/d is increased. In other words, the required rotation angle for the objective bolt stress increases when using longer bolts, and then the effect of Rzt relatively gets smaller, thereby reducing the scatter in bolt stress.

In the following, the types of bolted joints appropriate for the elastic angle control method are discussed, and some guidelines are given to take advantage of its tightening characteristics. The tightening accuracy is basically enhanced as larger rotation angle is required for the same target bolt stress.

Suitable application range

(1) Jointed portions clamped with high bolt stress

In the case of bolt stress σ_b being low, surface roughness shows a large effect on the relationship between bolt stress and nut rotation angle. Therefore, due to the difficulty

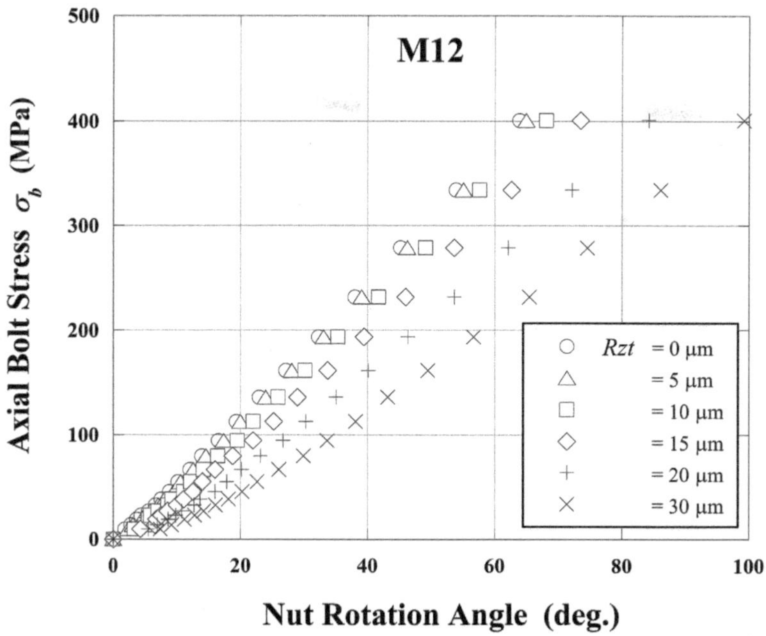

Fig. 3.26 Effect of surface roughness on the relationship between bolt stress and nut rotation angle.

in determining the appropriate snug torque T_{sng}, the tightening accuracy is necessarily lowered. The preferable objective bolt stress σ_b is, for instance, above 300 MPa.

(2) Jointed portions with large grip length

Since a large rotation angle is required for the objective bolt force F_b, the effect of engaged threads, whose stiffness is difficult to evaluate with high accuracy, becomes relatively small.

(3) Jointed portions clamped by threaded fasteners with fine thread

The rotation angle ϕ required to produce the same bolt force is inversely proportional to thread pitch P. Accordingly, when using fine screw threads, higher tightening accuracy is expected than using coarse screw threads, since larger ϕ is required.

(4) Jointed portions with small surface roughness

Since surface roughness has a small or negligible effect on the relationship between bolt force and nut rotation angle, high tightening accuracy is expected.

(5) Jointed portions composed of materials with low stiffness

When Young's modulus of joint materials is low, the required rotation angle becomes large, as explained in numerical example 3.8. Hence, high tightening accuracy is expected due to the same reason as in (2) and (3).

In relation to (2), it is predicted that bolt force shows a considerable scatter with smaller grip length because of the effect of surface roughness and geometric errors

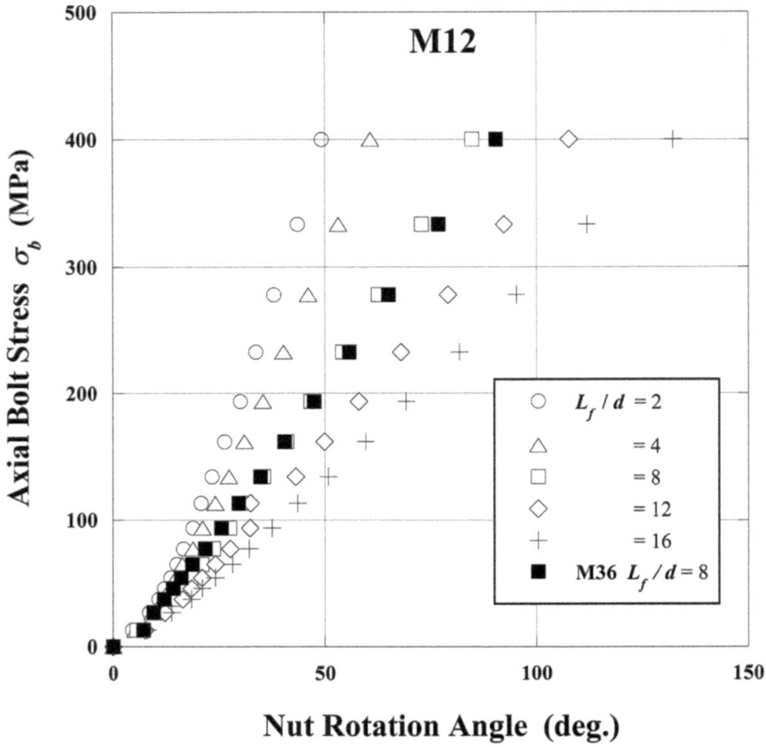

Fig. 3.27 Effect of grip length on the relationship between bolt stress and nut rotation angle.

of the jointed portion. Specifically, it is fairly difficult to evaluate the spring constant of engaged threads k_{th} with high accuracy, as shown in fig. 2.8. That is, the existence of only small geometric errors can significantly reduce the stiffness, and the amount of k_{th} may drop to the low value of stiffness comparable to the spring constant of fastened plates k_f. Therefore, the geometric errors can greatly affect the tightening accuracy of elastic angle control method. As a countermeasure, for instance, grip length L_f is increased by devising the joint geometry to lower the stiffness of fastened plates and then k_f is decreased. In any event, it is important that the elastic angle control method should not be applied to joints with very small grip length.

Guidelines for tightening operation

(1) Snug torque T_{sng} is selected to generate axial bolt stress σ_b, say, from 50 to 100 MPa. Then, the target bolt stress σ_b is converted to bolt force F_b. The magnitude of T_{sng} is determined by Eq. (3.15) or Eq. (3.16), with which the jointed portion is tightened. When the effect of surface roughness and geometric errors does not matter too much, snug torque can be lower than the above recommended value, because the accuracy in bolt stress achieved by nut angle control is essentially high.

(2) Next, the target bolt is tightened until the objective bolt force F_b is attained, based on the equation relating F_b to ϕ. The required rotation angle is calculated by Eq. (3.34). When surface roughness and geometric errors are both small, Eq. (3.33) can be used instead of Eq. (3.34). Attention must be paid that the amount of bolt force substituted into Eq. (3.33) or Eq. (3.34) is obtained by subtracting the bolt force generated by snug torque from the objective bolt force. The tightening guideline is explained in more detail in [54].

3.4 Direct tension method using hydraulic tensioner

3.4.1 Tightening principle

The direct tension method using hydraulic tensioners is widely applied to the joints in large diesel engines, nuclear power plants, chemical plants, etc., where bolt force needs to be controlled with high accuracy. It is said that by use of the direct tension method, the bolt forces can be controlled within an error of a few percent.

In the direct tension method, a hydraulic pump is connected to a tensioner, shown in Fig. 3.1, and the tensile force is directly applied to the target bolt. Fig. 3.28 illustrates the construction of a hydraulic tensioner. A grip nut is attached to the thread tip of the target bolt to apply tensile force. The grip nut is pushed up by a piston subjected to hydraulic oil pressure and generates an axial force in the target bolt. In the figure, the piston and the grip nut are movable parts. The tightening process using a hydraulic tensioner consists of four steps, as shown in Fig. 3.29.

Step 1: A pair of bolt and nut is attached to the jointed components and the nut is tightened finger tight. Next, a hydraulic tensioner is installed around the target bolt and the grip nut is attached to the tip of the bolt thread. No bolt force is generated in this state.

Step 2: The hydraulic pump is started. When a piston pushes up the grip nut by the action of hydraulic oil pressure, initial tension F_t is generated in the bolt. At that moment, the nut-bearing surface moves away from the plate surface due to bolt elongation.

Step 3: An appropriate amount of torque, which is termed nut rundown torque, is applied to the nut in order to bring the nut-bearing surface and plate surface into close contact. In the case of tightening not very large bolts, bolt force may be slightly increased due to nut rundown torque. However, when tightening large bolts, the increase in bolt force due to nut rundown torque is almost negligible compared to initial tension F_t.

Step 4: Removing hydraulic oil pressure, the tightening operation is completed. Then, the bolt force is decreased from initial tension F_t to final bolt force F_f, since the nut-bearing surface deforms downward by compressing the plate surface.

F_f corresponds to the objective bolt force F_b. Therefore, initial tension F_t is larger than F_f. The ratio of F_f to F_t is the most important value in the tightening operation by hydraulic tensioners. The amount of nut rundown torque, explained in Step 3, is also a major factor controlling the tightening accuracy. If the contact pressure on the nut-bearing surface is too low because of a small nut rundown torque, the deformation of microprojections existing on the surface significantly affects the tightening process

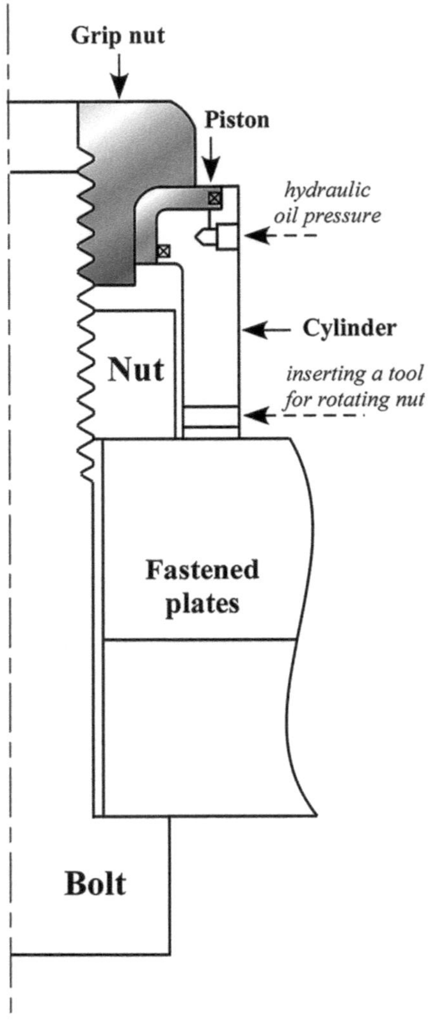

Fig. 3.28 Basic structure of a hydraulic tensioner.

in Step 4. As a result, the bolt force obtained is sometimes noticeably lower than the objective one.

3.4.2 Effective tensile coefficient

The ratio of objective bolt force F_f to initial tension F_t, which is applied in advance by using a hydraulic tensioner, is defined as effective tensile coefficient γ herein:

$$\gamma = \frac{F_f}{F_t} \tag{3.35}$$

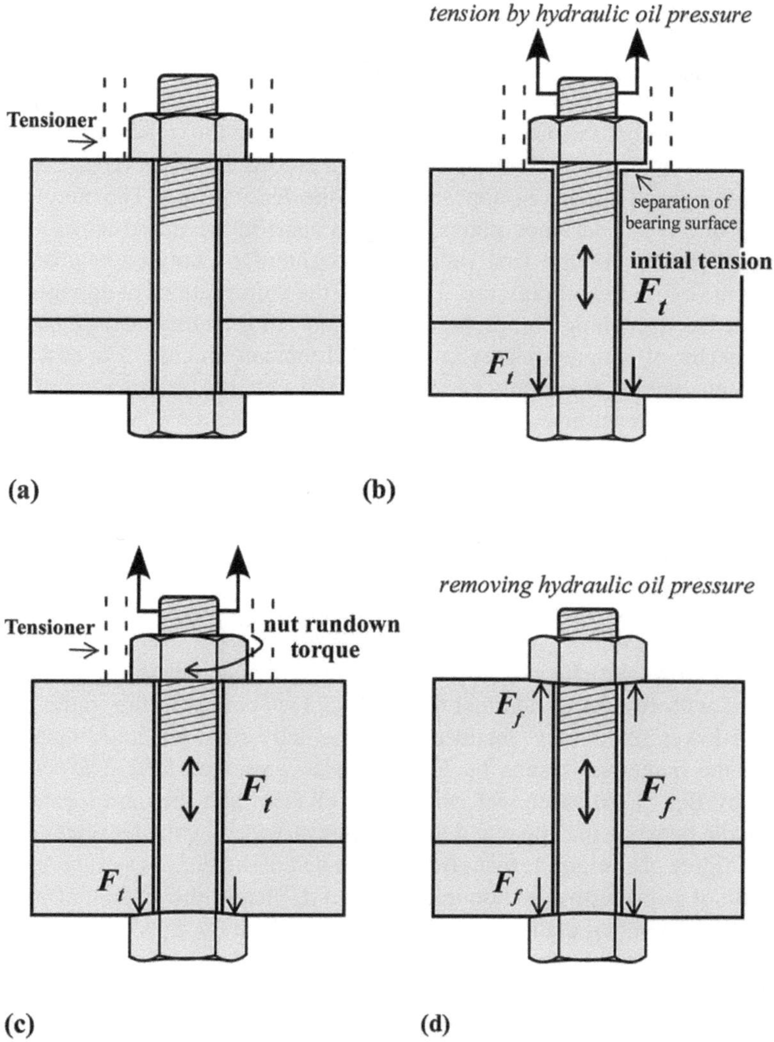

Fig. 3.29 Tightening process using a hydraulic tensioner. (A) Installation of tensioner, (B) generation of initial tension, (C) applying nut rundown torque, and (D) completion of tightening operation.

where γ is less than unity as a natural consequence. Deformations of each part in the initial tension process and the tightening completion process can be expressed by means of one-dimensional spring models, explained in fig. 2.5. The equation for effective tensile coefficient γ can be derived by the equilibrium of the deformations of each part, in Step 2 and Step 4, using spring constants composing the target bolted joint [34]. By use of the five spring constants introduced in section 2.2.2, k_{th}, k_s, k_{cyl}, k_{hd}, k_f, effective tensile coefficient γ is given in the following formula:

$$\gamma = \frac{\frac{1}{k_s} + \frac{1}{k_{cyl}} + \frac{1}{k_{hd}} + \frac{1}{k_f^*}}{\frac{1}{k_{th}} + \frac{1}{k_s} + \frac{1}{k_{cyl}} + \frac{1}{k_{hd}} + \frac{1}{k_f}} \tag{3.36}$$

In the above equation, the numerator and denominator in the right side are related to the initial tension and the tightening completion processes, respectively. The spring constant of engaged threads k_{th} appears only in the denominator. The mating surface consisting of nut and fastened plates, which is apart in the initial tension process, comes into contact for the first time in the tightening completion process after removing the hydraulic oil pressure. Therefore, the deformations of engaged threads only affect the tightening completion process. k_f^* in the numerator represents the spring constant of fastened plates in the initial tension process. The stiffness of a hydraulic tensioner is commonly very high. In addition, the deformation of fastened plates is strongly constrained by the ring-shaped tensioner base, whose area is considerably larger than that of the nut-bearing surface. As a result, it is assumed that k_f^* is related to k_f by $k_f^* = 2k_f$. Eqs. (3.36) and (3.37) in the next section are simple formulas derived under some assumptions. Although the effect of the outer diameter of fastened plates D_o on γ is thought to be limited, it has been demonstrated that experimental results and numerical ones by FEM, performed for the joints having comparatively large D_o, correlate fairly closely with the values of γ calculated by Eqs. (3.36) and (3.37).

Fig. 3.30 shows the effect of grip length L_f on effective tensile coefficient γ. L_f is normalized with respect to nominal diameter d. Two curves in the figure show the upper and lower limits of γ obtained experimentally by a tensioner manufacturer [79], and the numerical results by FEM are also presented [80]. The values of γ obtained by Eq. (3.36) agree well with those of FEM and they are located almost in the middle between the upper and lower limits provided by the tensioner manufacturer. The figure also suggests that effective tensile coefficient γ is simply determined by the ratio of grip length L_f to nominal diameter d. That is, the amount of γ is almost independent of nominal diameter. This can be explained by the statement in section 2.2.2 that the stiffness of bolted joints with similar form is almost proportional to nominal diameter. Threaded fasteners are not similar in most cases for different nominal diameters, as depicted in fig. 1.8. However, since the nonlinearity is limited to the engaged thread portion, nominal diameter d scarcely affects the amount of γ. It is determined by the ratios among the stiffnesses of each part of a bolted joint in the axial direction, as shown in Eq. (3.36).

3.4.3 Effects of surface roughness and nut rundown torque

As surface roughness increases, the effective tensile coefficient γ decreases even if the joint geometry is the same. Interface stiffness due to surface roughness can be expressed by nonlinear spring models whose stiffness increases with increasing bolt force. Expressing the interface stiffness on thread surface, nut-bearing surface, bolt head-bearing surface, and plate interface by four spring constants, K_{th}, K_{nu}, K_{hd}, and K_f, effective tensile coefficient γ is calculated by the following equation:

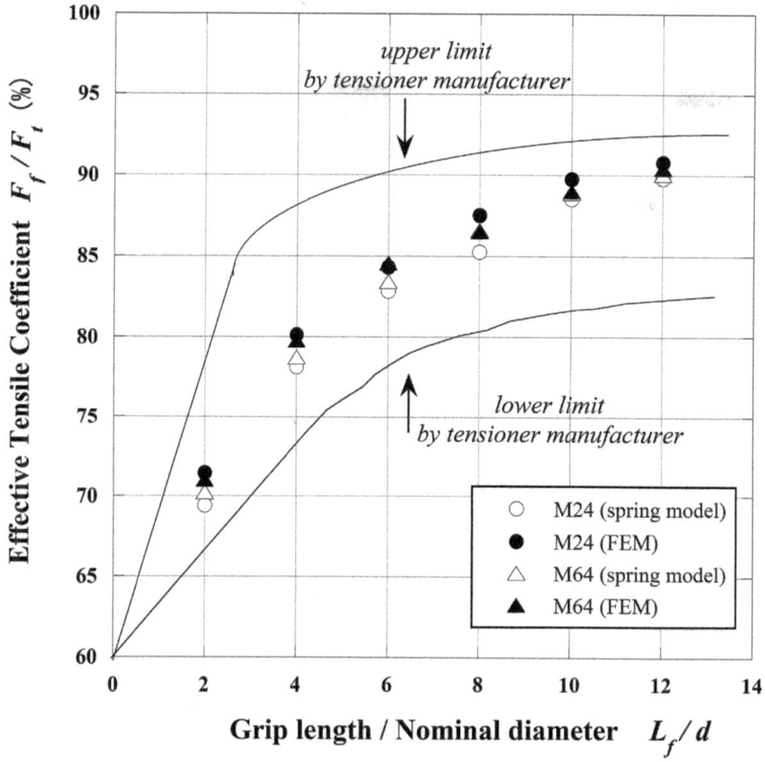

Fig. 3.30 Effect of grip length on effective tensile coefficient.

$$\gamma = \frac{\frac{1}{k_s} + \frac{1}{k_{cyl}} + \frac{1}{k_{hd}} + \frac{1}{k_f^*} + \frac{1}{K_{hd}} + \frac{1}{K_f}}{\frac{1}{k_{th}} + \frac{1}{k_s} + \frac{1}{k_{cyl}} + \frac{1}{k_{hd}} + \frac{1}{k_f} + \frac{1}{K_{th}} + \frac{1}{K_{nu}} + \frac{1}{K_{hd}} + \frac{1}{K_f}} \tag{3.37}$$

The derivation process of Eq. (3.37) is explained in detail in [36]. In Fig. 3.31, the effect of surface roughness on γ is presented, where bolt stress σ_b is set to be 200 MPa. Rzt is the sum of maximum height roughness Rz of the mating surfaces, as in the elastic angle control method. The amount of γ decreases as Rzt is increased. The tendency is more clearly observed with smaller grip length L_f and smaller nominal diameter d. Although not shown in the figure, the rate of decrease in γ gets larger as bolt stress σ_b is decreased.

The amount of γ in the actual application is largely influenced by the magnitude of nut rundown torque given in Step 3 of the tightening process, explained in Section 3.4.1. This suggests the necessity of a guideline for nut rundown torque to produce the bolt force with sufficient practical accuracy. In relation to the interface stiffness explained in section 2.6, a guideline has been proposed on the relationship between contact pressure p_n and approach of interface ζ under compression force,

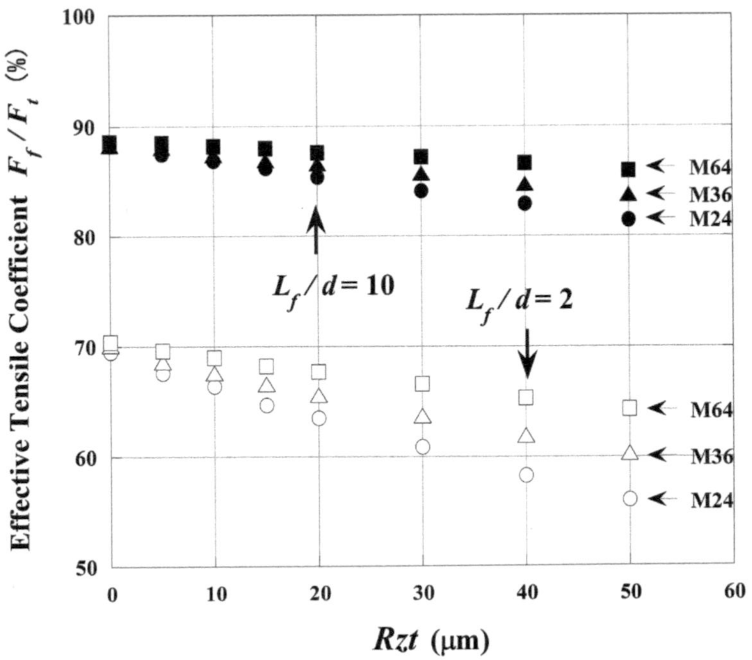

Fig. 3.31 Effect of surface roughness on effective tensile coefficient.

by conducting the experiments for varying nut rundown torque. The recommended guideline is that the amount of nut rundown torque should be selected so as to generate contact pressure of 10 MPa or higher on the nut-bearing surface [81]. This guideline is based on the view that there exists a critical contact pressure, above which surface roughness has little effect on the interface stiffness, because of the microprojections on the contact surface being crushed by the action of rundown torque. Explained below is a concrete method to determine the amount of nut rundown torque.

Bolt stress generated by nut rundown torque is calculated by the ratio of bolt cylindrical area A to nut-bearing surface area A_n. Fig. 3.32 shows the relationship between the ratio A/A_n and nominal diameter d. In the calculations, bolt holes of first grade to third grade are covered, in which the effect of chamfering around bolt hole is neglected. As a natural consequence, A/A_n shows the highest value in the case of third grade bolt hole. It is important that the effect of nominal diameter is quite noticeable. In the cases of first and second grade bolt holes, A/A_n reaches a peak approximately between M12 and M20, and then starts to decrease. Then, A/A_n increases again beyond around M48 and distributes roughly between 0.6 and 0.8. In the case of third grade, A/A_n becomes even larger, although showing a similar tendency. It is worth pointing out that A/A_n sharply decreases in the range below M10.

In the following, it is shown how to calculate the amount of nut rundown torque based on the foregoing consideration. The bolt force F_b generated by nut rundown

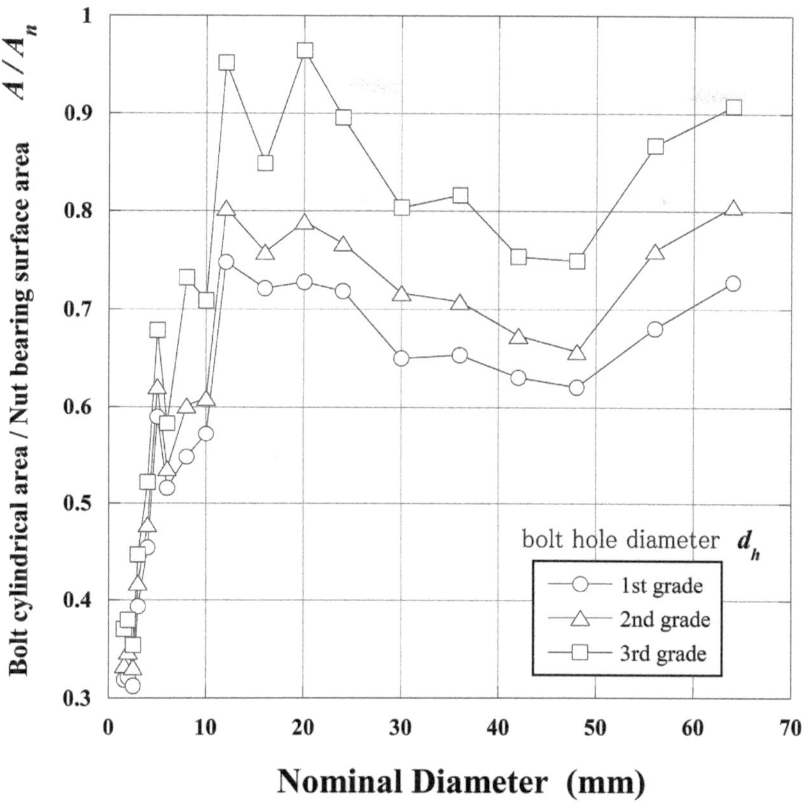

Fig. 3.32 Ratio of bolt cylindrical area to nut-bearing surface area.

torque can be calculated with use of Fig. 3.32. Assuming A/A_n as being 0.7, for instance, when applying the nut rundown torque that generates the contact pressure of 10 MPa for bolts with large nominal diameter, the corresponding bolt stress is found to be about 14 MPa by dividing 10 MPa by 0.7. Then, the bolt force F_b by nut rundown torque is obtained by multiplying 14 MPa by bolt cylindrical area A. Next, nut factor K is calculated by use of Eq. (3.18). Finally, the amount of nut rundown torque is evaluated by substituting F_b and K into Eq. (3.16). The area ratio A/A_n, presented in Fig. 3.32, is also important when considering the relationship between the bolt stress and the critical contact pressure allowable on the nut-bearing surface, as discussed in section 2.1.1.

Geometric errors of the jointed components are another factor influencing effective tensile coefficient γ besides surface roughness. Specifically, they are the flatness between nut-bearing surface and plate surface, pitch error existing in engaged threads, etc. More detailed discussions on geometric errors and an expression for calculating γ by taking account of geometric errors are given in [35].

3.4.4 Application range and tightening guidelines

When applying the direct tension method, care should be taken that the initial tension F_t is larger than the objective or final bolt force F_f. If initial tension F_t is too high, plastic deformations start from thread roots and spread widely. These may significantly impair the joint strength or lower the accuracy of final bolt force, due to the unloading after the completion of tightening process.

Suitable application range

(1) Jointed portions requiring high tightening accuracy

The direct tension method provides high accuracy in bolt force because its tightening process is basically not influenced by coefficients of friction on contact surfaces.

(2) Jointed portions with large grip length

Among the five spring constants in Eq. (3.36) for calculating γ, spring constant of engaged threads k_{th} is likely to be evaluated with insufficient accuracy, as shown in fig. 2.8. In the case of the grip length being large, however, the entire joint stiffness is small and the effect of k_{th} is relatively small. Hence, the amount of γ can probably be predicted with high accuracy.

(3) Jointed portions whose stiffness can be accurately evaluated

When the dimensional accuracy of the members composing a bolted joint is high and its surface roughness is small, high tightening accuracy is expected because the spring constants in Eq. (3.36) can be evaluated accurately.

(4) Jointed portions clamped with not too high bolt stress

Direct tension method is basically applied to the tightening operation in the elastic range, and the initial tension F_t is higher than the objective bolt force F_f. Therefore, it is favorable to apply the method to bolted joints, in which plastic deformations do not spread too much.

In relation to the recommended application rages from (1) to (4), there are cases in which the application of direct tension method should be avoided or particular attention must be paid. Regarding (2), this method is not suitable for joints with small grip length, because the evaluation accuracy becomes too small for all spring constants besides k_{th}. In particular, it should not be applied to joints with very small grip length. In the case of plate warping being a problem, the use of direct tension method should be avoided. It is explained in the following why the method is not suitable for joints with small grip length.

First, it is difficult to accurately evaluate the spring constant of fastened plates k_f. Second, the effect of spring constant of engaged threads k_{th} becomes relatively large. Finally, in the initial tension process, plastic deformations easily occur in a bolt-nut connection and fastened plates. That is because the effective tensile coefficient is fairly small, which leads to an excessive initial bolt force. In the case of clamping plates with fairly small grip length, although being thicker than so-called thin plates, the tightening accuracy could be improved by making the grip length larger through design ingenuity. Additionally, regarding (3), this method is not suitable for joints in which nonmetal materials with low stiffness, like gaskets, are inserted. Sheet gaskets, for example, show nonlinear and hysteresis characteristics in the stress–strain relationship.

Therefore, spring constant of the fastened plates k_f changes as bolt force is varied. It follows that the joint stiffness significantly changes between the initial tension process and the tightening completion process. When external forces are loaded in the running condition, the stiffness is further changed. Another important point is that the stiffness of sheet gaskets is extremely low compared to that of metals. Hence, it has a dominant influence on the accuracy of effective tensile coefficient, which results in a significant reduction of tightening accuracy. Mechanical behavior of gaskets is explained in more detail in Chapter 7, where pipe flange connections are treated. If direct tension method is unavoidably applied to those joints discussed above, the amount of effective tensile coefficient must be measured in advance by attaching strain gages to the target bolt.

Guidelines for tightening operation

(1) Using the dimensions of each part of the target joint, effective tensile coefficient γ is calculated by Eq. (3.36). The value of γ is located almost in the middle of the two lines obtained experimentally, as shown in Fig. 3.30. In actual application, the value may become slightly or much smaller due to the effect of surface roughness and other factors.

(2) The amount of initial tension F_t is calculated by dividing the objective bolt force F_f by γ. The tensile force of F_t is applied to the target bolt using a hydraulic tensioner.

(3) Calculating the amount of nut rundown torque that generates the contact pressure of 10 MPa or higher on nut-bearing surface, the target nut is firmly seated on the plate surface by applying the torque. Removing the hydraulic oil pressure, the tightening operation is completed.

Two equations, Eqs. (3.36) and (3.37), are proposed to calculate γ here. If surface roughness is small and geometric errors and warping existing in the joint are negligibly small, Eq. (3.36) can be used to calculate γ with practical accuracy. That is because the spring constants in Eq. (3.37) associated with interface stiffness, K_{th}, K_{nu}, K_{hd}, and K_f, can be regarded as infinite.

3.5 Thermal expansion method using bolt heater

3.5.1 Tightening principle [81]

The thermal expansion method is applied to large bolts that cannot be tightened by any other means. In addition, this method is commonly used to simultaneously tighten multiple bolts. Target bolts are hollow so as to insert a rod-shaped heater and are elongated by heating. Fig. 3.33 illustrates a bolt heater being inserted into a hollow bolt to perform tightening experiments. The target bolt is usually aligned vertically, with the engaged thread portion upwards, and the bolt heater is placed in the middle of the bolt hole. The tightening process using a bolt heater consists of four steps, as shown in Fig. 3.34.

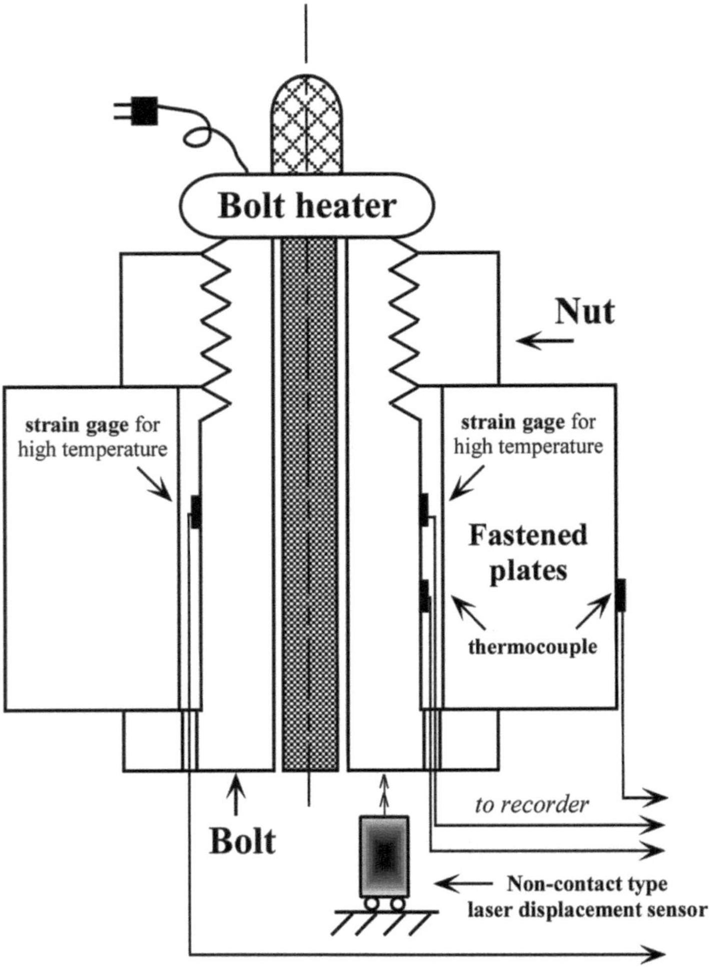

Fig. 3.33 Bolt heater inserted into a hollow bolt for tightening operation.

Step 1: The target hollow bolt is attached to fastened plates and a bolt heater is inserted.

Step 2: The bolt heater is turned on and the heating is continued until the objective elongation is reached. Then, a clearance equal to the bolt elongation is generated around the bolt head-bearing surface.

Step 3: Applying an appropriate amount of nut rundown torque, the bolt head-bearing surface is brought into close contact with the mating plate surface.

Step 4: By shrinking the bolt by natural cooling or forced cooling, the objective bolt force is generated.

Regarding the nut rundown torque, as in the direct tension method, the objective bolt force is produced with sufficient accuracy by applying the torque that generates contact pressure of 10 MPa or higher on the nut-bearing surface [81].

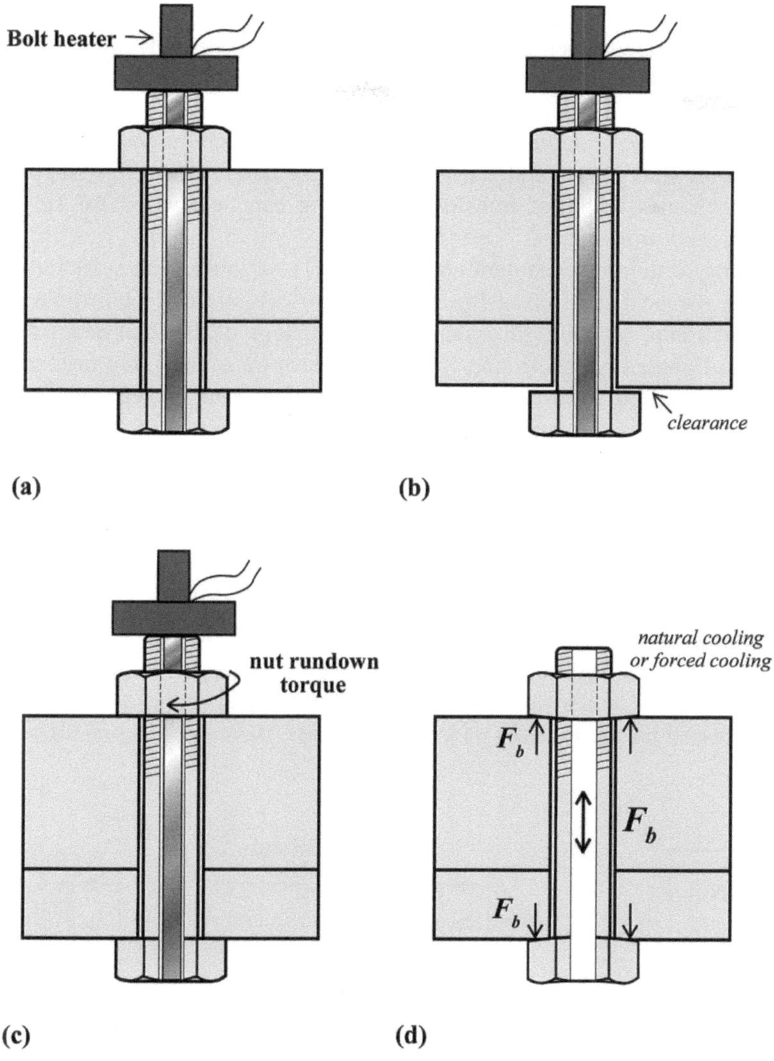

Fig. 3.34 Tightening process of thermal expansion method consisting of four steps. (A) Insertion of bolt heater, (B) elongation due to heating, (C) applying nut rundown torque, and (D) completion of tightening operation.

3.5.2 Simplified model for evaluating tightening process

Tightening process with a bolt heater is a complex problem that temperature and stress fields are coupled. With reference to Fig. 3.33, it is found that some percentage of the heat supplied to a hollow bolt by the heater is transferred to fastened plates through engaged threads and the nut-bearing surface or bolt head-bearing surface. In that case, the amount of heat transferred through a thin air layer, from bolt cylindrical surface to bolt hole surface of fastened plates, also cannot be ignored. By taking account of this phenomenon, a finite element approach, which can provide a fairly high accuracy, is proposed to evaluate the temperature distributions of bolted joint and the bolt force variations in the tightening process [81].

On the other hand, when the primary purpose is to simply obtain the required heating time and the relationship between heating temperature and bolt force to be generated with practical accuracy, a more simple method is effective—basic theories in solid mechanics and heat transfer engineering can be applied for solving the abovementioned problems.

According to the finite element analyses in [81], in joints with sufficiently large grip length, the heat transferred from bolt cylindrical surface to bolt hole surface, through small gap, accounts for a dominant percentage of the total heat transferred from bolt to fastened plates. Hence, ignoring the heat transferred into fastened plates through nut and bolt head-bearing surfaces, a bolted joint is replaced by a one-dimensional model with simple geometry [82], shown in Fig. 3.35. Model 1 consists of a hollow bolt and fastened plates. In Model 2, only a hollow bolt is modeled.

3.5.3 Equation relating axial bolt force to heating temperature [82]

Once the temperature distribution of a hollow bolt $T=T(r)$ in the radial direction is known, the amount of bolt elongation δ generated by a bolt heater can be obtained. Replacing a hollow bolt with a hollow cylinder with inner radius a and outer radius b, δ is expressed by the following formula using the theory of thermal stress:

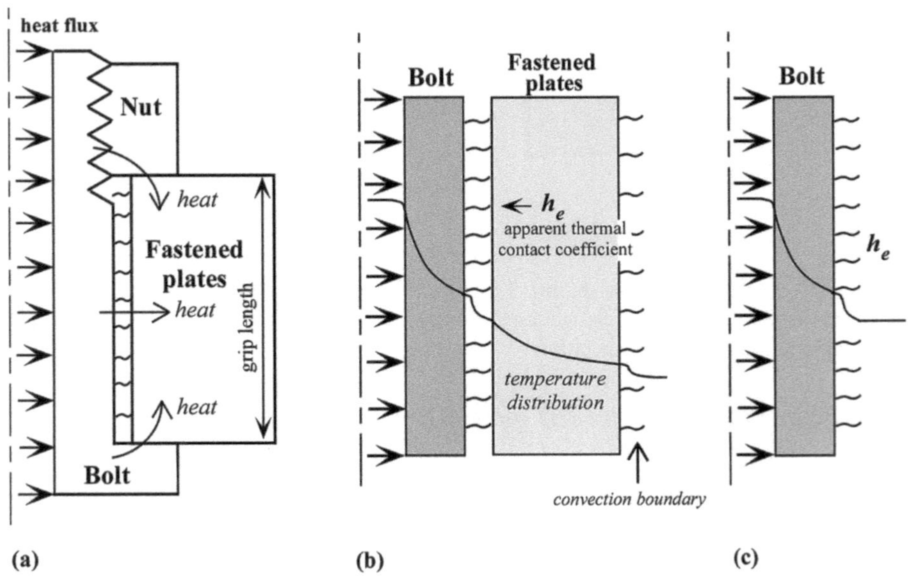

Fig. 3.35 One-dimensional numerical models for evaluating tightening process. (A) Heating by bolt heater, (B) bolt-fastened plates model (Model 1), and (C) bolt model (Model 2).

$$\delta = \frac{2\alpha_b L_f}{b^2 - a^2}\int_a^b Trdr \tag{3.38}$$

where α_b is coefficient of linear expansion of bolt materials.

Considering that the grip length L_f is the sum of the lengths of unengaged threads L_s and bolt cylindrical portion L_{cyl}, Eq. (3.38) is rewritten in the following form:

$$\delta = 8\alpha_b\left(\frac{L_s}{{d_S}^2 - 4a^2}\int_a^{d_S/2} Trdr + \frac{L_{cyl}}{d^2 - 4a^2}\int_a^{d/2} Trdr\right) \tag{3.39}$$

where b is set to be half the diameter of stress area d_s and nominal diameter d for unengaged threads and bolt cylindrical portion, respectively.

Eqs. (3.38) and (3.39) are applicable to an entirely hollow bolt.

In the case of long bolts, however, the target bolt is partially hollow because the maximum heater length is limited. The temperature distributions, bolt elongations, and bolt forces to be generated in those cases have been analyzed by FEM [81]. When heating an entirely hollow bolt, the temperature gradient in the radial direction is small and the bolt temperature can be regarded as wholly uniform, excluding the case that the diameter of heater insertion hole $2a$ is much smaller than the bolt nominal diameter. Let the mean temperature increase of the target bolt from the ambient temperature be ΔT_b; the bolt elongation δ for the grip length portion is the product of coefficient of linear expansion of the bolt material α_b, ΔT_b, and grip length L_f:

$$\delta = \alpha_b \Delta T_b L_f \tag{3.40}$$

When the heating is completed and the bolt is cooled down to the ambient temperature again, bolt force F_b is generated due to the bolt shrinkage equal to δ. Considering the relationship that the bolt elongation by heating is equal to the sum of axial displacements of each part, produced by bolt force F_b, and the approaches of interface on each contact surface, the following equation is derived [82]:

$$\begin{aligned}\alpha_b \Delta T_b L_f &= F_b\left(\frac{1}{k_{th}} + \frac{1}{k_s} + \frac{1}{k_{cyl}} + \frac{1}{k_{hd}} + \frac{1}{k_f}\right) + \zeta_{th} + \zeta_{nu} + \zeta_{hd} + \zeta_f \\ &= \frac{F_b}{k_{total}} + \zeta_{th} + \zeta_{nu} + \zeta_{hd} + \zeta_f \end{aligned} \tag{3.41}$$

where ζ_{th}, ζ_{nu}, ζ_{hd}, and ζ_f are the approaches of interface on thread surface, nut-bearing surface, bolt head-bearing surface, and plate interface, respectively.

The right side of Eq. (3.41) is equal to the equation that the term $2\pi/P$ is removed from Eq. (3.34), which represents the relationship between nut rotation angle and bolt force in the elastic angle control method [54].

Fig. 3.36 compares the bolt force F_b evaluated by the numerical analysis to the corresponding experimental results. In the numerical calculations, using the

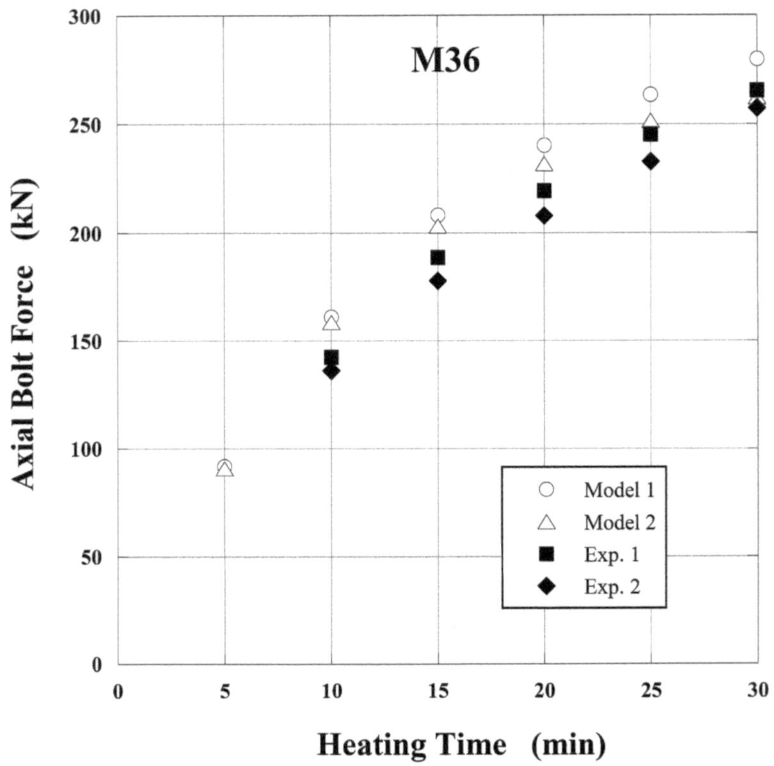

Fig. 3.36 Relationship between heating time and bolt force evaluated by numerical analysis and experiment.

temperature distributions obtained by one-dimensional finite difference method, the bolt elongation δ is calculated by Eq. (3.39), and then the bolt force is evaluated by substituting δ into the left side of Eq. (3.41) instead of $\alpha_b \Delta T_b L_f$. Nominal diameter of bolt and nut is M36. In [82], the joint geometry, heating conditions, and the numerical calculation procedure are explained in detail. The amount of heat transferred through bolt hole clearance between the bolt cylindrical surface and bolt hole surface is evaluated by apparent thermal contact coefficient h_e, discussed in section 2.7. Regarding the boundary conditions of Model 2, shown in Fig. 3.35C, the coefficient of heat transfer on the bolt cylindrical surface is equated to h_e, corresponding to the coefficient for the bolt hole clearance. The comparison between the numerical and experimental results shows that Model 2 gives a better approximation than Model 1, regardless of more assumptions being made.

- Learning by numerical example 3.9: Effect of grip length and nominal diameter on the tightening characteristics of the thermal expansion method

The effect of grip length and nominal diameter on the relationship between heating time and bolt stress to be generated is examined using Eq. (3.41), where the effect of approach of interface is not considered. For a pair of bolt and nut of M16 with coarse thread, changing the ratio of grip length L_f to nominal diameter d, L_f/d, to 2, 4, 6, 8, and 10, numerical calculations are performed to estimate the heating temperature ΔT_b required to produce a bolt stress of 100 MPa. Assuming that the other conditions are identical with those in numerical example 2.1, the heating temperatures are calculated to be 104, 87, 81, 78, and 76 °C for each case of L_f/d.

The calculation results indicate that when using bolts with identical nominal diameter, the heating temperature required to produce a constant bolt stress decreases as grip length is increased. The relationship between heating temperature ΔT_b and grip length L_f can be explained by the relationship between two spring constants k_b, k_f, and L_f in Eq. (3.41).

In order to examine the effect of nominal diameter d, heating temperature ΔT_b required to produce a bolt stress of 100 MPa is calculated for the case of $L_f/d=6$. The target joints are all geometrically similar and are clamped by a pair of bolt and nut with nominal diameter of M12, M16, M24, and M36. The calculation results of ΔT_b are 83, 81, 81, and 80 °C, respectively. It is interesting that the amount of ΔT_b is scarcely influenced by nominal diameter. The reason can be explained by considering the relationship among two spring constants k_b, k_f, grip length L_f, and nominal diameter d, discussed in section 2.2.

3.5.4 Application range and tightening guidelines

The most important information in the tightening operation by thermal expansion method is the relationship between heating time and bolt force to be generated. Using FEM, the relationship can be evaluated regardless of the joint geometry [81]. However, it is not practical to perform complicated finite element analysis for each bolted joint. Meanwhile, the bolts clamped by thermal expansion method are usually hollow over the entire length. Hence, simple formulas, derived in Section 3.5.3, are conveniently applied to those bolts. In the following, the suitable application range of the thermal expansion method and its tightening guidelines are summarized for entirely hollow bolts.

Suitable application range

(1) Jointed portions clamped with large bolts

Since no limitation exists for the maximum applicable bolt size, thermal expansion method is the only means to clamp very large or huge bolts. On the other hand, the lower limit of the applicable bolt size is determined by the heater diameter. In relation to the strength of hollow bolt used in the thermal expansion method, it has been reported that when the heater hole diameter is less than half the nominal diameter, no significant effect is observed in the load distributions along engaged threads [83]. Bolt heaters actually used generally satisfy the dimensional limitation.

(2) Jointed components to be clamped simultaneously by many bolts

Preparing the same number of bolt heaters as the target bolts, it becomes possible to clamp a lot of large bolts simultaneously. In that case, the scatter in bolt forces due to elastic interaction, which is a serious problem occurring when multiple bolts are sequentially tightened, can be excluded. The mechanism of elastic interaction is expounded in Section 3.7.

(3) Jointed portions with large grip length

The thermal expansion method utilizes the difference in the deformations of grip length portion between the elongation by heating and the shrinkage by cooling to ambient temperature. Therefore, high tightening accuracy is expected in joints with large grip length, for the same reason as in the elastic angle control method and direct tension method. In other words, this method is not suitable for joints with small grip length.

Guidelines for tightening operation

(1) Using the dimensions of each part of the target joint, heating temperature T_b required to produce objective bolt force F_b is calculated by Eq. (3.41). In the case of surface roughness being not so large, the terms relating to approach of interface in Eq. (3.41) can be omitted. From the viewpoint of working guidelines, it is necessary to obtain the heating time until ΔT_b is reached. In relation to the temperature distributions in a hollow bolt heated by a bolt heater, an exact solution exists for the transient temperature field of a hollow cylinder, in which the inner surface is subjected to uniform heat flux and the outer surface is exposed to convection heat transfer. However, the actual calculation is quite difficult. Therefore, if possible, it is recommended to calculate the approximate heating time by a simple numerical analysis such as one-dimensional finite difference method.
(2) Bolt elongation δ induced by heating temperature ΔT_b is calculated by Eq. (3.40). Dividing δ by thread pitch P and multiplying it by 360 degrees, the amount of nut rotation angle ϕ is determined as a tightening guideline.
(3) Attaching the bolt to the jointed portion, heating is started. Measuring the temperature of a specific point of the bolt, e.g., the tip of the bolt thread, nut rundown operation is started as the temperature increase of the selected point approaches ΔT_b.
(4) Keeping the heater in the bolt hole, the nut is rotated and firmly run down onto the plate surface until the rotation angle ϕ obtained in (2) is reached. The amount of nut rundown torque is determined so as to generate contact pressure of 10 MPa or higher on the nut-bearing surface.

If the nut rundown operation is started after removing the bolt heater, a nonnegligible amount of bolt shrinkage occurs due to rapid cooling, and the generated bolt force may be significantly smaller than the objective value. It should be pointed out here that ΔT_b is used for reference only, since a certain amount of error inevitably occurs depending on the measurement position. That is, the working guideline should be determined in terms of nut rotation angle ϕ or bolt elongation δ. Regarding the heating temperature, although larger bolt force is produced as ΔT_b is higher, ΔT_b should be selected so as not to cause material degradation. The above guideline is based on the nut rotation angle ϕ. On the other hand, when skilled workers are in charge, it is common practice to directly measure the bolt elongation using dial gages, etc.

The guidelines explained so far all correspond to the case of bolt heaters being placed vertically. To take advantage of the tightening characteristics of the thermal expansion method, it is necessary to establish a guideline for the cases of bolt heaters being placed horizontally or tilted to some extent. In [84], it is confirmed by experiments that the thermal expansion method is applicable to the case of bolt heaters being placed horizontally, in which the side surface of the bolt heater is in contact with the bolt hole surface. A concrete guideline for horizontal use is given in [84].

3.6 New tightening method utilizing real-time measurement of nut factor [85]

3.6.1 Development intention

Among the various tightening methods, torque control method is most widely used because of its easy tightening operation. However, the tightening accuracy of this method is unfortunately not so high. Therefore, the elastic angle control method is frequently introduced to clamp joints of critical importance; however, this method involves another problem of low working efficiency. In this section, a newly developed tightening method is introduced, aiming at simultaneously realizing easy operation and high tightening accuracy, in which the basic principle of torque control method is combined with that of elastic angle method. The fundamental characteristic of the proposed method is that the target torque is changed during tightening operation in order to achieve the objective bolt force with high accuracy. This is because the nut factor significantly varies depending on coefficients of friction, which cannot be predicted precisely prior to the tightening work.

3.6.2 Tightening principle

The relationship between bolt force F_b and tightening torque T_t is shown in Eq. (3.16).

$$T_t = KF_b d \tag{3.42}$$

Nut factor K can be calculated once coefficients of friction, μ_{th} and μ_{nu}, are given.

$$\begin{aligned} K &= 0.556\mu_{th} + 0.65\mu_{nu} + 0.019\,(\text{coarse thread}) \\ K &= 0.565\mu_{th} + 0.65\mu_{nu} + 0.011\,(\text{fine thread}) \end{aligned} \tag{3.43}$$

Meanwhile, when using the elastic angle control method, nut rotation angle ϕ is related to bolt force F_b by Eq. (3.33).

$$\phi = \frac{2\pi}{P}\frac{F_b}{k_{total}} \tag{3.44}$$

Using Eqs. (3.16) and (3.33), the ratio of T_t to ϕ is expressed by the following equation:

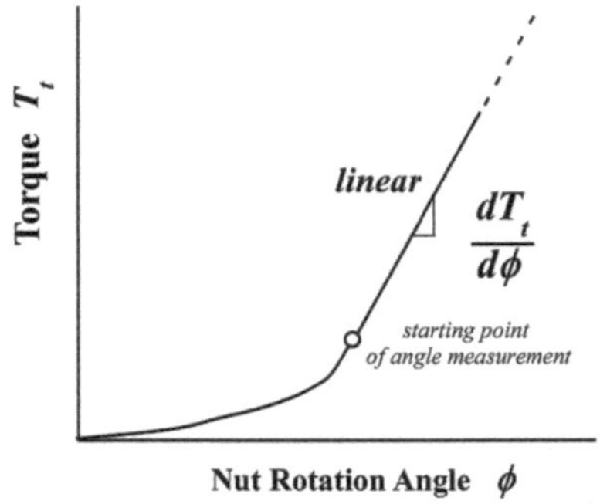

Fig. 3.37 Principle of proposed tightening method.

$$\frac{T_t}{\phi} = \frac{KPd}{2\pi} k_{total} \tag{3.45}$$

Fig. 3.22 shows the relationship between F_b and ϕ in the tightening process of elastic angle control method. A similar relationship holds between T_t and ϕ, as shown in Fig. 3.37. Replacing T_t/ϕ in Eq. (3.45) by the differential form, $dT_t/d\phi$, it represents the gradient of the linear portion in $T_t - \phi$ curve. Using $dT_t/d\phi$, the formula for true nut factor K_{tru} is derived from Eq. (3.45):

$$K_{tru} = \frac{2\pi}{Pdk_{total}} \frac{dT_t}{d\phi} \tag{3.46}$$

This equation implies that if $dT_t/d\phi$ can be measured during the tightening operation, K_{tru} that corresponds to the actual tightening condition is obtained. Then, the amount of tightening torque to accurately produce the objective bolt force F_b is calculated by Eq. (3.16), in which the value of K_{tru} is substituted into K. That is, the torque value is changed during the tightening operation to produce the objective bolt force with high accuracy.

3.6.3 Verification of proposed method using prototype tightening device

Fig. 3.38 shows a prototype tightening device developed following the principles explained above. This device is constructed on the basis of an electric-powered torque wrench and is equipped with a function of continuously measuring $dT_t/d\phi$ during the tightening operation. When $dT_t/d\phi$ reaches almost a constant value, true nut factor K_{tru} is calculated by Eq. (3.46) and the target torque is changed from the initial value to the corrected one calculated by Eq. (3.16).

Fig. 3.39 shows a set of test specimens composed of a pair of bolt and nut of M16 with 10.9 strength grade. Two flat washers are inserted under the nut- and bolt head-bearing surfaces. The grip length is 80 mm including the thickness of two washers.

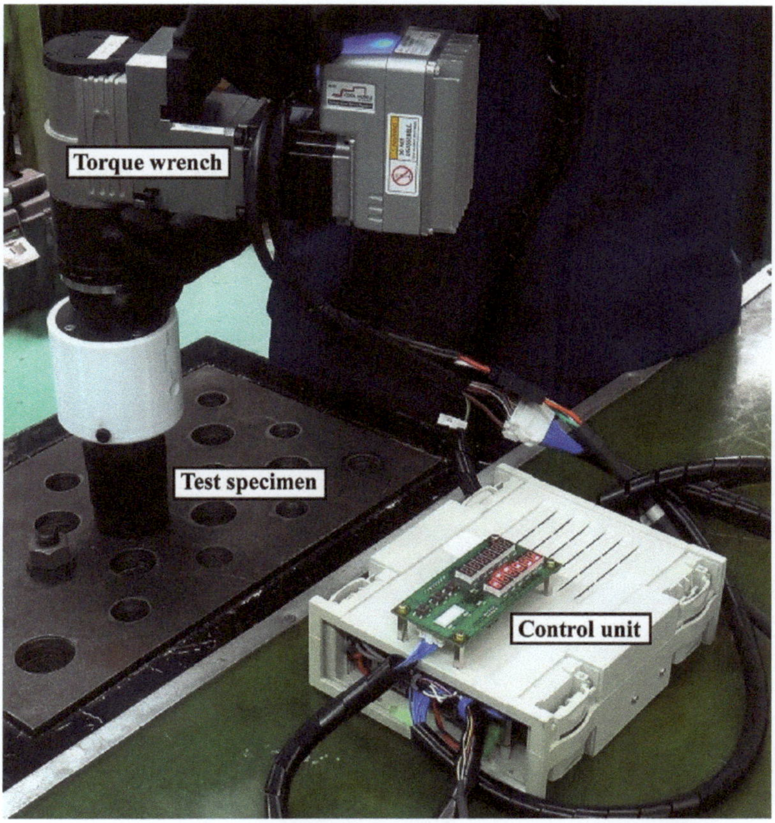

Fig. 3.38 Prototype tightening device developed following the proposed principle.

Tightening experiments are carried out for two types of test specimens with identical geometry, which are subjected to different surface treatment. For Type-A and Type-B, 20 sets of bolts, nuts, and flat washers are prepared, respectively. Average values of nut factor measured in advance are 0.15 for Type-A and 0.20 for Type-B. The objective bolt force is set to be 104 kN, which corresponds to a bolt stress of 517 MPa. The initial torques calculated using the above nut factors are 250 Nm for Type-A and 333 Nm for Type-B. In the experiment, the snug torque T_{sng} is set to be 150 Nm. Fig. 3.40 shows the experimental results. The ordinate and abscissa represent axial bolt force and the number of test specimen, respectively. It can be concluded that the scatter in bolt force is suppressed to less than 5%. The explanation of the statistical procedure is omitted here.

The developed device is a tentative one. To design a tightening device of practical use, it is necessary to prepare comprehensive data for spring constants of fastened plates k_f, since the amount of k_f significantly changes depending on various factors, such as the joint geometry, size, material, etc. In the above verification, the amounts of k_f are evaluated using fig. 2.13. Accordingly, if some database on k_f can be prepared

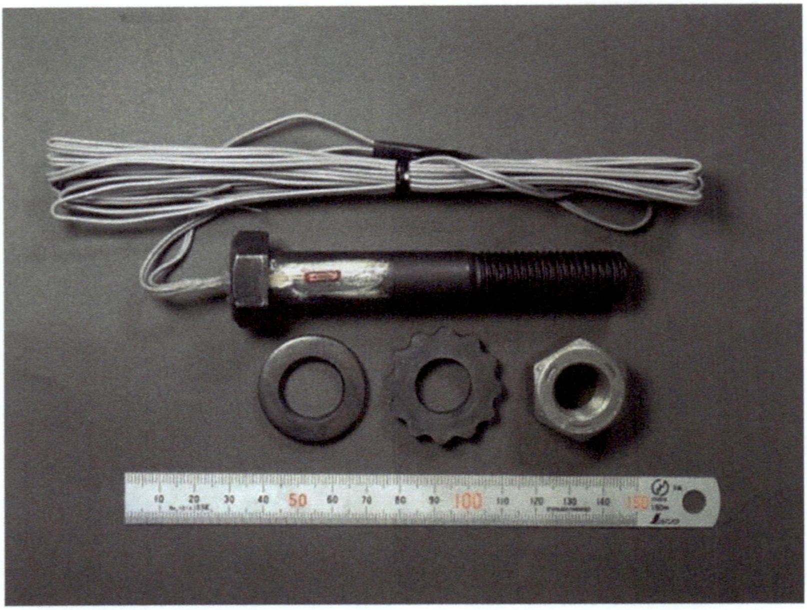

Fig. 3.39 A set of test specimens.

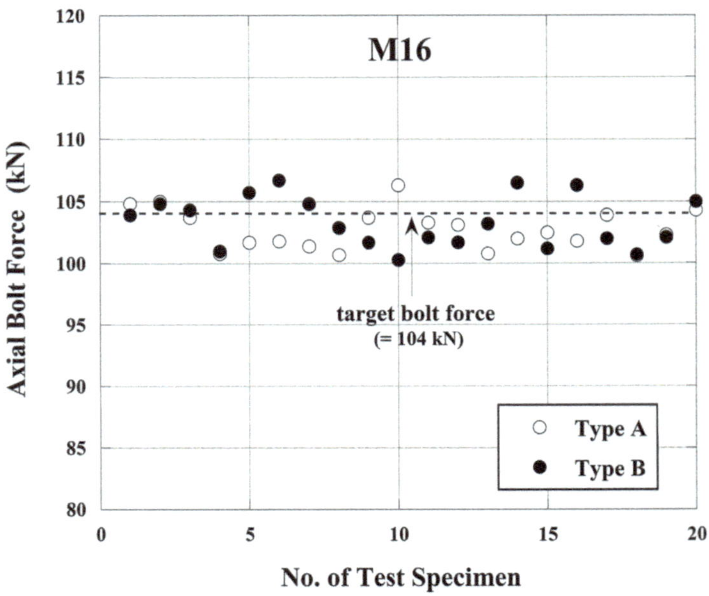

Fig. 3.40 Experimental results of the scatter in bolt force.

by a numerical method like FEM, corresponding to commonly used joints, the proposed device becomes a practical one and it can be expected to achieve a significant reduction of bolt force scatter.

3.7 Sequential tightening of multibolted joints and induced elastic interaction

3.7.1 Tightening operation of bolt-nut connections and elastic interaction

In general, multiple bolts are used to clamp the parts of machines and structures. In such multibolted joints, due to operational procedure or cost issues, it is often impossible to simultaneously tighten all the bolts close to each other. Therefore, if tightening the target bolts one by one in actual work, the amount of bolt force, whose tightening operation has already been completed, can change to some extent or sometimes significantly. This phenomenon is called elastic interaction.

In Fig. 3.41, for the case of tightening three bolts sequentially with identical bolt force F_b, the process is analyzed by FEM and the resulting deformations are illustrated on a larger scale [86]. Elastic interaction is explained focusing on bolt 1. All bolts are tightened with identical bolt force F_b in the order of bolt 1, bolt 2, and bolt 3. In the tightening process of each bolt, the scatter of bolt force traced to the characteristics of the tightening method, such as the scatter of coefficient of friction in torque control method, is not considered.

(1) Bolt 1 is tightened with the objective bolt force of F_b. At this point, axial bolt force of bolt 1, F_{b1}, is equal to F_b.
(2) Tightening bolt 2 with F_b, F_{b1} is changed because the tightening operation of bolt 2 affects the deformation around bolt 1. At this point, F_{b1} is no longer equal to F_b due to elastic interaction.
(3) Tightening bolt 3 with F_b, the bolt force of bolt 2, F_{b2}, is changed by the same mechanism as in (2). The bolt force of bolt 1, F_{b1}, is further changed from the state of (2).

The bolt forces of bolt 2 and bolt 3 when tightening bolt 1, or the bolt force of bolt 3 when tightening bolt 2, are not shown in Fig. 3.41. Those bolt forces are generally zero. However, in pipe flange connections, discussed later, it sometimes occurs that a certain amount of bolt force is generated, due to the flange deformation with an open-mouth shape, in the bolts located about 180 degrees apart from the bolt just tightened.

When tightening a lot of bolts sequentially, the bolt forces deviate from the target value due to elastic interaction, except for the bolt tightened last. The phenomenon of elastic interaction is always encountered, to a greater or lesser extent, when tightening multiple bolts sequentially. To avoid this interaction, all bolts must be tightened simultaneously. In the tightening operations performed in nuclear power plants, pressure vessels used under severe conditions, large-scale marine diesel engines, etc., in which high tightening accuracy is required, direct tension method using hydraulic tensioners is frequently employed to eliminate the effect of elastic interaction. In those

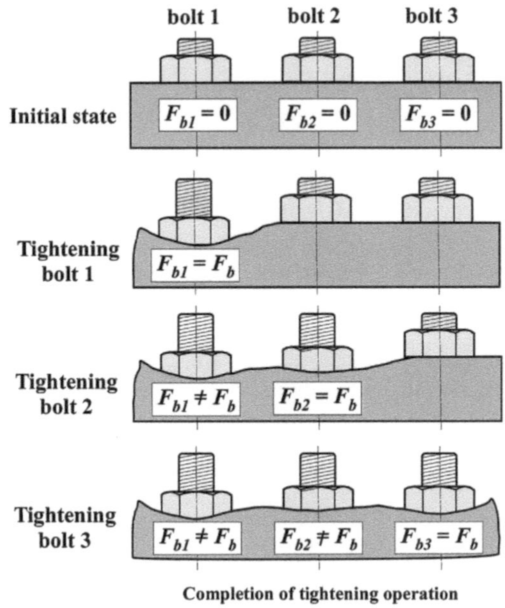

Fig. 3.41 Elastic interaction occurring when tightening multiple bolts one by one.

cases, as many hydraulic tensioners as there are target bolts are prepared for the simultaneous tightening. Furthermore, the effect of elastic interaction can be eliminated by using as many bolt heaters as the target bolts and tightening all the bolts simultaneously. This method is commonly employed when tightening the casing of large-scale steam turbines.

3.7.2 Effect of joint geometry on elastic interaction

The extent of bolt force scatter due to elastic interaction ranges widely from a negligible level to a considerable one that significantly influences the joint performance. The latter example is a pipe flange connection with a sheet gasket inserted. The extent of elastic interaction greatly varies depending on the joint geometry and stiffness. When clamping pipe flange connections, elastic interaction among multiple bolts frequently becomes an issue and the extent of interaction changes with flange geometry.

Fig. 3.42 schematically shows the shapes of commonly used pipe flanges. They are called flat face, large raised face, and small raised face flanges, respectively [87]. In general, flat face flanges are used for low pressure contained fluids. As the inner pressure increases, large or small raised face flanges come into use. Elastic interaction is more noticeable as the bearing surface area decreases. The primary cause is the phenomenon called flange rotation, shown in Fig. 3.43A. Since a certain clearance exists between the mating flange surfaces in large and small raised face flanges, the two surfaces deform so as to approach each other when clamping the bolt. In addition to adjacent bolts, such deformations also affect the axial bolt force of the bolts located about 180 degrees apart, as shown in Fig. 3.43B.

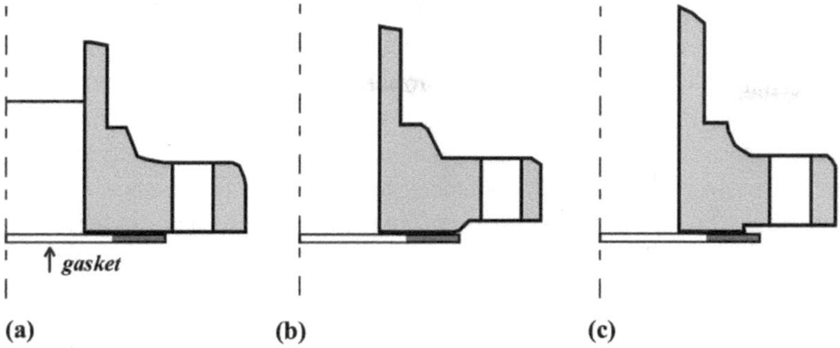

Fig. 3.42 Commonly used flanges with various geometries. (A) Flat face flange, (B) large raised face flange, and (C) small raised face flange.

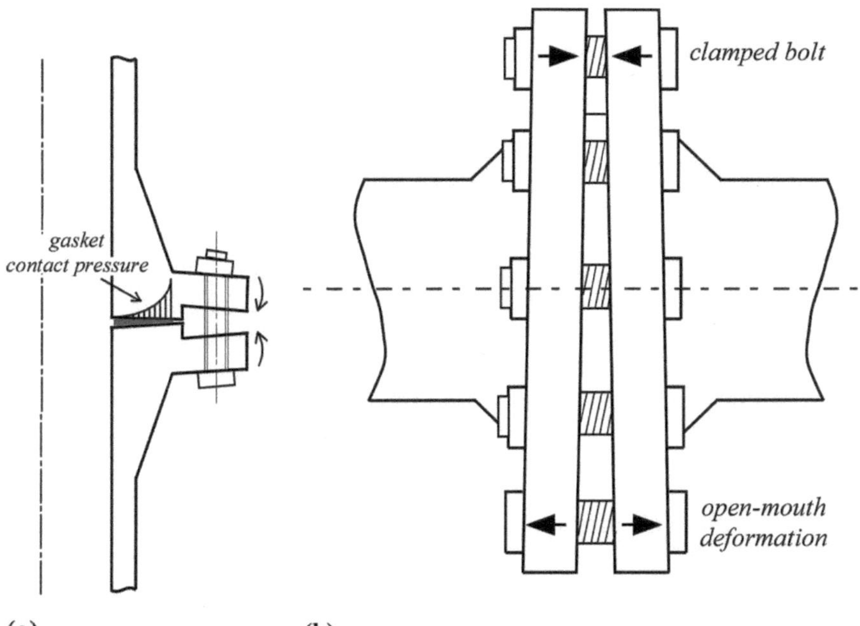

Fig. 3.43 Flange rotation and open-mouth deformation. (A) Flange rotation and (B) open-mouth deformation of pipe flange.

The elastic interaction in pipe flange connections is summarized as follows. When a certain bolt is clamped, bolt forces of the adjacent bolts are decreased and the forces of the bolts located about 180 degrees apart are sometimes increased, due to a specific deformation pattern called open-mouth deformation, in which two mating flange surfaces deform to be separated from each other. As a result, the bolt forces vary in a very complex manner during the tightening process. The above phenomenon causes the specific and complex mechanical behavior in the tightening process of pipe flange connections. Additionally, in case of clamping rectangular-shaped flanges, the effect of elastic interaction on bolt force scatter is considerably smaller comparing to ordinary round-shaped flanges, because open-mouth deformation is unlikely to occur.

In ordinary pipe flange connections, a gasket is inserted between two flanges, aiming at the sealing of contained fluids. The elastic interaction there is greatly influenced by the gasket stiffness. When using ring-shaped flat metal gaskets, the scatter in bolt forces due to elastic interaction is as much as the flange connections without gaskets [88]. In contrast, the effect of elastic interaction is markedly observed when using low stiffness gaskets. In particular, when using sheet gaskets or spiral wound gaskets with low stiffness, the scatter in bolt forces could become very large [89,90], because the gaskets show nonlinear and hysteresis characteristics in the stress–strain relationship.

Meanwhile, regardless of tightening multiple bolts sequentially, elastic interaction is hardly detected when tightening the rear wheel of large vehicles with dual-tire structure, as discussed in section 8.2. The bolts for clamping the wheel are arranged concentrically, and tires are attached to thin metallic wheels. The surfaces of two mating wheels are flat and closely in contact. Consequently, the deformation pattern, like flange rotation, does not occur because of the high stiffness due to the small grip length. Hence, elastic interaction is rarely observed [70]. When clamping thin fastened plates with high stiffness and negligible geometric errors, it is considered that the bolt force scatter due to elastic interaction can virtually be disregarded, including the wheel tightening operation of ordinary passenger cars.

Elastic interaction also becomes a problem in the disassembly process for inspection purposes, etc., in which each bolt is removed sequentially. In this process, when removing a certain bolt, bolt forces of the adjacent bolts can significantly increase. The elastic interaction occurring in the disassembly process sometimes becomes an issue in such joints where the bolt forces significantly vary in the tightening process due to elastic interaction. The bolt force variation in the disassembly process has been evaluated by FEM, for the cases of removing the clamped bolts in the order of star-pattern or a simple one-directional pattern [91].

3.7.3 Estimation of bolt force scatter and optimal tightening procedure

A pipe flange connection is a typical structure in which elastic interaction becomes a serious problem. In particular, in the case of a low stiffness gasket being inserted, the final bolt forces obtained show a significant scatter among the clamped bolts, even if each bolt is tightened with a specified force. Thus, there are two challenges to be solved:

Challenge 1: Prediction of the bolt force scatter occurring when tightening multiple bolts sequentially.

Challenge 2: Establishment of optimal tightening procedure that enables uniform bolt force distribution among all the clamped bolts at the completion of tightening operation.

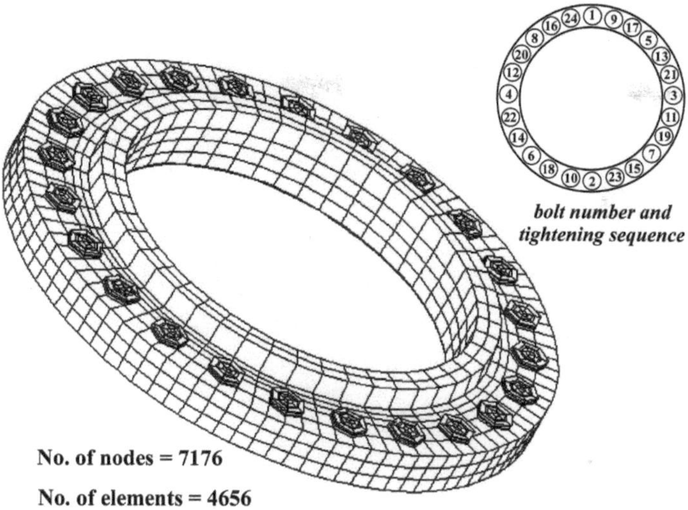

Fig. 3.44 Finite element model for pipe flange connection of 20-in. inner diameter.

In order to solve the above two challenges, numerical methods based on FEM have been proposed [89,90,92,93]. Detailed procedures are given in each reference paper. These methods are explained in the following with emphasis on bolt force scatter due to elastic interaction.

Fig. 3.44 shows an FE model for analyzing the tightening process of a pipe flange connection, whose inner diameter is 20 in., clamped by 24 bolts with nominal diameter of M33 [93]. A joint sheet gasket is inserted between the two flanges. Bolt numbers and tightening sequence are given in the figure. Fig. 3.45A shows the bolt force scatter when tightening 24 bolts sequentially in a diagonal pattern by applying an initial bolt force of F_i to each bolt, which is equal to the objective bolt force. The ordinate represents the ratio of bolt force in the final state F_b to F_i. The abscissa shows the bolt number corresponding to bolt-up sequence. It is found that experimental results and numerical ones by FEM are in fairly good agreement. Roughly speaking, final bolt forces F_b of the eight bolts, clamped in the first stage, are very small, those of the eight bolts clamped in the next stage are about 50% of the target value, and those of the eight bolts clamped in the final stage are almost equal to the target value.

The problem is that the degree of bolt force scatter greatly varies depending on the size and rating of the target flanges. That is, since the dimensions and geometries of standardized pipe flanges are not similar, the scatter pattern in bolt forces significantly changes depending on the target flange standard. In Fig. 3.45B, for a pair of large raised face flanges with a spiral wound gasket being inserted, bolt force scatters are compared between experimental results and numerical ones by FEM, where eight bolts are tightened sequentially in a diagonal pattern. The specification of the target flange is nominal pressure of 40 K and nominal diameter of 40 in JIS B 2238. As shown in the figure, at the completion of one turn tightening of the eight bolts, bolt forces of

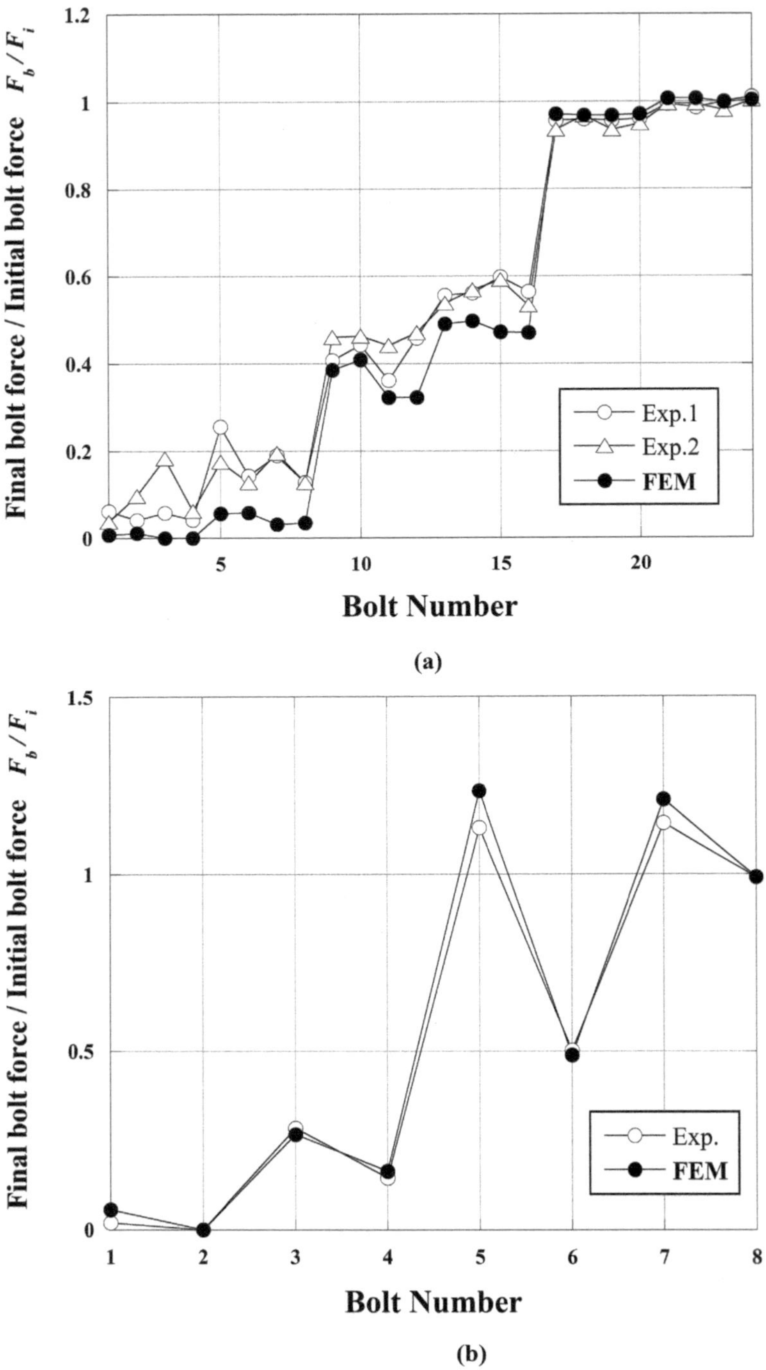

Fig. 3.45 Bolt force scatter due to elastic interaction when tightening bolts diagonally. (A) Pipe flange connection clamped by 24 bolts and (B) that clamped by 8 bolts.

No. 1 and No. 2 bolts are almost zero. Those of No. 3 and No. 4 bolts are around 20% of the target value, and bolt forces of No. 5 and No. 7 become about 20% higher than the target value. In No. 6 bolt, about 50% of the target value remains [89]. In the tightening experiments conducted here, in order to remove the bolt force scatter caused by factors other than elastic interaction, the bolt forces are given with reference to the output of strain gages. Therefore, the bolt force of No. 8 bolt is exactly equal to the target value.

In relation to Challenge 2, mentioned above, for a pipe flange connection of 20 in. in diameter, finite element analyses are performed to obtain the initial bolt forces, which make all the bolt forces equal to the objective one at completion of the tightening operation. Twenty-four bolts are used for clamping. The experimental results of bolt forces in the final state are shown in Fig. 3.46, in which the tightening experiment is carried out using the initial bolt forces obtained by FEM. It is found that almost uniform bolt forces are achieved with one-turn tightening operation, although some experimental errors are recognized. The procedure proposed for Challenge 2 is effective, e.g., for the joints of high value that need to be clamped with one-turn tightening operation. On the other hand, when this method is applied to a joint with a low stiffness insert, like a sheet gasket, it may cause a serious strength problem that very high initial bolt force needs to be applied to some bolts [90]. Moreover, it is fairly difficult, from the field work point of view, to tighten a lot of bolts with different bolt forces.

Therefore, a tightening guideline limited to use in pipe flange connection is published as JIS B 2251 (2008) "Bolt tightening procedure for pressure boundary flanged

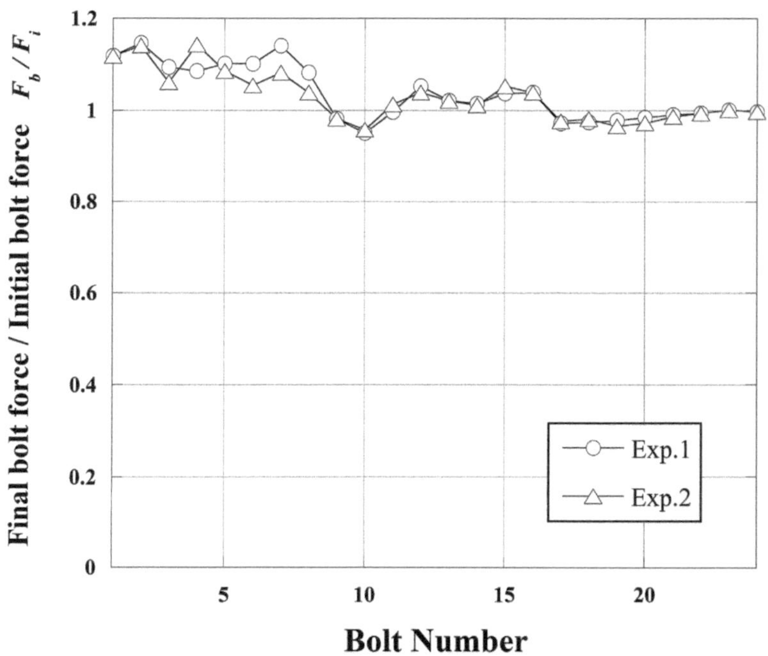

Fig. 3.46 Validation of the proposed analytical procedure aiming at uniform bolt force distribution.

joint assembly." The proposed guideline intends to comply with the on-site work, using the analytical method developed for Challenge 1. In pipe flange connections clamped by eight bolts or less, the tightening sequence is in the order of diagonal pattern as usual. In the flange connections clamped by 12 bolts or more, the target bolts are simply tightened in the order of one-directional pattern. The points of the guideline are as follows. First, four bolts 90 degrees apart from each other are tightened with the objective bolt force. Next, all the bolts are tightened in the order of one-directional pattern with the objective bolt force. The second process of one-directional tightening is repeated several times. Meanwhile, if tightening operation is done following the conventional diagonal pattern, the bolt force of a certain bolt is significantly changed when tightening the two neighboring bolts, which means that each bolt force greatly changes twice during the whole tightening operation. On the other hand, when tightening bolts with one-directional pattern, a large variation in bolt force occurs only once when tightening the adjacent bolt located in the tightening direction. Therefore, the proposed method can largely suppress the effect of elastic interaction. That is why JIS adopts a one-directional tightening pattern. Its detailed procedure is explained in JIS B 2251 for various shapes and ratings of pipe flanges.

3.8 Energy required for tightening threaded fasteners

3.8.1 Tightening energy consumed in the torque control method [94]

In the tightening operation, the required tightening energy increases with larger grip length, even if a constant bolt force is applied to threaded fasteners with identical nominal diameter. This is easily understood by considering that the bolt elongation increases as grip length is increased for a constant bolt force. In this section, a formula that calculates the energy required for tightening operation is derived for the torque control method, which is the most widely used tightening method. The effects of various factors, such as thread pitch, grip length, etc., are quantitatively clarified. When tightening a pair of bolt and nut by the torque control method, the required energy W_{trq} can be expressed in terms of tightening torque T_t and nut rotation angle ϕ:

$$W_{trq} = \frac{1}{2} T_t \phi \tag{3.47}$$

As is evident from Eq. (3.47), W_{trq} is proportional to ϕ when applying a constant torque. The relationship between nut rotation angle ϕ and bolt force F_b is given in Eq. (3.33).

$$\phi = \frac{2\pi}{P} F_b \left(\frac{1}{k_b} + \frac{1}{k_f} \right) = \frac{2\pi}{P} \frac{F_b}{k_{total}} \tag{3.48}$$

Furthermore, the relationship between tightening torque T_t and bolt force F_b is shown in Eq.(3.15).

$$T_t = \frac{1}{2} F_b \{ d_2 \tan (\rho_{th}' + \beta) + d_{nu} \mu_{nu} \} \tag{3.49}$$

Using Eqs. (3.33) and (3.49), the relationship is examined between tightening energy W_{trq} and thread pitch P. With reference to Eq. (3.33), nut rotation angle ϕ and thread pitch P are inversely related. In threaded fasteners with fine screw thread, however, pitch dimeter d_2 becomes a bit larger than that of coarse screw thread, thus, the amounts of k_{th} and k_s are slightly increased. Accordingly, the increasing rate in nut rotation angle ϕ caused by the decrease in P is a bit smaller than the value predicted by the inverse proportional relationship.

As shown in section 1.4.3, lead angle β is almost proportional to thread pitch P. Hence, β is decreased with decreasing P in Eq. (3.49). However, since the absolute value of β is considerably smaller than friction angle of thread surface ρ'_{th}, the effect of β is limited. As a result, the relationship between W_{trq} and P is summarized as follows. For screw threads with identical nominal diameter, the tightening energy W_{trq} increases almost inversely with thread pitch P. The increasing rate is somewhat smaller than the value predicted by the perfect inverse proportional relationship between W_{trq} and P.

Next, the cases of coarse and fine screw threads with identical nominal diameter are compared. Supposing that thread pitch of fine screw thread is half that of coarse screw thread, the fine screw thread needs nearly twice as much as the tightening energy required for the coarse one. Most of the tightening energy is consumed as friction works on thread surface and nut-bearing surface. The energy consumed for the deformations of a bolt-nut connection and fastened plates is a small amount. When tightening threaded fasteners with fine thread, they require larger energy than those with coarse thread. It is presumed, therefore, that coefficients of friction tend to be increased because the sliding length in fine threads is longer than that in coarse threads. The relationships between tightening energy W_{trq} and the influencing factors, other than thread pitch P, are summarized as follows:

(1) In reference to ϕ in Eq. (3.33), W_{trq} is inversely proportional to Young's modulus of plate materials.
(2) In reference to T_t in Eq. (3.49), W_{trq} increases roughly in proportion to coefficients of friction, μ_{th} and μ_{nu}, on thread surface and nut-bearing surface.
(3) In reference to ϕ in Eq. (3.33), spring constants, k_b and k_f, decrease almost in inverse proportion to grip length L_f. Thus, W_{trq} increases proportionally as L_f is increased.
(4) In relation to bolt force F_b in Eqs. (3.33) and (3.49), tightening energy W_{trq} calculated by Eq. (3.47) is proportional to the square of F_b.
(5) In relation to Eqs. (3.33) and (3.49), for a constant bolt stress σ_b, W_{trq} is nearly proportional to the cube of nominal diameter d.

The relationship between W_{trq} and d, explained in (5), is fairly complicated. In reference to ϕ in Eq. (3.33), although thread pitch P increases with d, the increasing rate decreases with larger nominal diameter. When bolt stress σ_b is constant, bolt force F_b is strictly proportional to the square of d. Meanwhile, five spring constants, k_{th}, k_s, k_{cyl}, k_{hd}, and k_f, composing the bolted joints with similar form, are almost in proportion to d. From the above, it is seen that the required nut rotation angle ϕ somewhat increases as d is increased. Next, in reference to T_t in Eq. (3.49), F_b is proportional to the square of d and the term in a brace is nearly proportional to d. Hence, tightening torque T_t is almost in proportion to the cube of nominal diameter d. As a result, W_{trq} is a bit larger than the value that is proportional to the cube of d. From the above results, the relationship between tightening energy and its influencing factors is summarized as follows:

(1) A large amount of tightening energy W_{trq} is required in such cases when Young's modulus of the joint material is small and/or the jointed portion with large grip length is clamped by threaded fasteners with fine thread.

(2) For the jointed portions clamped with identical bolt force by use of screw threads with identical nominal diameter, the required tightening energy is larger in fine screw threads than in coarse screw threads. Accordingly, if coefficients of friction are increased in fine screw threads, due to longer sliding length, the produced bolt force becomes smaller comparing to coarse screw threads.

In JIS, the maximum thread pitch P of coarse screw threads is 6 mm for M64. In actual joints, when using screw threads larger than M64, thread pitch is sometimes selected to be less than 6 mm. This means that when using large screw threads, the consumed tightening energy tends to increase at a greater rate proportional to nominal diameter. Accordingly, large bolts and nuts are susceptible to some nuisance problems, such as the seizure of threaded fasteners.

3.8.2 Calculation of tightening energy for various influencing factors

Tightening energy is calculated for bolted joints with various dimensions under varying tightening conditions.

(1) Calculation conditions

Bolt stress $\sigma_b = 100$ MPa, nominal diameter $d = 16$ mm, bolt hole diameter $d_h = 17.5$ mm, outer diameter of fastened plates $D_o = 128$ mm, thread pitch $P = 2$, 1.5, and 1 mm, grip length $L_f /d = 2$–20, coefficients of friction μ_{th}, $\mu_{nu} = 0.05$–0.3, length of unengaged threads $L_s = 1/5L_f$, Young's modulus of joint materials $E = 70$, 100, 150, and 200 GPa.

(2) Standard calculation condition

$P = 2$ mm, $L_f/d = 4$, μ_{th}, $\mu_{nu} = 0.12$, $E = 200$ GPa.

In Fig. 3.47A and B, the effects are shown of thread pitch P and coefficients of friction, μ_{th}, μ_{nu}, on tightening energy W_{trq}. It is assumed that μ_{th} is equal to μ_{nu}. W_{th} and W_{nu} represent the energies consumed in engaged threads and on nut-bearing surface. They correspond to the first and second terms in a brace of the right side of Eq. (3.15). From Fig. 3.47A, it is found that W_{trq} increases almost in inverse proportion to thread pitch P. The ordinate on the right side represents nut rotation angle ϕ. Fig. 3.47B shows that W_{trq} increases nearly in proportion to μ_{th} and μ_{nu}. In this case, ϕ is constant regardless of coefficients of friction, as evident from Eq. (3.33). In Fig. 3.48A and B, the effects of grip length L_f and Young's modulus E are depicted.

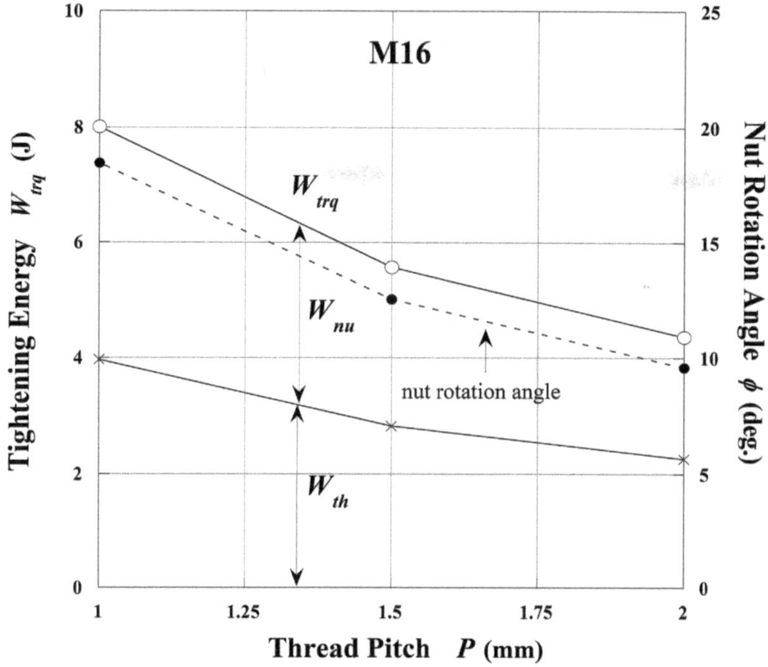

(a)

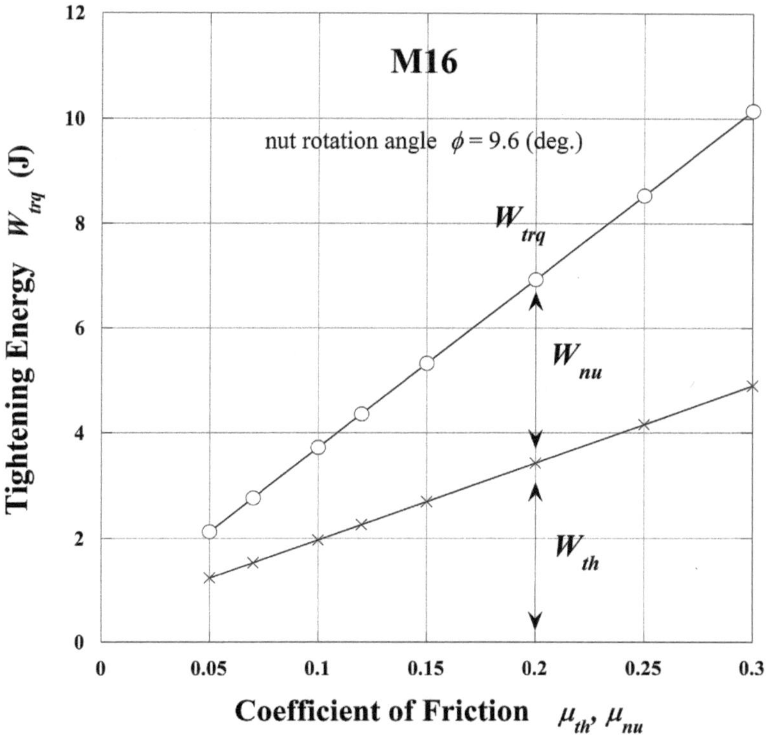

(b)

Fig. 3.47 Effect of thread pitch and coefficient of friction on tightening energy. (A) Thread pitch and (B) coefficient of friction.

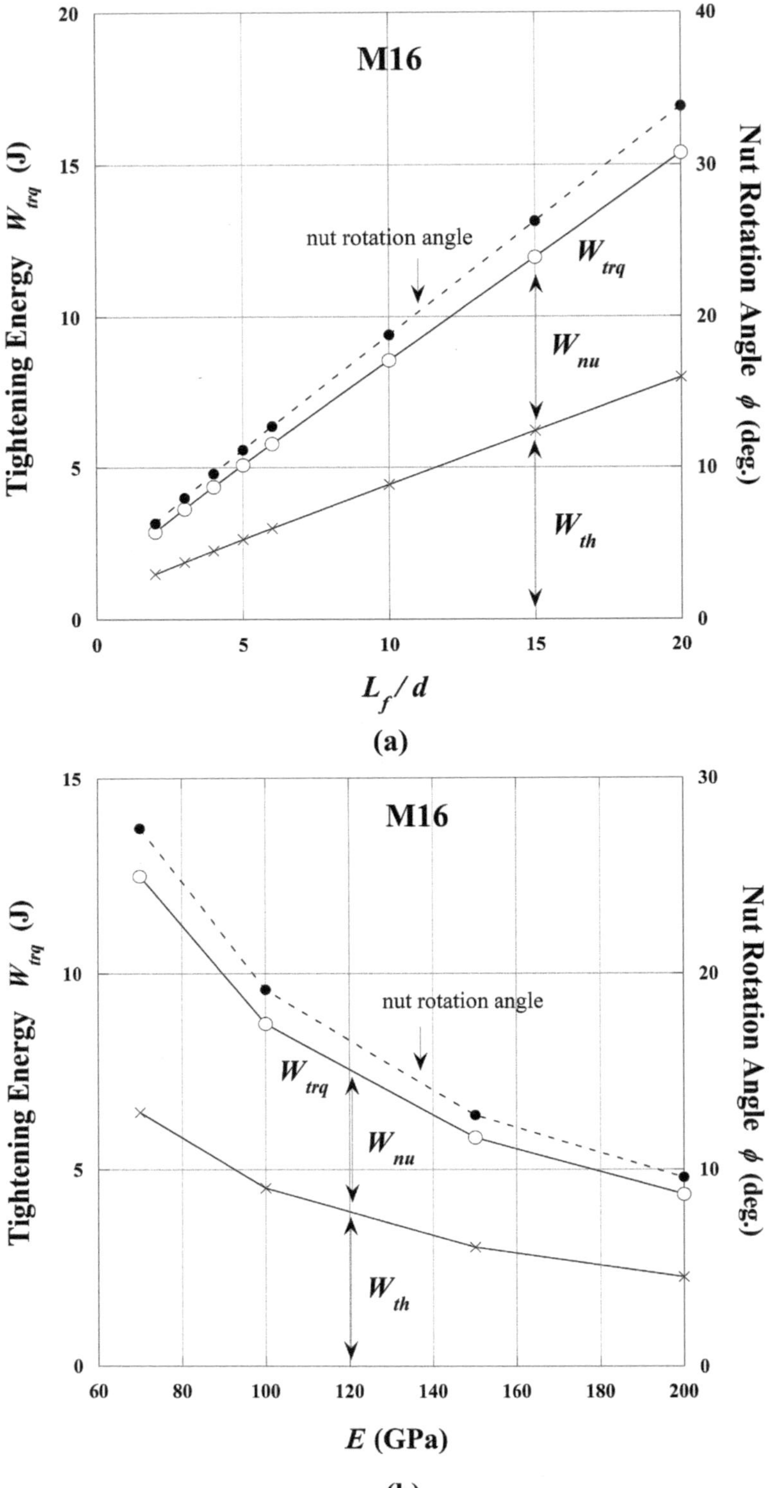

Fig. 3.48 Effect of grip length and Young's modulus on tightening energy. (A) Grip length and (B) Young's modulus.

Regarding grip length L_f, W_{trq} increases almost in proportion to L_f/d, as shown in Fig. 3.48A. Fig. 3.48B shows that W_{trq} increases nearly in inverse proportion to E, as in the case of thread pitch P. The above results are consistent with the considerations on the effect of each factor on the tightening energy, as discussed in Section 3.8.1.

Static and fatigue strengths of threaded fastener

4

4.1 Load distribution and ratio of flank loads of engaged threads

4.1.1 Load distribution in bolt-nut connections

When clamping fastened plates with a pair of bolt and nut, the bolt is subjected to tensile force and the nut is, in contrast, subjected to compression force due to bolt force F_b. Hence, thread ridges of triangular screw threads are in contact only on one side. Thread surfaces in contact and not in contact, shown in Fig. 4.1A, are termed pressure flank and clearance flank, respectively. Fig. 4.1B shows an example of von Mises stress distribution and the deformation pattern around engaged threads. The bolt is subject to pure tension. The numerical analysis is performed by axisymmetric FEM, i.e., thread geometry is assumed to be shaped like abacus balls. The magnitude of the stress is displayed with a 16-step gray scale. Focusing on the deformation of engaged threads, it is found that only the pressure flank side is in contact and the deformation becomes larger as the nut-bearing surface is approached. Regarding the thread ridge deformations, thread pitches of bolt and nut are substantially varied, because thread pitch of bolt increases and that of nut decreases due to tensile and compression forces acting on each part. When considering the mechanical behavior of a bolt-nut connection, it is important to notice that the top face of the nut is a free surface and the axial deformation around the nut-bearing surface is constrained by the plate in contact.

Because of the abovementioned phenomena, distinctive load distributions are observed along engaged threads. That is, the axial force acting on the pressure flank of engaged threads becomes largest at the first thread nearest to the nut-bearing surface, shown in Fig. 4.1A, and decreases toward the top face of the nut. Therefore, the maximum stress in a bolt-nut connection usually occurs at the first bolt thread root nearest to the nut-bearing surface.

Yamamoto derived a differential equation that represents the load distribution along engaged threads [95]. In Yamamoto's approach, a bolt is replaced by a cylinder with thread ridges being wound around its outer surface. As for the nut, it is assumed that thread ridges are wound around the inner surface of a hollow cylinder. Then, using the formulas that represent those deformations, the stiffness is expressed by spring constants for all parts of a bolt-nut connection. The resulting equation is shown below:

$$\frac{d^2F}{dx^2} - a_1F = 0 \tag{4.1}$$

The Mechanics of Threaded Fasteners and Bolted Joints for Engineering and Design.
https://doi.org/10.1016/B978-0-323-95357-3.00008-3

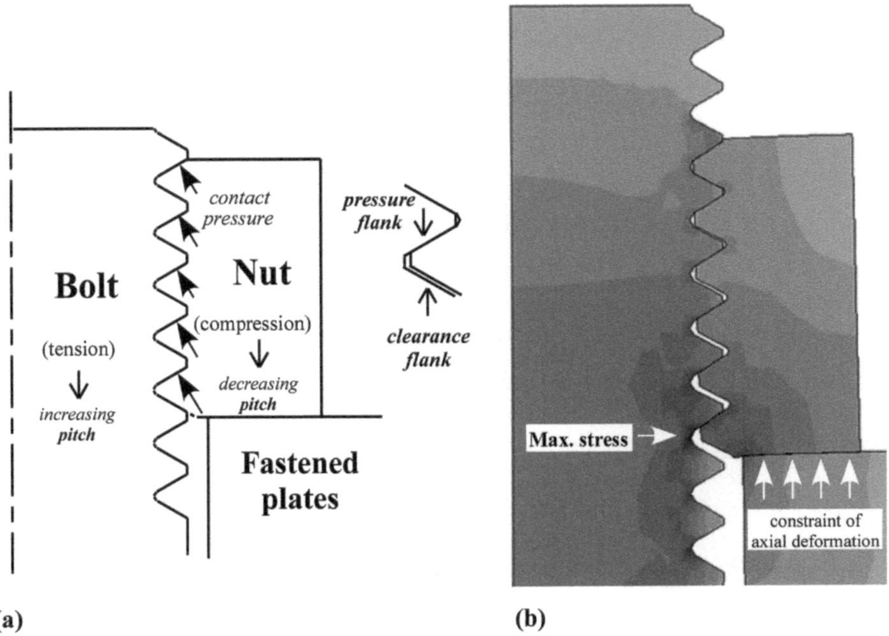

Fig. 4.1 Forces acting around engaged threads and von Mises stress distribution and deformation pattern. (A) Forces acting around engaged threads and (B) von Mises stress distribution and deformation pattern.

$$a_1 = \frac{\dfrac{1}{A_{ex}E_{ex}} + \dfrac{1}{A_{in}E_{in}}}{P\left(\dfrac{1}{k_{ex}} + \dfrac{1}{k_{in}}\right)}$$

where x designates the distance from top face of nut toward nut bearing surface; F is the magnitude of axial force acting on the cross section at coordinate x; A and P are cross-sectional area and thread pitch, respectively; male and female threads are distinguished by the subscripts ex and in; and k_{ex} and k_{in} represent the spring constants for the axial stiffness per one-pitch heights, respectively, of male and female threads.

Denoting the thickness of nut as H_{nu} and integrating Eq. (4.1), F can be expressed in the following form:

$$F = F_b \frac{\sinh\left(\sqrt{a_1}x\right)}{\sinh\left(\sqrt{a_1}H_{nu}\right)} \tag{4.2}$$

The above equation indicates that from top face to bearing surface of nut, the load distribution along engaged threads varies from zero to bolt force F_b following a function of $\sinh(x)$.

Then, how much bolt force is supported by each thread ridge? Assuming that thread geometry is axisymmetric, like abacus balls, the ratio of axial force supported by each thread ridge is termed ratio of flank loads herein. The sum of ratio of flank loads is necessarily unity. It is often used as an index for evaluating the strength of screw threads, because it provides the fundamental information on the load distribution along engaged threads. Ratio of flank loads of a bolt-nut connection can be calculated using Eq. (4.2). For example, in the case of M16 with thread pitch P of 2 mm and nut thickness H_{nu} of 12 mm, the magnitude of F can be calculated at six points from $x=0$ to $x=H_{nu}$ at intervals of 2 mm. That is, calculating the difference in F between the target point and at its adjacent point and dividing the difference by bolt force F_b, the ratio of flank loads is successively obtained for each thread ridge. As a result, all the ratios of flank loads can be evaluated from the first thread, which is nearest to the nut-bearing surface, to the sixth thread, next to the top face of the nut.

Distribution patterns of ratio of flank loads considerably vary depending on the thread geometry and the type of usage, e.g., bolts and nuts, pipe threads, eyebolts and eyenuts, etc. In bolt-nut connections, the first thread nearest to the nut-bearing surface supports the largest load and the ratio decreases toward the top face of the nut, since a bolt is under tension and a nut is under compression in the clamped state. In the next section, distribution patterns of ratio of flank loads in eyebolts and eyenuts are explained. In the former case, the stiffness of female threads is very high, and in the latter case, male and female threads are both under tension.

4.1.2 Load distribution in eyebolts and eyenuts

Eyebolts and eyenuts, shown in Fig. 4.2, are commonly used threaded fasteners for the purpose of lifting large structures or machines. When clamping ordinary bolt-nut connections, the bolt is subjected to tensile force, and the nut is subjected to compression force whose magnitude is equal to the tensile force. On the other hand, in eyebolts and eyenuts, male and female threads are both under tension due to gravity and inertial forces. Therefore, the load distribution pattern along engaged threads is different from that in bolt-nut connections and it can be obtained by solving the following differential equation [96]:

$$\frac{\mathrm{d}^2F}{\mathrm{d}x^2} - a_1F + a_2F_b = 0 \tag{4.3}$$

$$a_2 = \frac{1}{PA_{in}E_{in}\left(\dfrac{1}{k_{ex}} + \dfrac{1}{k_{in}}\right)}$$

The symbol F in Eq. (4.3) represents the magnitude of axial force acting on the cross section, a distance x away from the tip of male thread. Comparing to Eq. (4.1) for bolt-

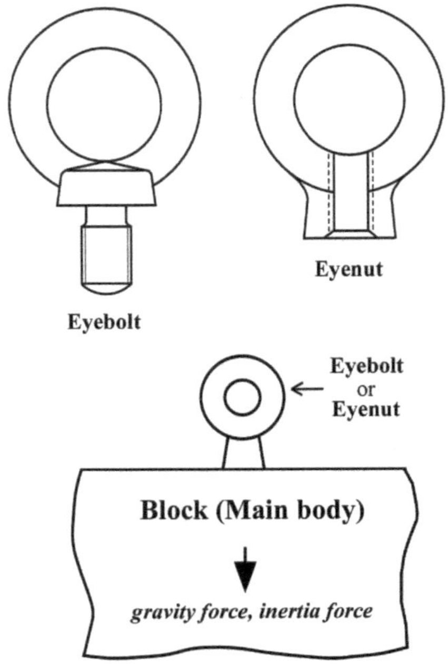

Fig. 4.2 Eyebolt and eyenut subjected to gravity and inertial forces.

nut connections, a constant term a_2F_b is added in Eq. (4.3). The ratio of flank loads in eyebolts and eyenuts can be calculated by integrating Eq. (4.3) in a similar manner to the case of bolt-nut connections.

Fig. 4.3A shows the ratio of flank loads in eyenuts, in which the calculated results by Eq. (4.3) are compared to those obtained by axisymmetric FEM. Fig. 4.3B depicts the finite element model of an eyenut. The ratio of flank loads in eyenuts becomes high around both ends. There are two main reasons for this phenomenon. First, male and female threads are both under tension. The second reason is the small volume of the eyenut that is engaged with the male threads. In the case of eyebolts, although not shown in the figure, the ratio of flank loads shows a steady decrease from the tip of the male thread toward the ring portion. The distribution pattern is similar to that in body-side engaged threads clamped by tap bolts or studs. In this case, the body-side female threads can be regarded as a high stiffness nut because of its very large volume, thereby producing a similar distribution pattern to the joints clamped by tap bolts or studs.

Pipe thread connections are another example of a joint in which male and female threads are both under tension or compression. In parallel pipe threads used for mechanical joining of two pipes, the ratio of flank loads becomes high around both ends of the engaged threads, and its distribution pattern is nearly symmetric about the center position. Additionally, as the pipe thickness is decreased, the radial deformation gradually increases due to the stiffness reduction, and this produces a characteristic load distribution pattern, i.e., the appearance of high value around the center position.

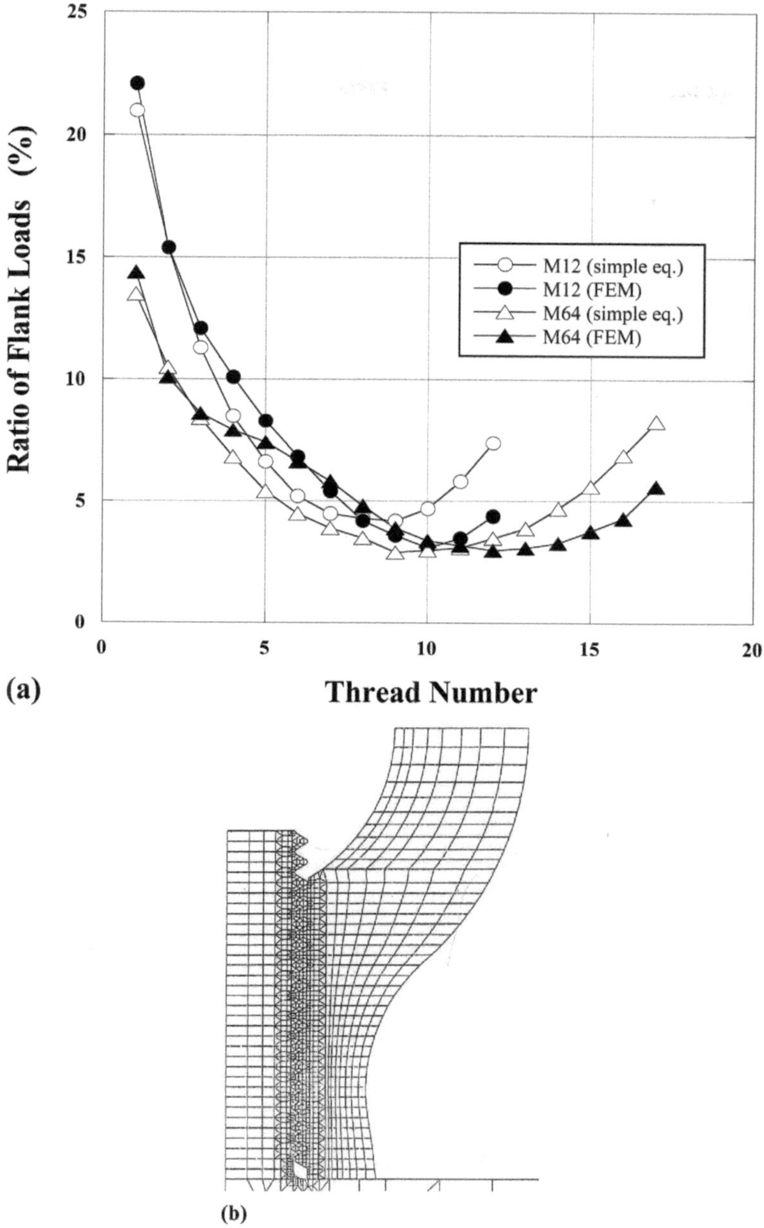

Fig. 4.3 Ratio of flank loads and FE model of eyenuts. (A) Ratio of flank loads in eyenut and (B) FE model of eyenut.

4.1.3 Analysis of ratio of flank loads by FEM

Assuming the shape of thread ridge as being axisymmetric, like an abacus ball, the ratio of flank loads can be evaluated with comparative ease by FEM. Applying bolt force F_b to bolt cylindrical body, the axial components of contact forces acting on the pressure flank are calculated for each combination of bolt and nut threads with one-pitch height. Then, the ratio of flank loads along the engaged threads can be obtained by dividing each axial component by F_b successively.

Fig. 4.4A shows an example of the numerical results of the ratio of flank loads for M16 with coarse thread. The number of engaged threads is six, since thread pitch and nut thickness are set to be 2 and 12 mm, respectively. Coefficients of friction on thread surface and nut bearing surface, μ_{th} and μ_{nu}, are assumed to be equal and changed to 0.1, 0.15, and 0.2. The ratio of flank loads shows a maximum value on the first thread and then decreases from nut-bearing surface toward top face of nut. Meanwhile, a slight increase is observed on the sixth thread. This is mainly due to the assumption of axisymmetric shape of thread ridge. That is, the axial stiffness of the FE model becomes higher than that of the actual thread ridge with helical shape. It is found that ratio of flank loads of the first thread somewhat increases as μ_{th} and μ_{nu} are increased. This phenomenon can be explained as follows. Unlike the bolt head-bearing surface, the pressure flank is inclined by 30 degrees to the plane perpendicular to the bolt axis. Therefore, a slip in the radial direction is liable to occur rather easily when μ_{th} is small. As μ_{th} is increased, however, the relative slip between male and female threads becomes more strongly restrained, which changes the deformation pattern around the engaged threads. Thus, the ratio of flank loads of the first thread is increased.

What has been discussed so far raises the question: how effective is increasing the number of engaged threads in enhancing the strength of threaded fasteners? To answer this question, comprehensive finite element analyses have been performed, in which only the number of engaged threads is changed using the foregoing FE model of M16. In Fig. 4.4B, the numerical results of ratio of flank loads are shown, where the number of engaged threads is changed from 6 to 40. It shows that even if the engaged number is increased, the decrease in the ratio of flank loads of the first thread is not substantial. It also indicates that there is no contact in the 18th and subsequent engaged threads when the engaged number exceeds 18. In JIS, the standard nut thickness for M16 is 13 mm, which means that the engaged thread number is 6.5. Accordingly, it can be said that the results shown in Fig. 4.4B demonstrate the effectiveness of the generally accepted standardization for nut thickness.

4.2 Static strength and stress concentration in threaded fasteners

4.2.1 Stress concentration and stress concentration factor

Stress concentration is a phenomenon in which the stress becomes locally high around a component with discontinuous shape, like a notch. In general, machines and

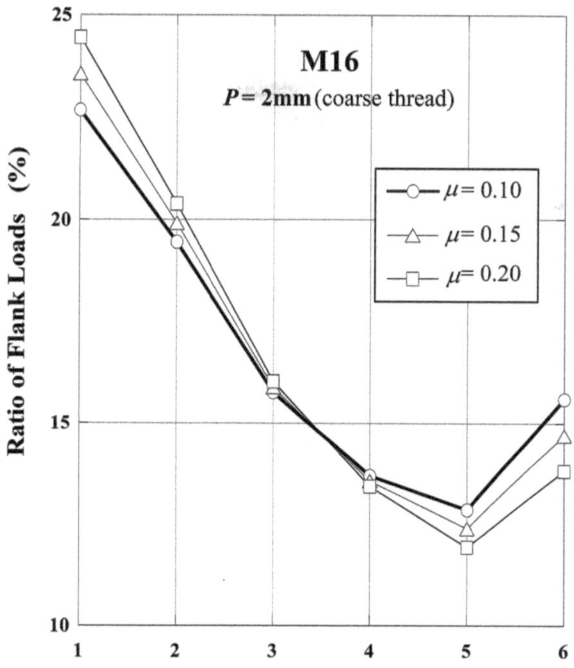

(a)

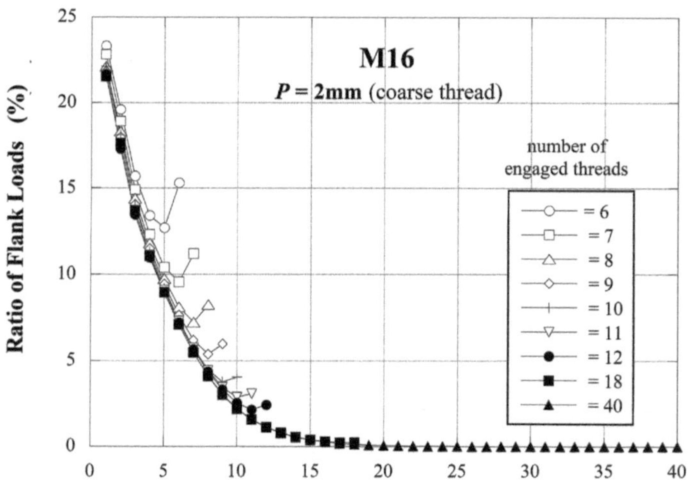

(b)

Fig. 4.4 Ratio of flank loads of bolt-nut connection of M16 with coarse thread. (A) Effect of coefficient of friction and (B) effect of number of engaged threads.

structures have cross sections of various complex shapes to achieve a specific function, such as arc- and U-shaped notches. Therefore, some level of stress concentration inevitably occurs. The extent of stress concentration is evaluated by the ratio of the maximum stress $\sigma_{\max}$ occurring at notch root to the mean stress σ_n at notched cross section, which is termed stress concentration factor or shape factor:

$$\alpha = \frac{\sigma_{\max}}{\sigma_n} \tag{4.4}$$

Concrete values of stress concentration factor α are given in the textbooks or handbooks [97] related to machine design, in which such members as plates and shafts with various notched shapes subjected to tension, compression, bending moment, and twisting moment are discussed. When the shape of notch and/or the way of applying external loads are too complicated to evaluate using the aforementioned handbooks, numerical analysis by FEM would be an effective alternative method.

Mechanisms of the generation of stress concentration and the variation of stress concentration factor can be explained using a simple example. Fig. 4.5A depicts the stress concentration phenomenon of an infinite plate with a single hole subjected to tensile force. Stress concentration factor in this case is 3, as is well known. In the case of the plate width being finite, as shown in Fig. 4.5B, the stress concentration factor becomes less than 3. The reason is as follows. When subjected to tensile load, the circular hole deforms into a vertically long shape along the load direction, while the stiffness due to the plate material existing lateral to the hole prevents such a deformation pattern. Accordingly, as the plate width is decreased from infinite to finite, the stiffness of the plate that prevents the hole deformation is decreased, hence, the stress

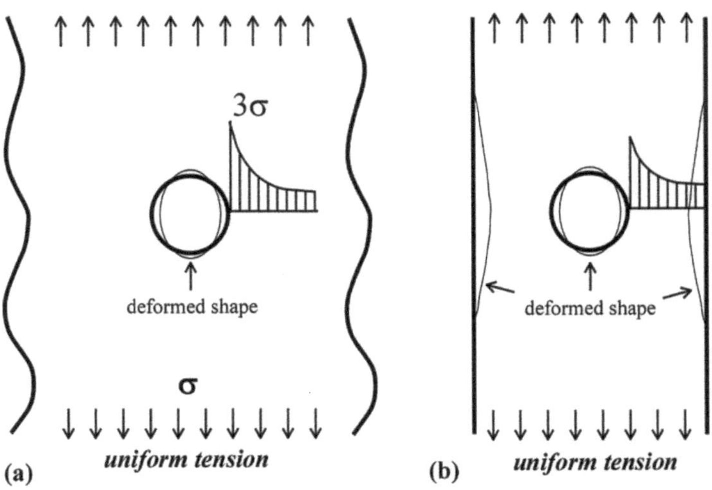

Fig. 4.5 Stress concentration of plates with a single hole. (A) Infinite plate and (B) plate of finite width.

concentration factor is also reduced. It follows that the stress concentration factor is maximum in the case of infinite plate and the factor decreases as the plate width is decreased.

Another example is a comparison between a plate with a single notch and a plate with multiple notches, each of which is under tension. Notch shape is supposed to be semicircular with identical size. In the case of a single notch, as shown in Fig. 4.6A, the open-mouth deformation around the notch is largely constrained by the adjacent portion. Accordingly, stress concentration factor α_I is fairly large and is about 3.06. In the case of having two notches, as shown in Fig. 4.6B, stress concentration factor α_{II} is smaller than α_I, because the deformation around the notch becomes less constrained. In the case of having three notches, stress concentration factor for the center notch α_{IV} is smaller than that for the end notch α_{III}. This suggests that the stress concentration factor for the notch located at the end of a row of continuous notches is larger than the factor for the center notch, because in the former case one side is heavily constrained due to the absence of an adjacent notch. Accordingly, the order of stress concentration factors is as follows:

Single notch > end notch of continuous notches > center notch of continuous notches.

The above statement is summarized in the more generalized expression that stress concentration factor is low when the stiffness around the target notch is low, which produces large deformation in its vicinity. Regarding the notch shape, stress concentration factor is increased with smaller notch root radius. It is also important that the magnitude of stress concentration factor is varied depending on the loading condition. In general, for plates or bars having identical notches, stress concentration factor is largest when subjected to tension or compression, followed by bending and twisting in descending order [98].

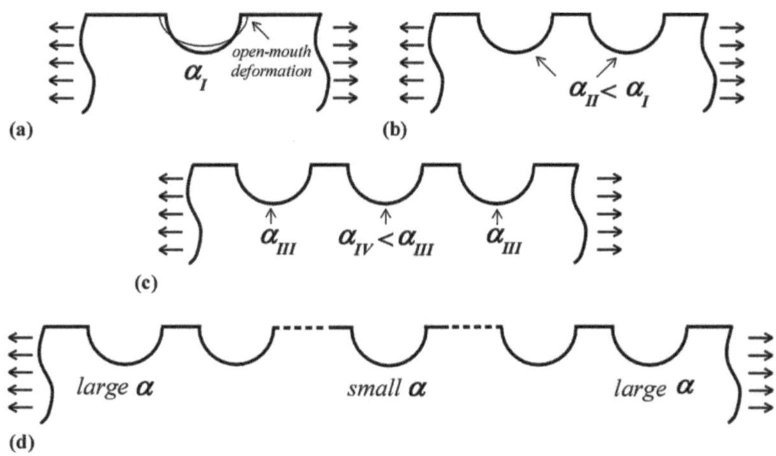

Fig. 4.6 Comparison of stress concentration between plates with a single notch and multiple notches. (A) Single notch, (B) two notches, (C) three notches, and (D) continuous notches.

4.2.2 Stress concentration in threaded fasteners

In bolted joints, high stress concentration usually occurs in three locations, i.e., the first bolt thread root adjacent to the nut-bearing surface, the thread runout of bolt, and the root radius under the bolt head. These are coincident with the locations where failures or ruptures are likely to occur, illustrated in fig. 2.1. Among them, the strength of the first bolt thread root is the most critical issue. The root radius of male thread ρ recommended by JIS is given in eq. (1.7). As for thread runout, high stress concentration occurs because it is located at the end of a series of continuous notches, as explained in Fig. 4.6. If the root radius gets smaller due to processing problems, the stress concentration factor is further increased.

Meanwhile, in the case of large bolts under repeated loading, like the crank pin bolts of large diesel engines, a slender bolt, which is a bolt with reduced shank, is often used, aiming at the improvement of fatigue strength. In slender bolts, thread runout portions are sometimes removed in the manufacturing process to avoid high stress concentration occurring there. Regarding the root radius under the bolt head, a minimum value is given in JIS B 1180 to avoid high stress concentration. If the bolt head-bearing surface is not perfectly parallel to the plate surface, some amount of bending moment is inevitably produced when applying bolt force. A practical example is where a plate surface being slightly tapered due to processing problems. Bending moment, induced by the nonparallelism between mating surfaces, also becomes an issue around the nut-bearing surface.

In the foregoing cases, stress concentrations in the three locations stated above could be larger than predicted, due to the additional bending moment. On the other hand, in thin plates with some residual distortion caused by plastic working, loosening of threaded fasteners becomes a problem more often than stress concentration.

Geometric errors involved in threaded fasteners also affect the stress concentration. It is stipulated in JIS B 1180 that the errors in parallelism at the nut-bearing surface and perpendicularity between bolt head and bolt cylinder must be less than one or two degrees, as shown in fig. 1.11. In large machines and structures, assembly errors sometimes cause a problem in addition to geometric errors. Due to the synergistic effect of both errors, there is a case in which stress concentration becomes considerably higher than the design value. It is quite difficult to quantitatively evaluate the stress concentration considering the effect of geometric and assembly errors. In such cases, numerical methods like FEM would be an effective means.

4.2.3 Mechanism of stress concentration at thread root

Stress concentration factor at thread root increases with larger nominal diameter. For identical nominal diameters, larger stress concentration is produced in fine screw threads than in coarse ones. Threaded fasteners do not have similar figures in most cases, as shown in fig. 1.8. In the case of bolts with coarse thread, the ratio of thread pitch P to nominal diameter d, P/d, decreases with increasing d. Here, the stress concentration at the bolt thread root is compared to the case of a plate of finite width under tension with continuous notches on one side. In the case of bolts, the depth of notch gets relatively smaller as nominal diameter increases, as found from fig. 1.8. Hence, it approaches the state of a semiinfinite plate with continuous notches, thereby increasing the stress concentration factor [97]. It has been reported that the stress concentration factor at the first bolt thread root exceeds 4, according to the numerical results established by FEM [9,99]. Such high stress concentration is produced by the following mechanism.

First, in relation to the ratio of flank loads, shown in Fig. 4.4, the thread ridge next to the nut-bearing surface supports a large amount of axial force, which exerts a large tensile force around the thread root. In that case, since male and female threads are in contact only on the pressure flank side, the compression force acting on the pressure flank generates open-mouth deformations around the thread root, as in the case of applying bending moment. This promotes stress concentration at the first thread root, coupled with the action of tensile force explained above. The mechanism of stress concentration at bolt thread root is illustrated in Fig. 4.7.

In summary, it is concluded that the superposition effect of tensile force and bending moment generates high stress concentration at the bolt thread root. The stress distributions and deformation pattern around engaged threads, shown in Fig. 4.1, confirm the foregoing explanation that very high stress concentration is produced around the first bolt thread root. Deformations of engaged threads become larger toward the nut-bearing surface and a salient open-mouth deformation pattern is observed around the first bolt thread. Regarding the effect of bending moment, even if the same amount of compression force acts on the pressure flank, the stress concentration at the bolt thread root is increased when the center of contact pressure approaches the tip of the thread ridge because of the longer moment arm.

Meanwhile, it has been reported that in a hollow bolt whose hole diameter is half the nominal diameter, the ratio of flank loads of the first thread is higher compared to a solid bolt. However, the stress concentration at the first thread root is somewhat

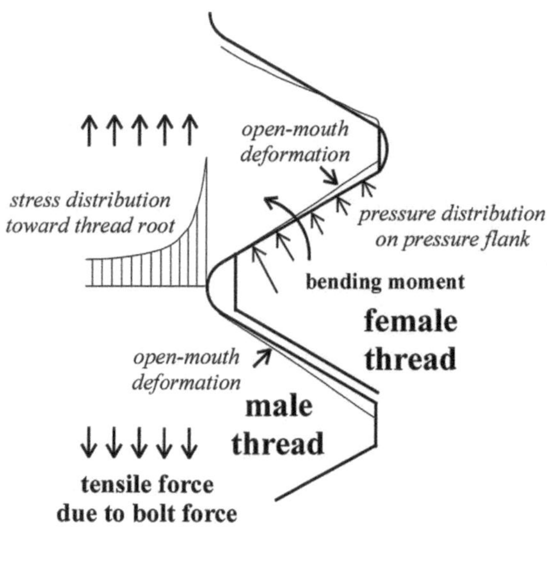

Fig. 4.7 Mechanism of stress concentration occurring at thread root.

smaller [83]. This phenomenon can be explained as follows. By adding a hole to a solid bolt, the center of contact pressure on the pressure flank moves toward the thread root, and the moment arm length is shortened.

4.2.4 Evaluation of stress concentration at thread root

The root radius of the thread root has a dominant effect on the stress concentration. Stress distributions along the thread profile from thread root to flank are evaluated by Otaki [7,8], in which the geometry around the thread root is represented by a continuous function having the same curvature and then the exact solutions of complex stress functions are applied. Using the analytical results obtained, stress concentration factors are quantitatively evaluated for metric threads and Whitworth threads. As a result, it has been clarified that stress concentration at the thread root increases as nominal diameter is increased and/or thread root radius is decreased, and stress concentration factor of fine screw threads is higher than that of coarse screw threads with identical nominal diameter. Using a copper plating method, Seika, et al. measured the stress distribution along the thread root [16]. Their experimental results show that the maximum stress occurs at a bolt thread root located 2/3 pitch away from the nut-bearing surface.

Fukuoka et al. conducted axisymmetric FE analyses for the case of bolts being under pure tension, from which it was clarified that stress concentration factor at thread root exceeds 4 [99], in which the meshes placed around the thread root are changed in various sizes to ensure numerical accuracy. In their analyses, the maximum length of triangular elements, placed around thread root, is selected to be less than one-tenth of the root radius. Then, by changing thread root radius and coefficients of friction on thread surface and nut-bearing surface into various levels, the stress concentration factor at thread root has been evaluated with two significant digits of precision. Fig. 4.8 depicts the whole mesh pattern of the analytical model and the detailed mesh patterns around the thread root. The analytical object is M24 with coarse thread and thread pitch is 3 mm. Root radius of male thread ρ is changed to 0.4, 0.3, and 0.2 mm. Additionally, the minimum value of ρ, recommended by JIS, is calculated to be 0.375 mm by using eq. (1.7). Numerical results are shown in Fig. 4.9.

In order to compare the numerical results to Maruyama's experimental ones obtained by the copper plating method, the ratio of the maximum principal stress to mean stress at thread root is selected as the ordinate. The abscissa represents thread root radius. This indicates that the numerical results are sufficiently effective, considering the scatter in experiments. Also, the numerical results confirm the validity of the size of root radius ρ recommended by JIS. In the above analyses, the constraint on the axial deformation of thread ridge is higher compared to actual screw threads, since helical thread geometry is not considered. It is therefore predicted that stress concentration factors thus obtained are somewhat high compared to the ones evaluated by using helical thread models, as explained later.

When bolts are tightened by the torque control method, the magnitude of von Mises stress produced in engaged threads becomes higher than the case of applying pure tension, as found from eq. (3.31), due to the effect of twisting moment. In [47], the stress concentration factor in tightening by the torque control method is evaluated by using

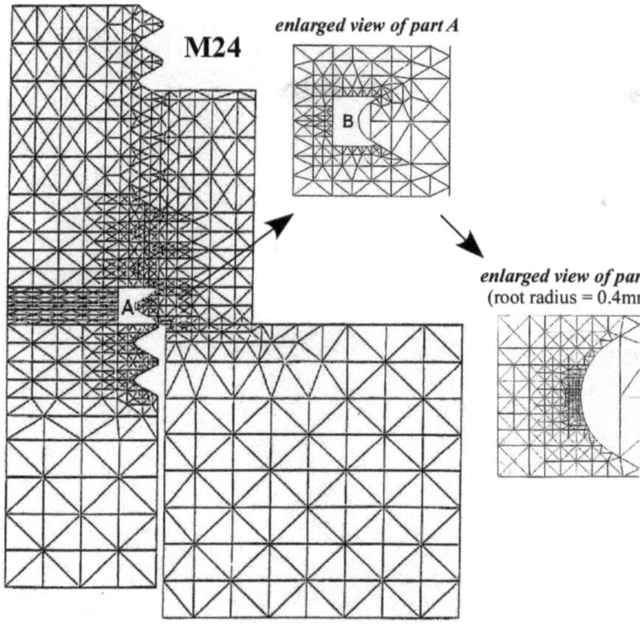

Fig. 4.8 FE model for analyzing stress concentration at thread root.

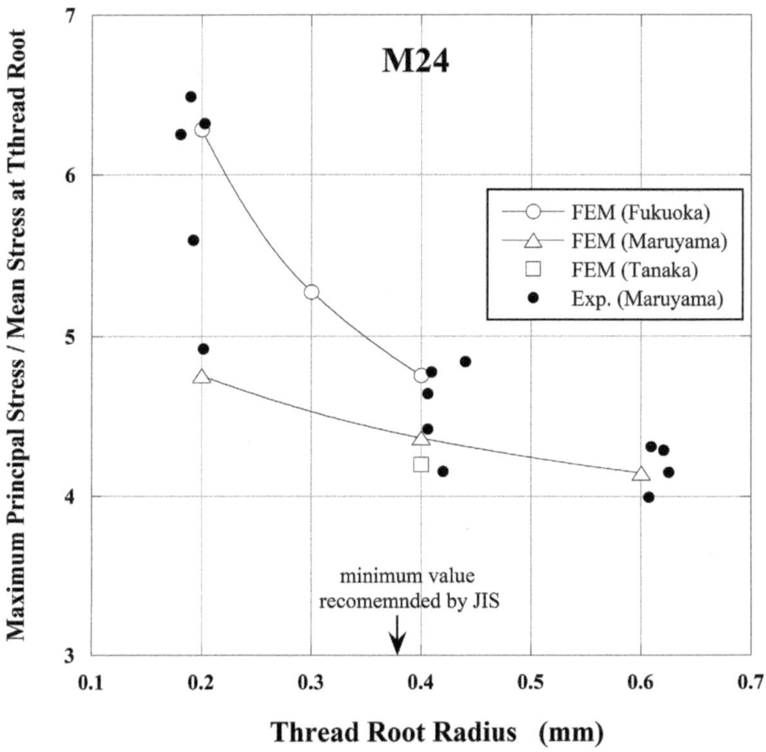

Fig. 4.9 Effect of thread root radius on stress concentration factor.

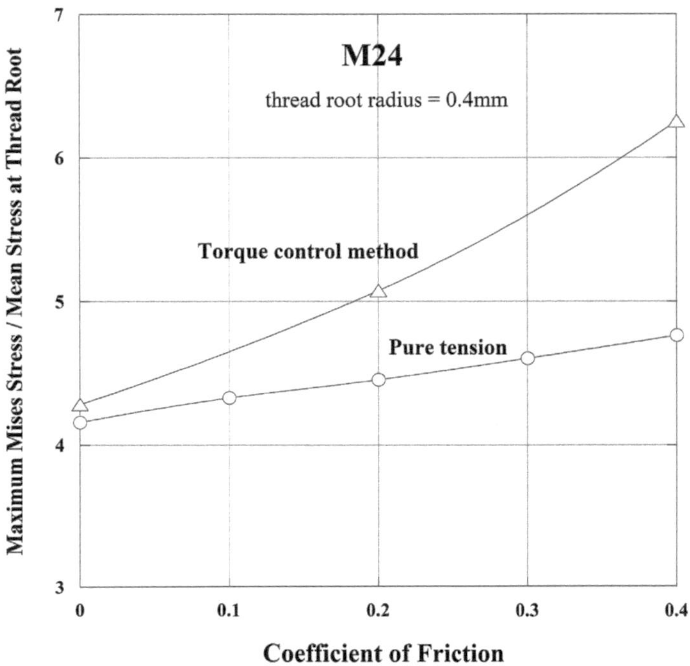

Fig. 4.10 Stress concentration factors produced by the torque control method and under pure tension.

two-dimensional FE models, in which the model shape is axisymmetric, but each node has three degrees of freedom. That is, in addition to r and z coordinates, deformations in the circumferential direction θ are considered. Fig. 4.10 compares the numerical results for M24 between maximum von Mises stresses generated by applying tightening torque and those under pure tension. When tightened by the torque control method, higher stress concentration occurs owing to the shear stress induced by the twisting moment. In addition, as coefficients of friction are increased, stress concentration factors show a rather steep increase because larger torque is required to produce the same amount of bolt force. This result is consistent with the relationship between the stresses in engaged threads and the coefficient of friction, discussed in numerical example 3.7 in section 3.2.9.

4.2.5 Stress concentration and plastic deformation of threaded fasteners

Stress concentration factors at thread root are generally higher than those of the notches frequently encountered in commonly used machines and structures, as discussed in the preceding sections. Therefore, plastic deformations are likely to occur around the thread root, even if threaded fasteners are clamped with relatively low bolt force. Such plastic deformations are usually limited to small regions, since the stress concentration factor is very high. That is, the occurrence of plastic deformation does not necessarily lead to joint failure.

Fig. 4.11 depicts how the plastic region expands with increasing bolt stress σ_b, when M12 and M16 bolts with coarse thread are subjected to pure tension. Areas shown in black are under plastic deformation. The target joint material is supposed to be elastic-perfectly plastic body, i.e., stress values are assumed to be constant after yielding. Young's modulus and Poisson's ratio are 200 GPa and 0.3, and coefficients of friction on contact surfaces are set to be 0.15. The magnitude of bolt stress σ_b is expressed as the ratio to the yield stress σ_Y of the joint material. Plastic deformations initiate around the first bolt thread root and gradually spread to the other thread roots. Finally, plastic deformations spread over the entire cross section around the center of unengaged threads. The expansion of plastic regions, shown in Fig. 4.11, is consistent with the problem cases in which bolt rupture due to over torquing is likely to occur around the center of unengaged threads. Regarding the magnitude of bolt stress σ_b, it is found that plastic regions suddenly start to propagate between $0.7\sigma_Y$ and $0.8\sigma_Y$. Focusing on the effect of nominal diameter, the plastic region shows a rapid spread in M16 when exceeding $0.76\sigma_Y$, while the cross section of M12 is almost in the plastic region around the center of unengaged threads, even under the condition of $0.72\sigma_Y$.

In Fig. 4.12A, the spread of plastic region is illustrated when tightening M12 bolts by the torque control method [48]. Numerical analyses are conducted using axisymmetric FE models, in which each node has three degrees of freedom, as explained in [47]. Coefficient of friction is 0.2. Initial yield stress is 790 MPa, and tangent modulus in the plastic region is assumed to be 2700 MPa to consider the effect of strain hardening. It is difficult to strictly compare the results between Figs. 4.12A and 4.11 because of the difference in the analytical conditions. To put it simply, even if bolt stress σ_b is low, the plastic regions spread over a considerably larger area than the case of pure tension, owing to the action of twisting moment. Fig. 4.12B shows the relationship between the spread of plastic regions and nominal diameter of bolt.

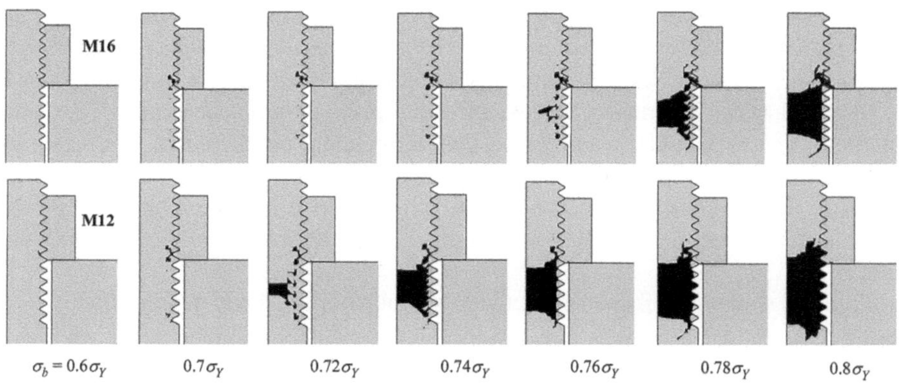

Fig. 4.11 Expansion of plastic regions for varying bolt stress and nominal diameters.

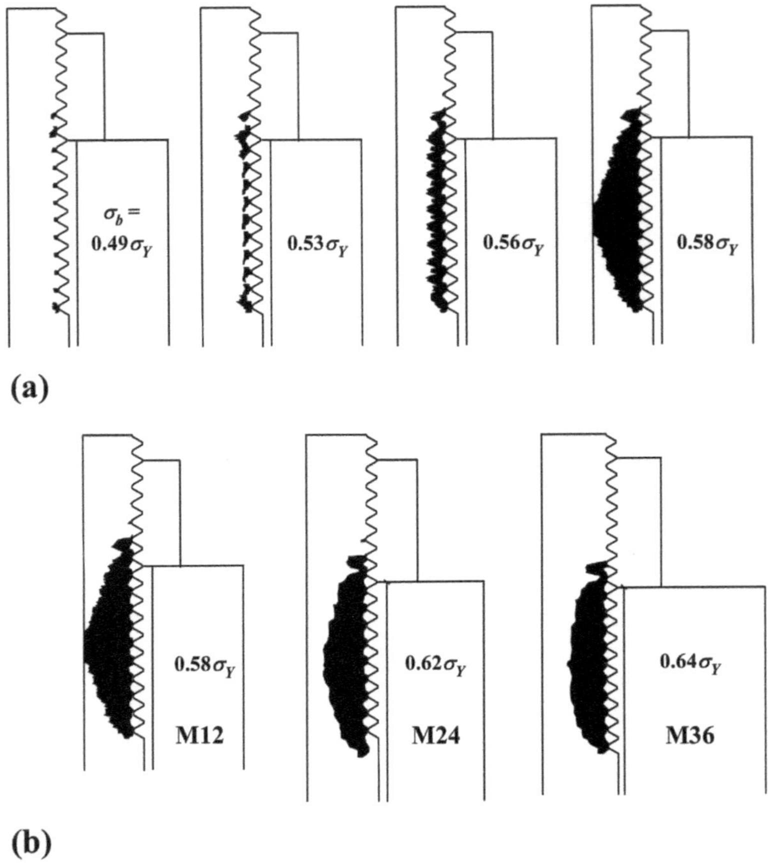

Fig. 4.12 Expansion of plastic regions when tightening by the torque control method. (A) Effect of bolt stress (M12) and (B) effect of nominal diameter.

The amounts of tightening torque applied to M24 and M36 are determined so that the spread of plastic regions becomes nearly the same as the rightmost result of M12, $\sigma_b = 0.58\sigma_Y$, in Fig. 4.12A. The amount of bolt stress σ_b of M12 is lower than those of M24 and M36. This means that the spread of plastic regions becomes larger with smaller nominal diameter, as in the case of being under pure tension. The reason is the nonsimilar figures of threaded fasteners, shown in fig. 1.8. That is, as nominal diameter d decreases, the ratio of thread pitch P to d increases, and the cross-sectional area at male thread root A_r becomes relatively smaller than bolt cylindrical area A.

Fig. 4.13 shows the relationship between the ratio of A_r to A and nominal diameter d, where A_r is calculated using the maximum value of thread root diameter, $d_{r|\max}$, shown in numerical example 2.2.

$$d_r|_{\max} = d - 1.227P \tag{4.5}$$

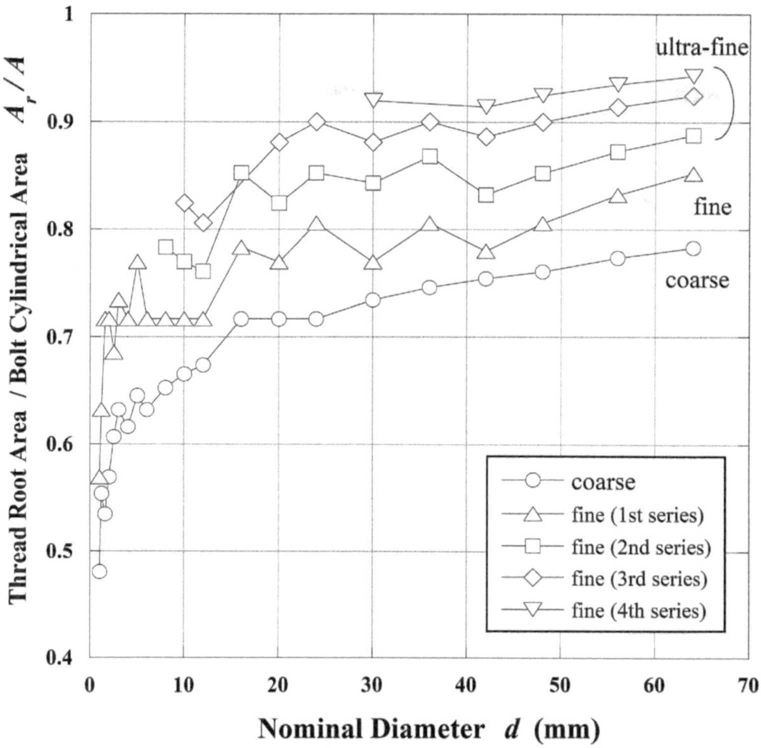

Fig. 4.13 Ratio of cross-sectional area at male thread root to bolt cylindrical area.

There are several specifications of fine screw threads for identical nominal diameter. In this book, screw threads whose thread pitch is one size smaller than that of coarse screw thread are simply called fine screw threads, and the ones with even smaller pitches are ultrafine screw threads. A_r/A is necessarily smallest in coarse screw threads. Fig. 4.13 shows that A_r/A becomes considerably smaller, particularly for coarse screw threads of M16 and smaller. It suggests that with smaller nominal diameter, plastic deformations are likely to occur due to the rise in the mean thread root stress, even if clamped with identical bolt stress. It is considered that the relationship between A_r/A and nominal diameter d, shown in Fig. 4.13, is one of the major causes for a phenomenon that plastic deformations or ruptures are liable to occur when tightening small bolts by the torque control method.

The above considerations suggest that when determining the magnitude of bolt stress σ_b, the amount of A_r/A varying with nominal diameter should be taken into account, in addition to the ratio of σ_b to yield stress σ_Y of the joint material, such as $0.7\sigma_Y$ or $0.6\sigma_Y$. Furthermore, it is also worth noting here that the progress of plastic deformation is largely influenced by coefficients of friction when tightening by the torque control method.

In Fig. 4.14, the relationship between bolt stress σ_b and nut rotation angle ϕ is shown, in which coefficients of friction μ on thread surface and nut bearing surface

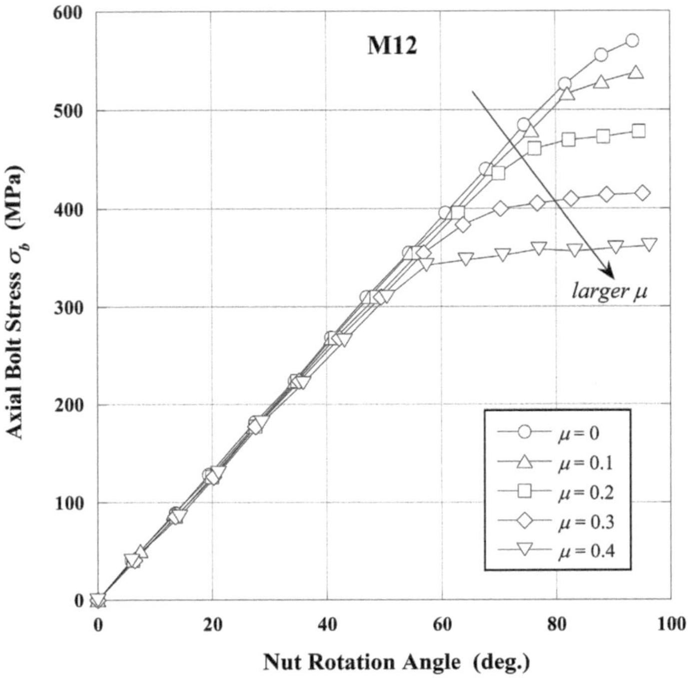

Fig. 4.14 Effect of coefficients of friction on plastic deformations.

are assumed equal. When μ is small, the relationship between σ_b and ϕ is almost linear until reaching the point of fairly high bolt stress. In contrast, when μ is large, a nonlinear relationship is observed between σ_b and ϕ from the range of relatively low bolt stress. That is because larger tightening torque T_t is required to produce the same amount of bolt force F_b as coefficient of friction is increased. It can be understood by eq. (3.15) which represents the torque-bolt force relationship. As T_t is increased, thread torque T_1 also increases, hence, shear stress and von Mises stress exerting on the bolt cylinder are raised correspondingly, thereby causing an expansion of plastic regions. Plastic deformation phenomena of bolts are summarized as follows:

1) Plastic regions tend to expand with smaller nominal diameter and larger thread pitch, which also suggests that plastic regions are more likely to expand in coarse screw threads than in fine ones.
2) When applying an identical bolt stress by torque control method, plastic regions are more likely to expand with increasing coefficient of friction in the circumferential direction.

In recent years, the use of the plastic-region tightening method has expanded [100], in which plastic deformations are allowed to some extent. An important point in the actual application is the selection of appropriate bolt stress σ_b. In the previous study [101] dealing with plastic-region tightening method, σ_b is set to be $0.8\sigma_Y$ or higher. Referring to Figs. 4.11 and 4.12, it is found that plastic deformations are considerably progressed from engaged threads to unengaged threads when tightening with this amount of bolt stress.

4.2.6 Reduction strategy of stress concentration at thread root

A large number of studies have been published so far aiming at reducing the stress concentration at bolt thread root. Incorporating those research results, many threaded fasteners with special shapes have been proposed. Listed below are some representative methods for reducing the stress concentration.

1) Increasing thread root radius

The radius of bolt thread root is increased as much as possible within the range that no interference occurs between male and female threads. This method is effective for high value-added bolts, in which high accuracies are required for the dimensions and geometry of thread root radius. Eq. (1.7) and the numerical results shown in Fig. 4.9 serve as a good reference.

2) Reducing axial stiffness of engaged threads

Stress concentration becomes noticeable when the target portion is difficult to deform. Hence, the geometry of threaded fasteners is designed so as to make the surrounding region of the first bolt thread root easy to deform, at which the maximum stress concentration occurs. For example, it is proposed to chamfer the nut-bearing surface toward the bolt axis with straight lines [102] or curved lines [103], as shown in Fig. 4.15. Chamfering with straight lines is more effective for reducing stress concentration. However, curved lines are more desirable to avoid damage to the bearing surface. It is shown in Fig. 4.16 how much the stress concentration factor can be reduced by introducing curved chamfering for the case of bolts under pure tension. The abscissa represents the chamfering angle. Experimental results obtained in [102] are also shown in the figure. Although the experimental results show a fairly large scatter for varying chamfering angle, it suggests the effectiveness of the chamfering processed on the nut bearing surface. Comparing the numerical results by FEM between M64 and M12, it is found that chamfering is more effective in reducing the stress concentration with larger nominal diameter. In the case of M64, the stress concentration factor can be reduced by up to 27%. When the chamfering angle exceeds a certain level, the stress concentration factor becomes almost constant. This result indicates that the stress concentration can effectively be reduced only by

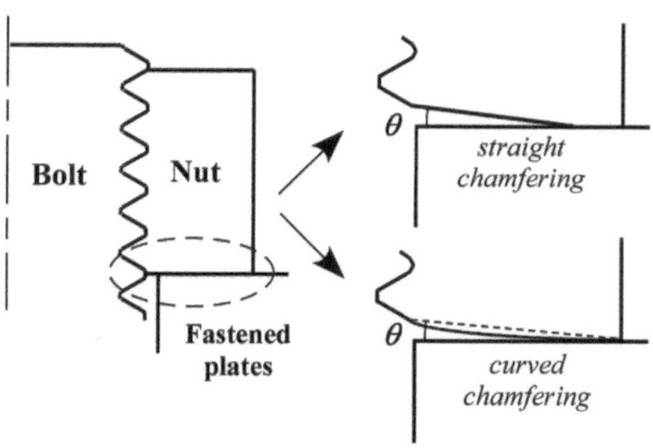

Fig. 4.15 Strategy for reducing stress concentration by chamfering nut-bearing surface.

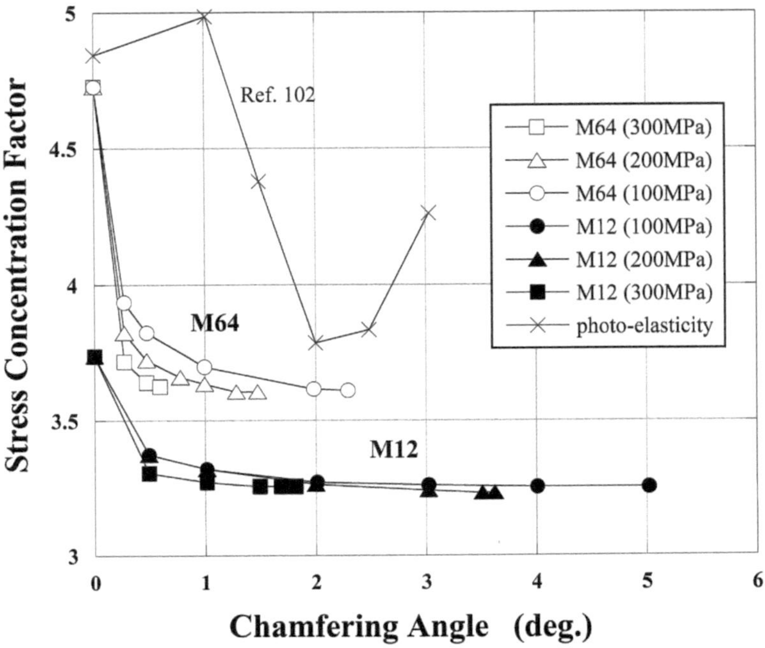

Fig. 4.16 Effect of chamfering on the reduction of stress concentration [102].

chamfering the nut bearing surface with some curvature. It also shows the robustness of this method. Incidentally, regarding the specifications of bolt hole diameter d_h, as d_h increases from first grade to third grade, the stress concentration factor decreases due to the decrease in axial stiffness.

3) Processing thread pitch of nut slightly larger than that of bolt

When tightening bolt-nut connections, the ratio of flank loads of the first thread becomes the largest, because the bolt elongates and the nut shrinks, thereby causing a difference in thread pitch between bolt and nut. To cope with this problem, the distribution of ratio of flank loads is averaged by processing the thread pitch of the nut to be slightly larger than that of the bolt. As a result, stress concentration at the first bolt thread root can be reduced. This method is theoretically effective, but it involves a difficulty in processing accuracy [103], because the amount of thread pitch modification is very small.

4) Reducing thread overlap between male and female threads toward nut-bearing surface

Thread ridges of bolts or nuts are processed with a certain taper angle along the axis; or, the tips of thread ridges are slightly cut off along the taper angle. This operation lowers the ratio of flank loads of the first thread. There are some examples in the actual threaded fasteners of diesel engines, in which the tips of male threads are cut off along a certain taper angle.

5) Processing the shape of thread ridge of nut to be subjected to tensile force

Nut shape is processed as shown in Fig. 4.17A [14] so that both male and female threads are subjected to tensile force, as in the case of pipe threads. In [14], the performance of various-shaped threaded fasteners aiming at reducing stress concentration, as shown in Fig. 4.17B, has been examined by photo-elastic experiments. It is considered that the chamfering method shown in Fig. 4.15 belongs to the type shown in Fig. 4.17B.

6) Making engaged thread length longer

In a bolt-nut connection, nut thickness H_{nu} is about 0.8 times the nominal diameter d. As H_{nu} is increased, stress concentration factor decreases. However, the effect is limited, as predicted from the numerical results of ratio of flank loads, shown in Fig. 4.4. As for studs, engaged thread lengths of 1.25d, 1.5d, and 2d are recommended for the body-side female threads made of carbon steel, cast iron, and light alloys, respectively, as stated in section 1.5.1. If the engaged thread length is changed in this way, according to the stiffness of the body-side material, stress concentration factors at the first bolt thread root are relatively close to each other among those three cases [64]. Accordingly, the engaged thread lengths recommended in JIS B 1173, shown above, are considered reasonable.

4.3 Stress distribution along thread root

4.3.1 Stress concentration in bolt-nut connections

The stress concentration factor at thread root can be evaluated by means of axisymmetric FEM. However, it is impossible to obtain the stress distribution along a helical thread root [99]. Furthermore, since the thread ridge geometry is assumed to be axisymmetric, the mutual constraint between male and female threads becomes larger than in actual threads. Additionally, the effect of slip along the thread helix cannot be considered.

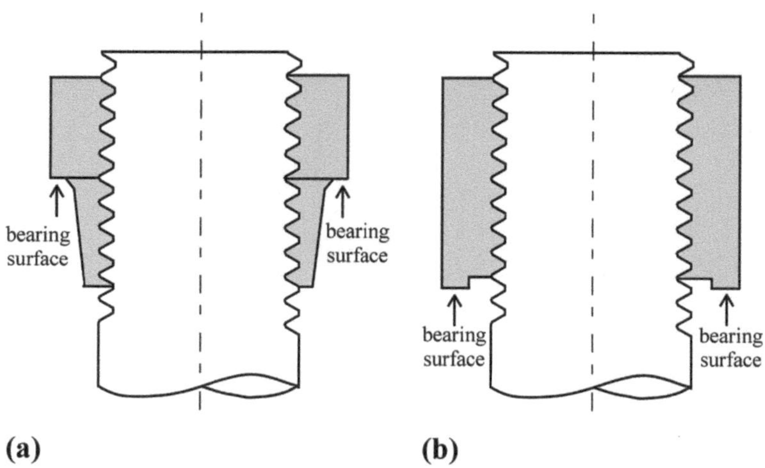

Fig. 4.17 Strategy for reducing stress concentration by modifying nut shape. (A) Shape of nut subjected to tensile force and (B) shape of nut aiming at reducing stress concentration.

From the above considerations, it is necessary to use helical thread models for the rigorous evaluation of the stress concentration around the thread root. In this section, stress concentrations along bolt thread roots are quantitatively evaluated by using the finite element models explained in section 2.5, in which helical thread geometry is rigorously reproduced.

Numerical analysis has been conducted using an FE model of M16 with coarse thread, shown in section 2.5. Uniform tensile stress of 100 MPa is applied to the cross section of bolt cylinder. Thread pitch P is 2 mm. Coefficients of friction on thread surface μ_{th} and nut-bearing surface μ_{nu} are assumed to be equal, and they are changed to 0.05 and 0.2. Supposing the jointed components as being a linear elastic body, Young's modulus and Poisson's ratio of joint materials are set to be 200 GPa and 0.3. Bolt hole dimeters are changed to 17, 17.5, and 18.5 mm, corresponding to first, second, and third grades. Fig. 4.18 depicts an example of von Mises stress distributions along bolt thread root [49]. The ordinate represents a dimensionless number defined by dividing von Mises stress $\bar{\sigma}$ by bolt stress σ_b. The abscissa shows the distance z from the nut-bearing surface. Assuming the number of engaged threads to be five, $z = 10$ mm corresponds to the top face of the nut. It is shown that the maximum stress occurs at the first bolt thread root, which is 1 mm or half a thread pitch away from the nut-bearing surface. This position corresponds to the right side of the mesh pattern of the nut, as shown in fig. 2.26B, where the nut thread height engaged with the first bolt thread is equal to one thread pitch.

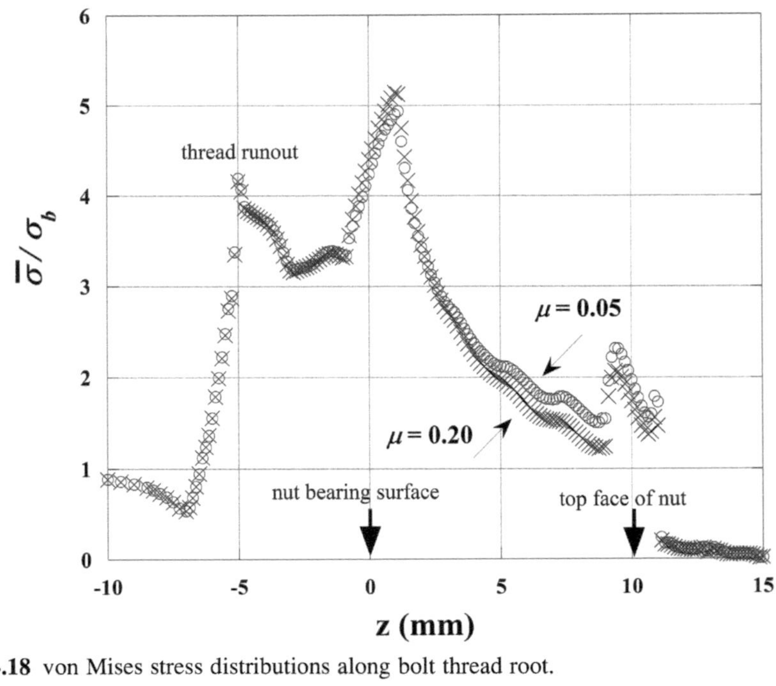

Fig. 4.18 von Mises stress distributions along bolt thread root.

It is found from these results that the thread root stress becomes maximum at the point which the deformation around the first bolt thread root is most strongly constrained. The effect of coefficients of friction reveals itself less remarkably compared to the case of axisymmetric analysis. The reason is that the mutual constraint between male and female threads becomes small, because the helical thread geometry is taken into account. Bolt thread root stress shows a rapid decrease from the first thread root toward the tip of bolt threads, and then it shows a small peak around the top face of the nut. This phenomenon can be explained as follows. Nut threads gradually vanish in the vicinity of the end of engaged threads. Then, the sign of σ_z, the primary component of von Mises stress, suddenly changes from positive to negative around the top face of the nut. Although not shown here, von Mises stresses along the nut thread root also decrease from the first thread root toward the top face of the nut. Then, the maximum stress at the nut thread root is about 60% of the maximum value at the bolt thread root [20].

Fig. 4.19 shows the effects of coefficient of friction μ, nominal diameter d, and bolt hole dimeter d_h on the stress concentration [49]. As for nominal diameter, the results

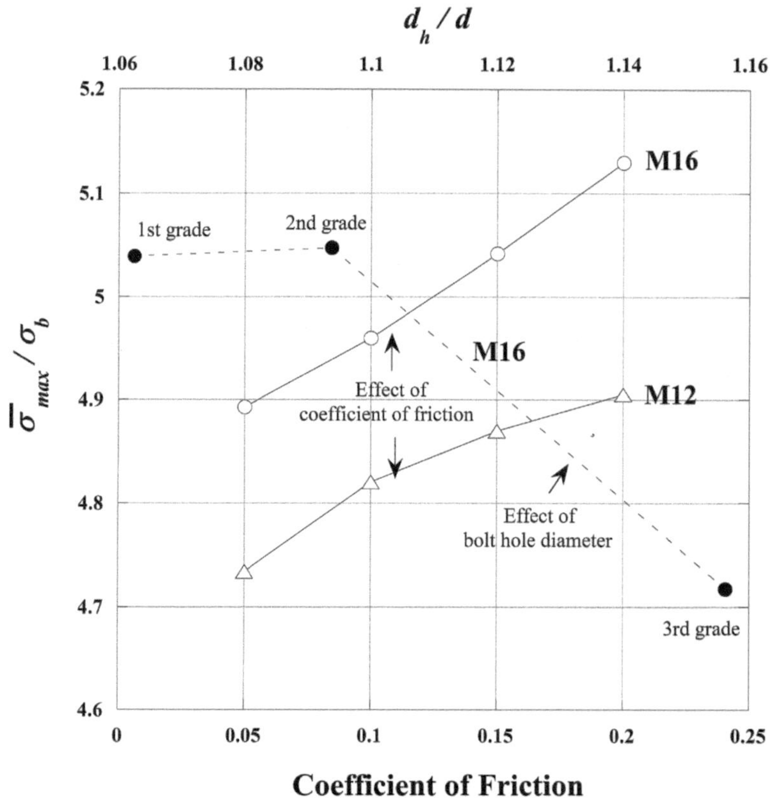

Fig. 4.19 Effects of coefficient of friction, nominal diameter, and bolt hole dimeter on stress concentration.

of M16 with coarse thread are compared to those of M12 with coarse thread, whose thread pitch P is 1.75 mm. Bolt hole diameter of 2nd grade is chosen as the standard. The ordinate represents the maximum von Mises stress $\bar{\sigma}_{max}$ occurring at the first bolt thread, divided by bolt stress σ_b. The abscissa shows coefficient of friction μ and bolt hole diameter d_h. The latter is given as a dimensionless quantity by dividing it by d. The result shown in Fig. 4.19 that stress concentration increases with larger nominal diameter agrees with the previous studies [7,8] using exact solutions of the theory of elasticity. Regarding the coefficient of friction μ, stress concentration slightly increases as μ is increased. As for bolt hole dimeter d_h, stress concentration decreases with increasing d_h due to the decrease of the axial stiffness around the first thread root.

Bolt hole diameter also affects the contact pressure distribution on the nut-bearing surface. Fig. 4.20 shows the contact pressure distribution in the circumferential direction for bolt hole diameter of second grade [49]. The symbol r represents the distance from the bolt axis. Highest contact stress is produced along the bolt hole. Contact pressure distributions show a slight variation in the circumferential direction due to the effect of helical thread geometry. Although not shown in the figure, the contact pressure along the bolt hole is higher in the second grade than in the third grade of bolt hole diameter. Incidentally, the mean contact pressure is necessarily higher in the third grade due to smaller nut-bearing surface area A_n. This means, therefore, that the contact pressure decreases more sharply in the radial direction in the second grade than in the third grade.

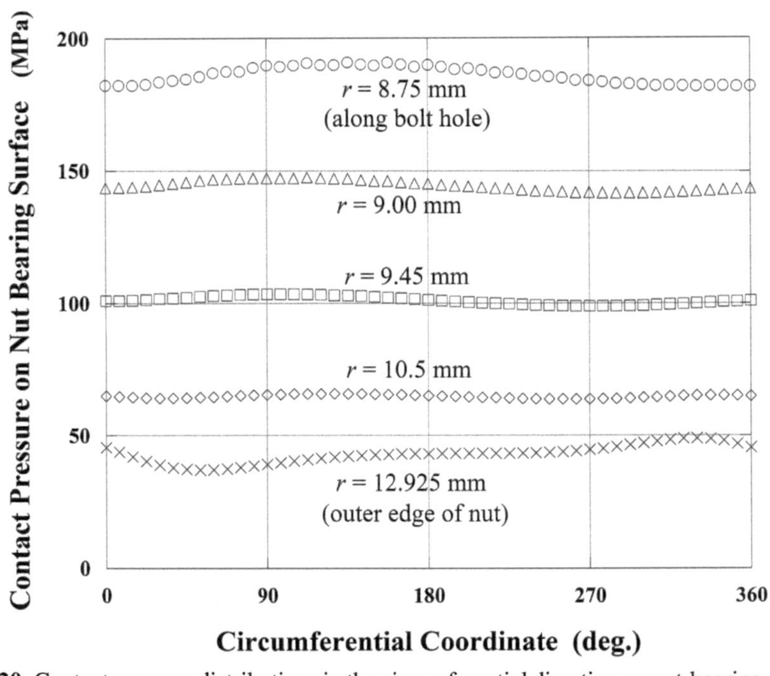

Fig. 4.20 Contact pressure distributions in the circumferential direction on nut-bearing surface.

When tightening bolt-nut connections by the torque control method, bolts are subjected to a twisting moment in addition to axial bolt force, by which von Mises stress at the bolt thread root is raised [47]. Comparison of stress concentration under pure tension to the case of applying tightening torque is made in Section 4.3.3, in which the stress concentrations in studs, tap bolts, and body-side female threads are discussed.

4.3.2 Effect of thread pitch and number of threads

Stress distributions along thread roots are compared for bolt-nut connections of M16, where thread pitch P is changed to 2, 1.5, and 1 mm. Since nut thickness H_{nu} is set to be 12 mm, the number of engaged threads is 6, 8, and 12, respectively, for each thread pitch. Bolt stress, material constants, and coefficient of friction are the same as those in the previous section. Fig. 4.21A and B depicts the mesh patterns of the cross section of FE models for fine screw threads [94]. Thread pitch P is 1.5 and 1 mm, respectively. Fig. 4.22A shows the distributions of von Mises stress $\bar{\sigma}$ along bolt thread root when applying pure tension, in which $\bar{\sigma}$ is normalized with respect to bolt stress σ_b. Stress concentration at the first bolt thread root is slightly increased with decreasing thread pitch. The phenomenon that smaller pitch raises the stress concentration appears more

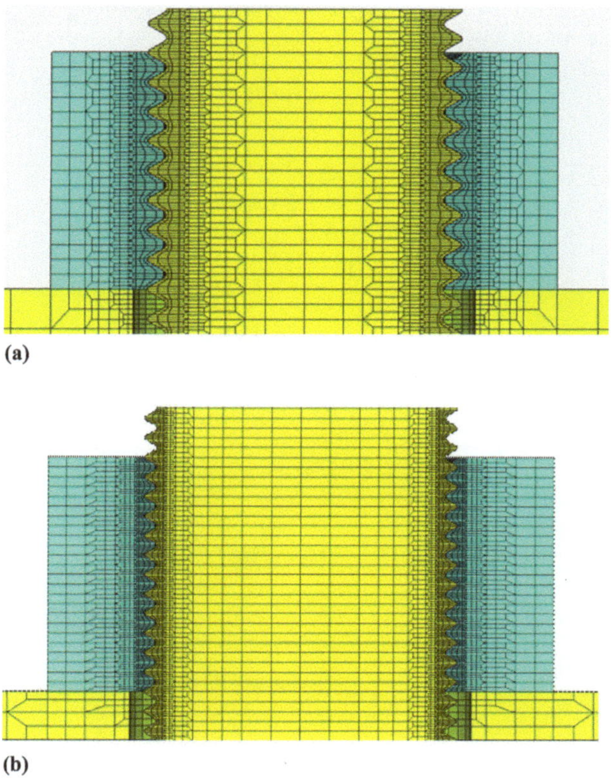

Fig. 4.21 FE models for fine screw threads. (A) $P=1.5$ mm and (B) $P=1$ mm.

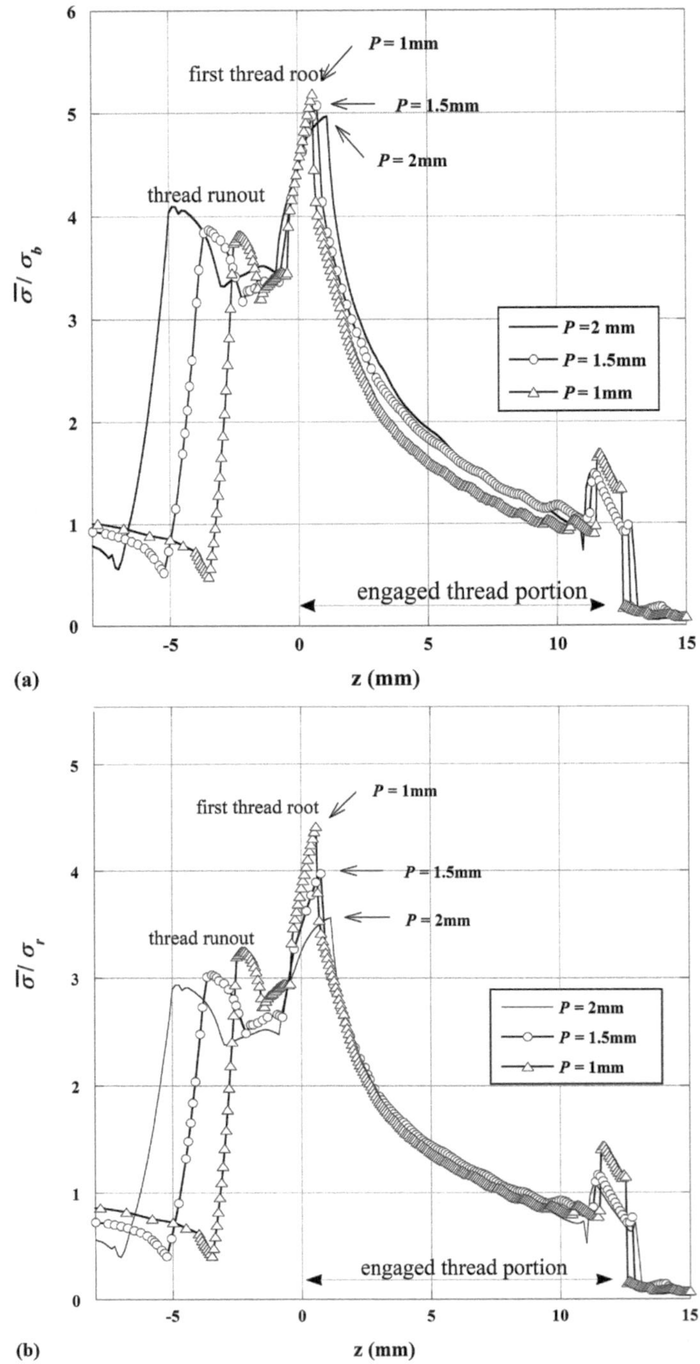

Fig. 4.22 von Mises stress distributions along bolt thread root under pure tension. (A) Normalized with respect to bolt stress σ_b and (B) mean stress at thread root σ_r.

distinctly by dividing $\bar{\sigma}$ by mean stress at thread root σ_r, as found from Fig. 4.22B. Shown in Fig. 4.23 are the numerical results for the case of tightening fine screw threads by torque control method under the same conditions. von Mises stress $\bar{\sigma}$ is divided by σ_b, as in the case of Fig. 4.22A. Comparing to the case of applying pure tension shown in Fig. 4.22A, the stress concentration phenomenon in fine screw threads appears more distinctly.

In fig. 2.27, helical thread FE models of double- and triple-thread screws are shown for M16 bolts with coarse thread. Supposing that bolt stress σ_b is 100 MPa and material constants and coefficient of friction are identical with those in the previous section, the distributions of von Mises stress $\bar{\sigma}$ along the bolt thread root are compared between the two cases of applying pure tension and using torque control method. The number of threads is changed to single-, double-, and triple-thread screws. The numerical results are shown in Fig. 4.24A and B, which represent the cases of using torque control method and applying pure tension, respectively. Since nut thickness H_{nu} is set to be 10 mm, the numbers of spiral turns are 5, 2.5, and 1.67, respectively. In both cases, the maximum stress increases as the number of threads is increased. This tendency is somewhat more noticeable in the torque control method.

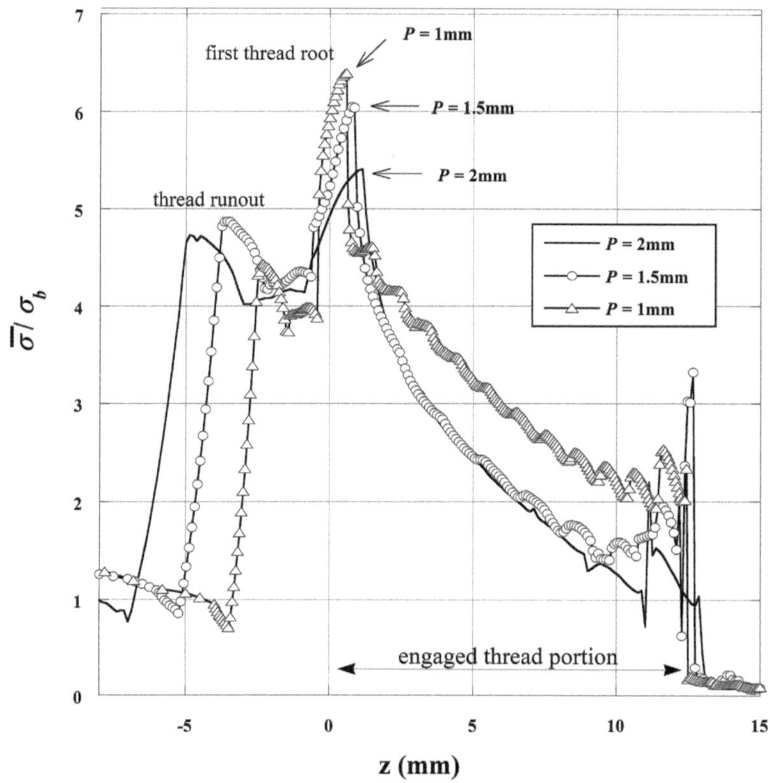

Fig. 4.23 von Mises stress distributions along bolt thread root when tightening by torque control method.

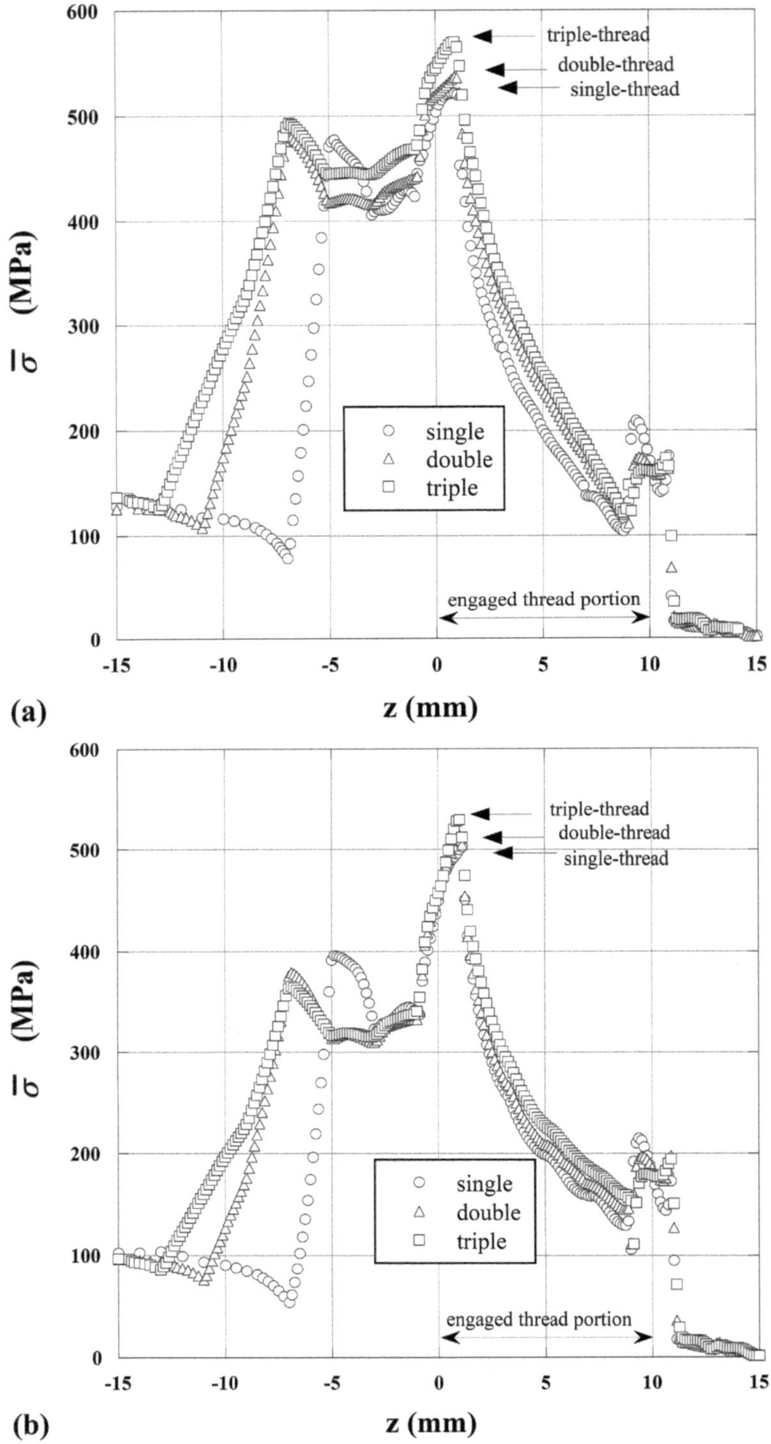

Fig. 4.24 Stress concentration occurring in multiple-thread screws. (A) Torque control method and (B) pure tension.

4.3.3 Stress concentration of engaged threads in main body side

When clamping multiple parts using threaded fasteners, the most general way is to use a combination of bolts and nuts. However, when the target jointed components have specific shapes or dimensions, female threads are machined along drilled holes of the body side instead of using nuts. Then, the tightening operation is carried out using hexagon bolts or studs. Since body-side female threads can be regarded as a nut with infinite thickness and width, its stiffness is higher compared to an ordinary nut. Accordingly, body-side engaged threads exhibit some specific mechanical behaviors different from bolt-nut connections.

Fig. 4.25 depicts an example of FE models of body-side engaged threads [20]. The target bolt is M16 and thread pitch is 2 mm. Figs. 4.26A and B show the distributions of von Mises stress along male and female thread roots, obtained by FE models. The numerical results are compared between the two cases of applying pure tension and using torque control method. Material constants and other conditions for FE analyses are the same as in the previous section. As for male thread root, the magnitudes of von Mises stress produced by torque control method are all higher than the other, over the region from thread runout to the first thread root. In contrast, regarding female thread root, the case of applying pure tension wholly produces higher von Mises stress. The mechanism of the latter phenomenon is estimated as follows. When tightening by torque control method, female threads are subjected to twisting moment in addition to the compression stress due to bolt force. Then, the amount of axial stress σ_z, which is the primary component of von Mises stress, may be slightly decreased due to the effect of helical thread geometry.

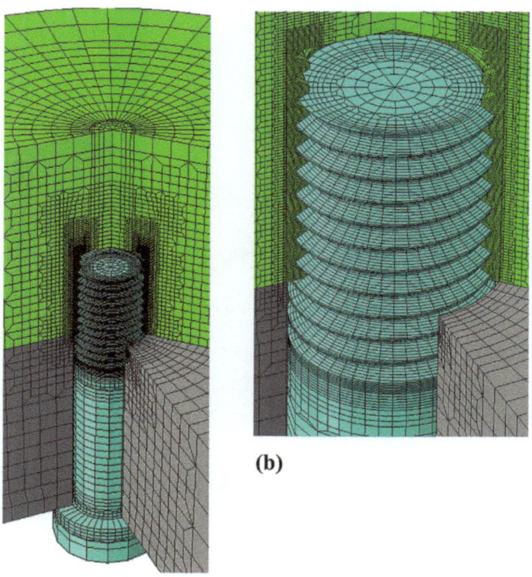

(a) (b)

Fig. 4.25 FE model of body-side engaged threads. (A) Entire model and (B) detail view of engaged thread portion.

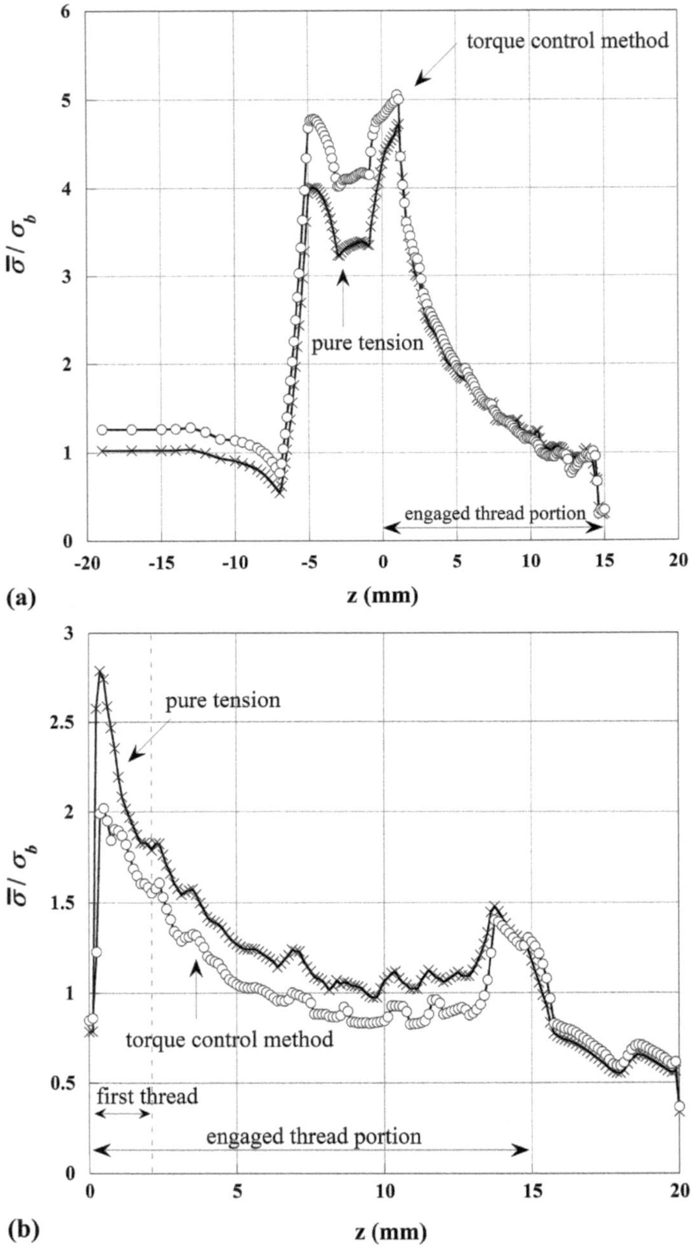

Fig. 4.26 von Mises stress distributions in body-side engaged threads. (A) Male thread side and (B) female thread side.

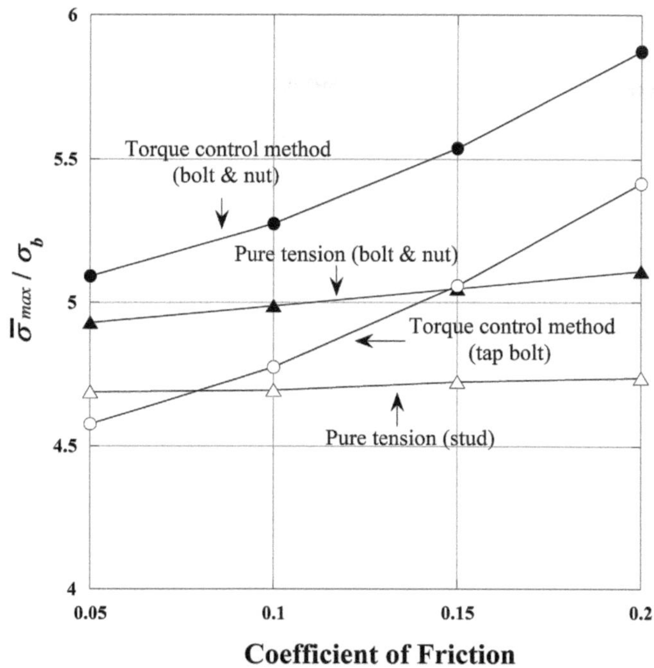

Fig. 4.27 Comparison of stress concentrations occurring in various clamping states.

Next, comparison is made between the maximum values of von Mises stress $\bar{\sigma}_{max}$ occurring at the bolt thread roots of bolt-nut connections and body-side engaged threads. In Fig. 4.27, the ordinate represents the ratio of $\bar{\sigma}_{max}$ to bolt stress σ_b and the abscissa shows coefficient of friction. In either case, stress concentration becomes more noticeable with increasing coefficient of friction. The tendency is more markedly observed when tightening by torque control method. For body-side engaged threads, comparison is made between the case of a stud under pure tension and that of a hexagon bolt being tapped into the main body. It is an interesting phenomenon that the maximum stress occurring in the tap bolt is relatively lower in the region of small coefficient of friction.

4.4 Fatigue failure of screw threads

4.4.1 Relationship between metal fatigue and stress amplitude

When machines and structures are subjected to repeated or fluctuating loads, cracks are sometimes initiated at the spot where high stress amplitude occurs, even if the magnitude of the applied load is considerably smaller than its static allowable load. Then, the cracks gradually propagate and finally lead to a rupture. This phenomenon is called fatigue failure, the most troublesome issue for the safety of machines and structures. Since fatigue failure of metals is a very complicated phenomenon, even today there is no end to the occurrence of accidents and problems caused by metal fatigue [104,105]. Especially in threaded fasteners, it is quite difficult to estimate the fatigue strength with high accuracy because they have sharp notches and are under complicated loadings, comparing to the case of parts or structures having smooth configurations.

Fatigue strength of machines and structures is influenced by a variety of factors, such as loading conditions, joint materials, processing method and treatment, stress amplitude occurring at notch root, mean stress exerted on the component, stress gradient in the depth direction at notch root, etc. The effect of stress amplitude is the most dominant among them. S–N curve, which relates stress amplitude to the number of repetitions of applied loads until rupture occurrence, is the most important diagram to evaluate fatigue strength. An example of S–N curve is given in Fig. 4.28. The ordinate represents stress amplitude σ_a. The abscissa shows the number of repetitions of applied loads N. The data are depicted by a semilog graph. The drawing process of S–N curve is as follows. Stress amplitudes are changed variously. Then, sets of the stress amplitude and number of repetitions at which the fatigue failure occurs are plotted one after another, and thus, the S–N curve is completed. For carbon steel materials, the S–N curve becomes a straight line parallel to the horizontal plane when the number of repetitions of applied loads exceeds a certain number, say, ranging from 10^6 to 10^7. The stress amplitude corresponding to the straight line is termed fatigue limit. It is widely used as the standard value to predict the fatigue strength of machines and structures. Recently, the research on ultrahigh cycle fatigue, in which fatigue failures occur after about 10^9 repetitions, has been conducted intensively. Meanwhile, nonferrous metals like aluminum alloys do not show a distinct fatigue limit. It is customary to define the fatigue limit as the stress amplitude at a certain number of repetitions, which is determined by the working conditions of the target machines and structures.

In general, fatigue strength increases with increasing tensile strength, which is the index for static strength. However, the increasing rate of the fatigue strength decreases when the tensile strength exceeds a certain level. For the relationship between fatigue

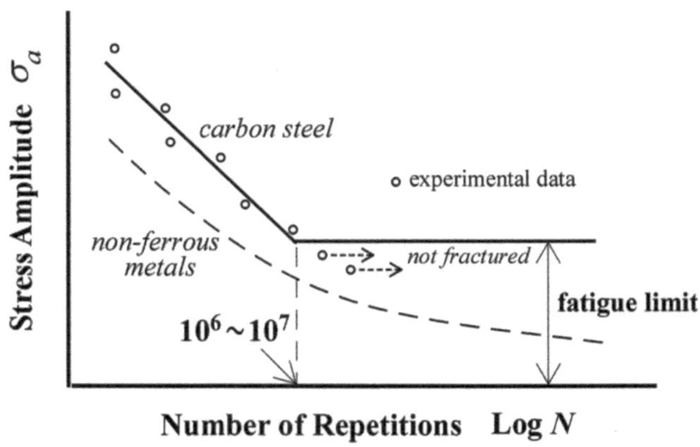

Fig. 4.28 S–N curve.

limit σ_w and tensile strength σ_B, the following formulas are proposed for smoothed test specimens subjected to alternating loads [106]:

$$\begin{array}{ll} \text{carbon steel} & \sigma_w \cong 0.5\sigma_B \\ \text{cast iron} & \sigma_w \cong 0.4\sigma_B \\ \text{non}-\text{ferrous metal and its alloy} & \sigma_w \cong 0.3\sigma_B \end{array} \tag{4.6}$$

Supposing the stress amplitude to be half the case of alternating loads, the fatigue strength for pulsating loads is lower than that for alternating loads. For instance, the value for carbon steel is slightly less than $0.4\sigma_B$. The ratio, σ_w/σ_B, varies depending on the loading conditions. In the case of alternating loads, for materials with identical tensile strength σ_B, the fatigue strength for the repetition of tension and compression is lower than that for rotary bending fatigue test [98]. In that case, the fatigue strength for rotary bending σ_w is about 50% of the tensile strength σ_B, as given in Eq. (4.6), and the value becomes slightly less than 40% when subjected to the repetition of tension and compression. Furthermore, it has been reported that the fatigue strengths for rotary bending and plane bending are almost equal [106]. Fatigue limit σ_w is not linearly related to tensile strength σ_B. The increasing rate of σ_w decreases as σ_B is increased. Especially for alternating torsional loads, when σ_B exceeds about 800 MPa, the fatigue limit of shear stress τ_w shows almost no increase from about 200 MPa [98].

The values shown above are all for the fatigue limit of smoothed test specimens. On the other hand, the fatigue limit of the components with sharp notches like screw threads cannot be improved very much, even if the tensile strength of the material is raised [107]. Furthermore, when tightening high strength bolts into body-side female threads made of high strength metals, it is sometimes impossible to achieve the desired fatigue strength, because the fatigue limit does not increase very much compared to the increase in tensile strength, and the processing accuracy of female threads is inevitably low.

Fatigue limit of notched portion σ_w is even lower than that of smoothed test specimen σ_{w0} with cross-sectional area equal to the notch root. The ratio of σ_{w0} to σ_w is termed the notch factor β:

$$\beta = \frac{\sigma_{wo}}{\sigma_w} \tag{4.7}$$

According to the definition, the notch factor is necessarily larger than unity. The difficulty in estimating the fatigue limit is traced to the relationship between fatigue strength and stress concentration. Notch factor β is smaller than stress concentration factor α, defined in Section 4.2.1. The relationship between β and α is expressed by notch sensitivity factor η, defined below:

$$\eta = \frac{\beta - 1}{\alpha - 1} \tag{4.8}$$

In general, the difference between α and β becomes larger with increasing α. This phenomenon suggests that if fatigue limit σ_w for components under high stress concentration is estimated based on stress concentration factor α, it may lead to an unduly conservative design. Meanwhile, stress concentration factors can be evaluated for most cases by means of the latest technologies of numerical analysis, even if complex geometry and boundary conditions are involved. In contrast, although various evaluation methods are proposed for the amount of notch factor β, it is quite difficult to accurately predict β without conducting fatigue tests, because β is influenced by many factors, as stated above.

When estimating the fatigue limit σ_w of actual machines and structures, size effect must be taken into account. Size effect indicates the phenomenon that the fatigue limit is decreased as member size is increased among multiple target members with different dimensions and similar figures. Size effect is rarely observed when subjected to the repetition of tension and compression, but it appears rather distinctly under bending and torsional loads. Applying the above phenomenon to threaded fasteners, it can be said that the fatigue limit is decreased with larger nominal diameter. In addition, a variety of alternating loads are applied independently or in combination to threaded fasteners, and their shapes are not similar for varying nominal diameters, as shown in fig. 1.8. Therefore, the fatigue strength of threaded fasteners shows very complex behaviors.

4.4.2 Fatigue failure of threaded fasteners

The locations in threaded fasteners at which fatigue failures are likely to occur, are illustrated in fig. 2.1. In the case of bolt-nut connections, the failure occurs in most cases around the first bolt thread, where the maximum stress concentration is generated. Sometimes it occurs around the second or third root, which are located somewhat closer to the top face of the nut. Next to the first bolt thread root, fatigue failure is apt to occur around thread runout. Since the thread runout is located at the end of continuous notches, stress concentration there is higher comparing to any other unengaged threads. In addition, the notch at thread runout may have too small a radius due to the processing problem. Another critical location is the root radius under the bolt head. Stress concentration factor there is smaller than those of the other two locations, but the fatigue strength may be lower due to metallurgical problems occurring in the manufacturing process.

In contrast, fatigue failures occurring around body-side engaged threads show considerably different behaviors from the case of bolt-nut connections. There are many cases of failures occurring in bolts. However, it has also been reported that a considerable number of failures occur in the body-side female threads. In that case, the crack initiates at the female thread root engaged with the far end bolt thread, which is the opposite side of the first bolt thread. This type of fatigue failure has been reported as a critical problem occurring around connecting rods of large diesel engines for quite some time. It has been proved by photo-elastic experiments [108] that large stress amplitudes are generated around the body-side female thread root engaged with the far end bolt thread.

Fatigue failures of threaded fasteners are closely related to stress concentration. However, as evident from the foregoing actual examples, fatigue failures are not necessarily initiated at the location where the maximum stress concentration occurs. Hence, in Section 4.7, stress amplitude distributions along thread root are evaluated by FEM using helical thread models. Then, the relationship between the magnitude of stress amplitude and the occurrence position of fatigue failure is discussed.

4.4.3 Influencing factors on fatigue strength of screw threads

To estimate the fatigue limit of threaded fasteners, the effect of the tightening and loading conditions must be considered, in addition to nominal diameter, thread pitch, thread root radius, and joint materials. Therefore, it is impossible to use the data of fatigue limit of smoothed test specimens or notched test specimens without any modification. Of course, the most reliable are fatigue test data obtained by using actual threaded fasteners.

Yamamoto collected reliable experimental data of fatigue limit performed by many researchers [109]. In Fig. 4.29, the relationship between fatigue limit σ_w and tensile strength σ_B is drawn using the collected data. Bolts and nuts are made of carbon steel and the processing methods are cutting and grinding. Based on the above experimental data, Yamamoto insists that the effect of mean stress exerted on the bolt can be negligible. Another important conclusion is that the fatigue limit is not influenced very

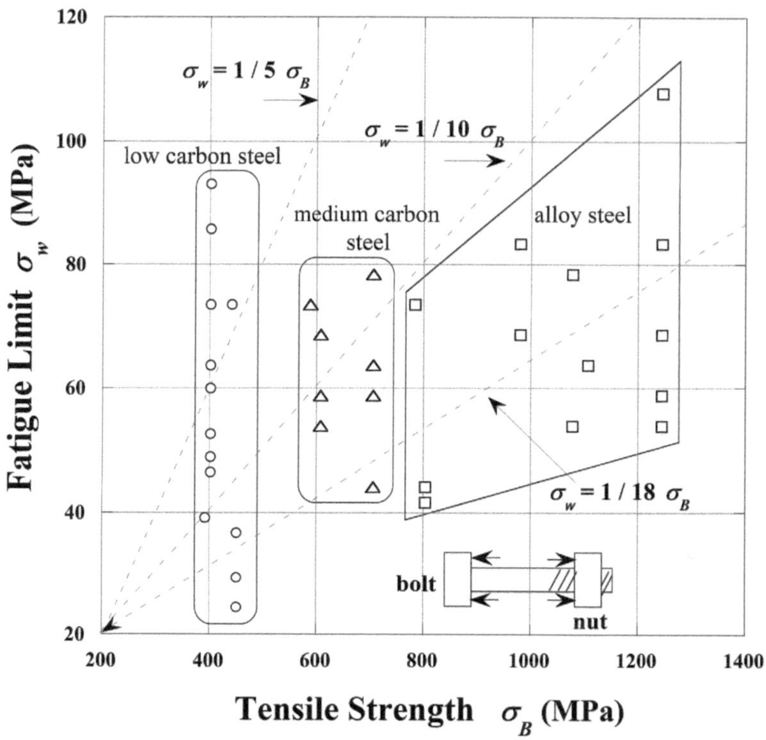

Fig. 4.29 Relationship between fatigue limit and tensile strength in bolt-nut connections.

much by thread pitch and the type of screw threads. However, the effect of thread pitch P appears by changing the way of data display. It is explained in detail later. As shown in the figure, external loads are applied to bolt head- and nut-bearing surfaces so that test bolts are subjected to pulsating tensile load. Fatigue limit σ_w is evaluated by use of the cross-sectional area of minor diameter d_1. Experimental results of σ_w are distributed in the range from 25 to 105 MPa, most of which are between 40 and 80 MPa. Fatigue limit σ_w increases with increasing tensile strength σ_B, and the ratio of σ_w to σ_B is much smaller than the 0.5 given in Eq. (4.6). For each bolt-nut material, the scatters in fatigue limit and σ_w/σ_B are summarized as follows:

Low carbon steel: σ_w *ranges from 25 to 90 MPa, in which* σ_w/σ_B *shows a large scatter.*

Medium carbon steel: σ_w *ranges from 40 to 80 MPa, in which* σ_w/σ_B *is distributed around 1/10.*

Alloy steel: σ_w *ranges from 40 to 105 MPa, in which* σ_w/σ_B *is distributed around 1/18 with a large scatter.*

The cause of the scatter in each material is presumably due to the difference in the influencing factors listed at the beginning of this section. In the fatigue tests performed in Fig. 4.29, three kinds of screw thread standard are used, i.e., metric, Whitworth and unified screw threads. The ratio of thread root radius ρ to thread pitch P is as follows:

Metric screw thread (abbreviated as M): $\rho/P = 0.108$

Whitworth screw thread (abbreviated as W): $\rho/P = 0.137$

Unified screw thread (abbreviated as U): $\rho/P = 0.144$

Threaded fasteners with different nominal diameter are not similar and the non-similarity can be expressed by the ratio of thread pitch P to nominal diameter d, P/d, as shown in fig. 1.8. Then, the data in Fig. 4.29 are redrawn into the relationship between fatigue limit σ_w and P/d, which is shown in Fig. 4.30. The newly drawn figure indicates that the fatigue limit σ_w is more closely related to P/d than tensile strength σ_B. The effects of thread pitch P, nominal diameter d, and thread root radius ρ are listed below:

1) In both coarse and fine screw threads, fatigue limit σ_w is clearly decreased with decreasing P/d. In other words, fatigue limit tends to decrease with increasing nominal diameter or as thread pitch decreases for bolt-nut connections with identical nominal diameter.
2) When P/d is reduced to about 0.1, fatigue limit σ_w shows a remarkable increase and ultrafine pitch screw threads form another group. In this group, σ_w also decreases as P/d is decreased and its reduction rate is quite large.
3) No significant effect of thread root radius ρ, which varies depending on the screw thread standard, is observed.

It is worth noting that when fatigue limit of screw threads σ_w is evaluated by use of P/d, the data on σ_w are divided into two groups, i.e., coarse and fine threads and ultrafine

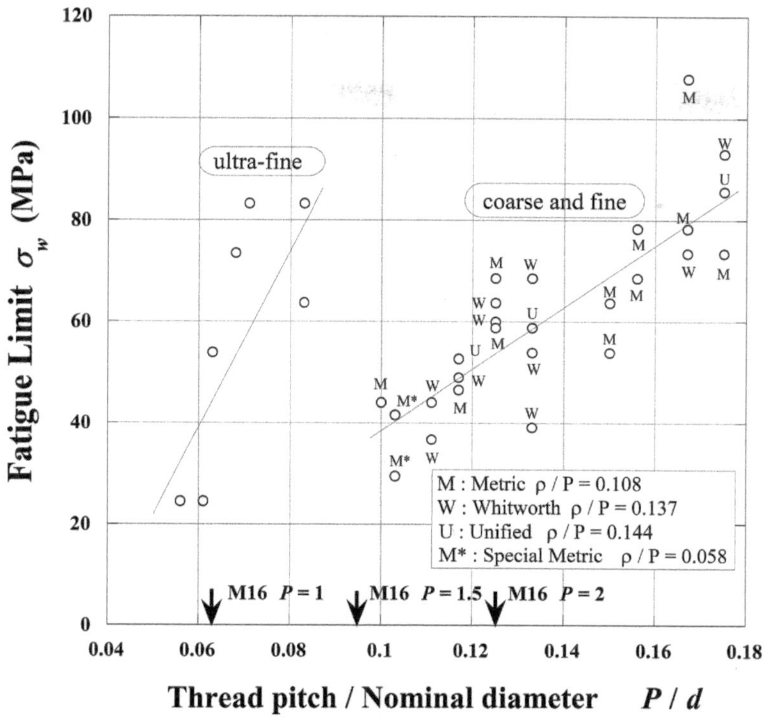

Fig. 4.30 Relationship between fatigue limit and ratio of thread pitch to nominal diameter.

threads, as found from Fig. 4.30. The relationship between σ_w and P/d is approximated by the least squares method. The resulting equations are given below for each group:

$$\begin{aligned} \sigma_w &= 617(P/d) - 23.1(\text{MPa}) \quad \text{for } P/d > 0.1 \\ \sigma_w &= 1839(P/d) - 72.3(\text{MPa}) \quad \text{for } P/d < 0.083 \end{aligned} \tag{4.9}$$

The experimental results in Fig. 4.30, in which fatigue limit σ_w shows a rapid change in the region of P/d, being from around 0.08 to 0.1, demonstrate the difficulty in evaluating the effect of thread pitch P on σ_w, in addition to the size effect appearing with varying nominal diameter. Meanwhile, the effect of the ratio of thread root radius ρ to thread pitch P on σ_w is smaller comparing to P/d. The experimental results support an important statement that stress concentration is not necessarily a decisive factor on fatigue limit.

Regarding the effect of nominal diameter d, the relationship between fatigue limit σ_w and d is graphically shown in VDI 2230(1986) [110] for high strength bolts of 8.8, 10.9, and 12.9 grades. Fatigue limit σ_w (MPa) of bolts heat treated after rolling and d (mm) are related by the following equation:

$$\sigma_w = 0.75\left(\frac{180}{d} + 52\right) \tag{4.10}$$

Eq. (4.10) shows that σ_w is decreased in an inverse proportional manner as d is increased. The results in Fig. 4.30 show that fatigue limit of screw threads σ_w is quite closely related to the ratio of thread pitch to nominal diameter, P/d, which is consistent with the above VDI formula on the effect of nominal diameter d, including size effect. Fatigue limit σ_w (MPa) of bolts rolled after heat treatment varies with initial bolt force F_b. Supposing that $F_{0.2}$ is the bolt force corresponding to 0.2% proof stress, σ_w is given as a function of the ratio of F_b to $F_{0.2}$.

$$\sigma_w = 0.75\left(\frac{180}{d} + 52\right)\cdot\left(2 - \frac{F_b}{F_{0.2}}\right) \tag{4.11}$$

Since $F_b/F_{0.2}$ is generally less than unity, it follows that the fatigue limit of bolts rolled after heat treatment is higher than that of bolts heat treated after rolling. It is found from these equations that the fatigue limit σ_w of the bolts rolled after heat treatment is decreased with increasing initial bolt force F_b, while the σ_w of the bolts heat treated after rolling is not influenced by F_b. According to Eqs. (4.10) and (4.11), the fatigue limit σ_w is to be determined only by nominal diameter d and $F_b/F_{0.2}$, regardless of the detailed dimensions such as thread pitch, thread root radius, etc. Nonetheless, those equations are useful to estimate the approximate value of σ_w for high strength bolts. The equations are applicable in the range of $0.2F_{0.2} < F_b < 0.8F_{0.2}$.

Otaki combined his original analytical method using complex stress functions with the experimental results by Yakushev and the proposition by Ishibashi. He concluded that fatigue limit σ_w is decreased with increasing nominal diameter and there exists a specific pitch size that makes the fatigue limit minimum [8]. Considering the above proposition together with the other research data, the statement that fatigue limit σ_w is decreased with larger nominal diameter can be highly trusted. As for the existence of thread pitch that minimizes σ_w, it can be explained by the experimental data on the fatigue limit shown in Fig. 4.30. That is, σ_w shows a minimum value in the range of P/d being about between 0.08 and 0.1, and σ_w increases again when P/d is further decreased. In the following, it is summarized why the fatigue limit of actual bolted joints is difficult to evaluate with high accuracy:

1) Published data on fatigue tests are mostly targeted at smoothed test specimens without notches. In order to apply the data to threaded fasteners, it is necessary to accurately estimate the notch factor.
2) Fatigue tests in the past have mostly been carried out under the condition that the external force is applied in the axial direction. On the other hand, in the actual machines and structures, bending, twisting, and shearing are commonly exerted besides axial loading. Also important is that those loads are often exerted in combination on the bolted joints.
3) When large external forces are applied to actual threaded fasteners, separation of the contact surface progresses as their magnitude is increased, except for fastened plates having small outer diameter. Hence, the area under contact pressure is reduced. Due to the interface separation, load factor, which is the ratio of bolt force increase ΔF_b to applied external load W, varies nonlinearly against the magnitude of applied load. Accordingly, it is difficult to evaluate the effect of interface separation with practical accuracy using a simple equation.

In order to evaluate the fatigue limit with high accuracy, the best way is to perform a fatigue test under the same conditions as the target joint. However, it is usually impractical except for the special case of the target joint being small. In addition, when joint geometry and/or loading conditions are complex, it is sometimes impossible to evaluate the stress amplitude with sufficient accuracy by means of the theories of solid mechanics or elasticity. In the following sections, various methods for evaluating the fatigue strength of bolted joints are introduced, the problems involved in each method are discussed, and the effectiveness of numerical analysis by FEM is also demonstrated.

4.5 Evaluation method of fatigue strength of threaded fasteners

4.5.1 Summary of the bolted joint diagram

When clamping bolt-nut connections, the bolt is subjected to tension and the fastened plates are compressed due to the reaction force. In actual bolted joints commonly encountered, the elongation of bolt-nut connection and the shrinkage of fastened plates are both proportional to bolt force. Fig. 4.31 schematically shows a bolted joint clamped with bolt force F_b. It is subjected to external force W. For simplicity, the external force is assumed to be exerted annularly on the plate surface in the axial direction.

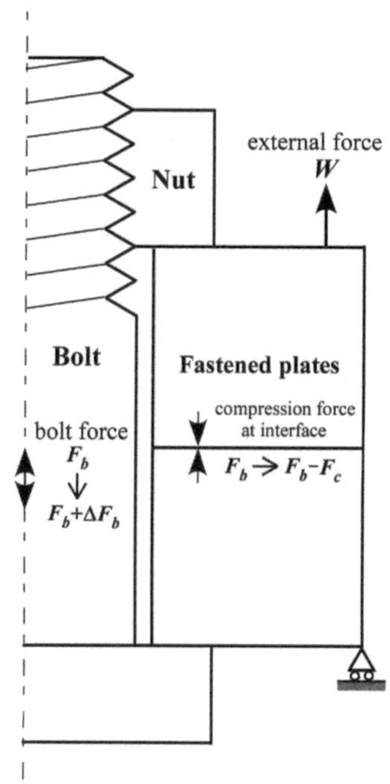

Fig. 4.31 Bolted joint subjected to external force.

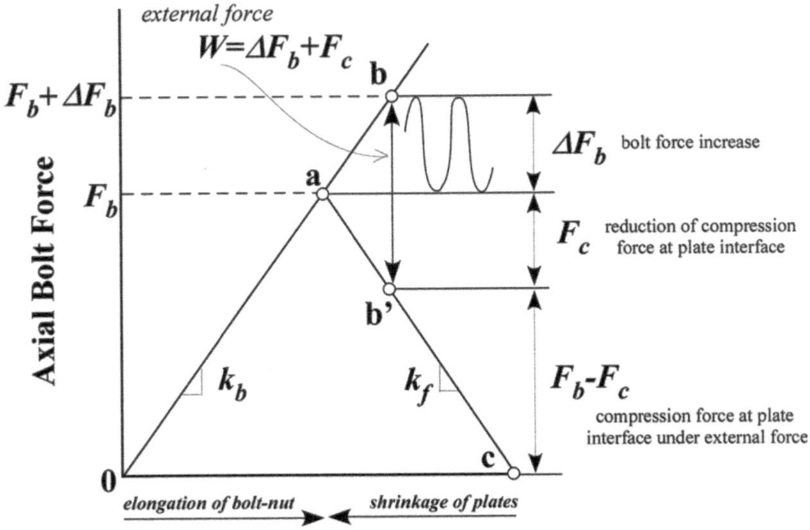

Fig. 4.32 Bolted joint diagram.

The bolted joint diagram shown in Fig. 4.32 represents the relationship among the related quantities. It has commonly been used to estimate the fatigue strength of bolted joints. In this section, a summary of the bolted joint diagram is given. Fundamental problems with the diagram are described in detail in the next section. The ordinate represents the bolt force and the abscissa shows the elongation of bolt-nut connection δ_b and the shrinkage of fastened plates δ_f. The triangular part 0-a-c in the figure is called the joint triangle herein. It is also used to calculate the bolt force reduction due to the embedment occurring at the interface, which is explained in section 6.4. The gradients of the two straight lines composing the joint triangle, inclined upward right and downward right, represent the spring constant of bolt-nut connection k_b for tension and the spring constant of fastened plates k_f for compression, explained in section 2.2. In the figure, F_b represents the initial bolt force before external force W is applied.

When the external force of tensile load W is applied against the initial state, the bolt force is increased along the line inclined upward right and the compression force exerted on fastened plates is decreased along the line inclined downward right. By the action of the external force, the bolt force is increased as much as ΔF_b from the initial bolt force F_b. In contrast, the compression force acting on fastened plates is decreased by amount F_c from F_b. Hence, the following equation holds among W, ΔF_b, and F_c:

$$W = \Delta F_b + F_c \tag{4.12}$$

When the external force is applied as pulsating load that changes between 0 and W, stress amplitude σ_a exerted on the bolt can be evaluated by dividing half the bolt force increase ΔF_b by the stress area A_s of threaded portion:

$$\sigma_a = \frac{\Delta F_b}{2A_S} \tag{4.13}$$

The ratio of ΔF_b to W is termed the load factor. It is used as a basic index when evaluating the fatigue strength of threaded portions:

$$\phi_u = \frac{\Delta F_b}{W} \tag{4.14}$$

When subjected to a constant external force, ϕ_u should be as small as possible because smaller ϕ_u produces smaller stress amplitude. Considering the similarity among the triangles in the figure, load factor ϕ_u can be calculated in terms of spring constants of bolt-nut connection k_b and fastened plates k_f, or using the elongation of bolt-nut connection δ_b and the shrinkage of fastened plates δ_f:

$$\phi_u = \frac{k_b}{k_b + k_f} = \frac{\delta_f}{\delta_b + \delta_f} \tag{4.15}$$

The above equation was first proposed by Thum. It indicates that ϕ_u becomes small when bolts are easily able to elongate and fastened plates difficult to deform.

4.5.2 Essential problems involved in the bolted joint diagram

According to Fig. 4.32, load factor ϕ_u is constant regardless of the amount of external force W. However, in actual bolted joints, ϕ_u is not constant except in special cases. This is because the spring constant of fastened plates under tensile external force is different from the spring constant under pure compression in the initial tightening process. It has been reported that the stiffness of fastened plates is different between the initial tightening state and in the state under tensile external force. Sawa et al. evaluated the amount of load factor ϕ_u for bolted joints with various configurations and loading conditions [111]. They defined a spring constant of fastened plates k_{pt} when subjected to tensile force. The proposed equation for evaluating ϕ_u is as follows:

$$\phi = \frac{k_b}{k_b + k_f}\left(\frac{k_f}{k_{pt}}\right) \tag{4.16}$$

The term in the parenthesis corresponds to the correction factor proposed by Junker. Using the basic theory of solid mechanics, Zhang et al. derived an equation for the spring constant of fastened plates under tensile force, and they show that the spring constant is nonlinearly related to the magnitude of external load [112,113]. The nonlinearity is caused by the interface separation. In fig. 1.12, fastened plates are classified into thin cylinder, thick cylinder, and plate, according to the distribution pattern of contact pressure at the interface. In the case of plate, e.g., when external force is applied, the separation starts at the outer end of pressure cone, at which the contact pressure is zero, then expands with increasing external force. In the case of thick cylinder, the contact pressure is higher than in plate, hence, the separation is less likely to occur; as external force is increased, however, the separation progresses in a similar manner to the case of plate.

The separation is more unlikely to occur in thin cylinder due to its small contact area and high contact pressure, which is almost uniformly distributed. However, when the external force exceeds a certain value, the whole interface suddenly loses contact. Accordingly, as far as thin cylinder subjected to axisymmetric external load is concerned, a conventional bolted joint diagram seems to be effective for evaluating load factor ϕ_u and stress amplitude σ_a generated in actual bolted joints. Unfortunately, it is a rare case that the external load is applied in the axisymmetric manner to the jointed components of a thin cylinder. Actual bolted joints are usually subjected to more complex loads, i.e., bending, shear, and twisting loads, or the combination of those loads. In such cases, it is almost impossible to accurately evaluate the stress amplitude exerted on the bolt and the compression force acting on the fastened plates by means of the bolted joint diagram shown in Fig. 4.32. The problems involved in the conventional bolted joint diagram are listed in the following:

1) Basically, the target of the diagram is bolted joints subjected to axisymmetric external forces.
2) Spring constant of fastened plates when subjected to tensile force is different from the spring constant k_f under pure compression in the initial tightening process. In most cases, it is nonlinearly related to the magnitude of external load.
3) When large tensile force is applied, a separation occurs at the interface of fastened plates. It starts from the outer edge of pressure cone. The effect of separation on the spring constant is not considered in the diagram, despite being a critical issue.
4) For bending, shear, and twisting loads applied to actual jointed portions, the diagram cannot properly evaluate the load factor, stress amplitude exerted on the bolt, or the compression force acting on the fastened plates.

When fatigue failure occurs in threaded fasteners, a crack initiates from the bolt thread root in many cases. Regarding this phenomenon, only the mean stress amplitude in the engaged thread portion or bolt cylinder is evaluated by the bolted joint diagram. At the thread root, larger stress amplitude is generated. In addition, when bending moment is applied to the joint, the tensile stresses occurring at the thread root due to initial tension are increased or decreased at two thread roots 180 degrees apart from one another. To cope with such phenomena, it is effective to introduce numerical methods represented by FEM.

4.5.3 Verification of bolted joint diagram by FEM

In this section, a bolted joint diagram is drawn using the numerical results obtained by axisymmetric FEM that takes account of the effect of plate interface separation, thereby clarifying the problems involved in the conventional diagram. Fig. 4.33 depicts the FE model together with the boundary and loading conditions. External force W is applied to the plate surface as annular pulsating tensile load. Nominal

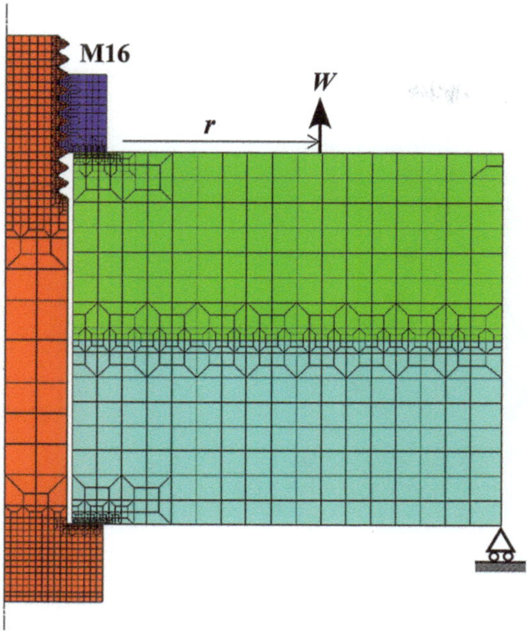

Fig. 4.33 FE model for verifying the validity of bolted joint diagram.

diameter of the bolt-nut connection is M16 and the bolt stress is set to be 100 MPa. Fastened plates are composed of two hollow cylinders with identical height. Grip length L_f is 48 mm, bolt hole diameter d_h is 17.5 mm, and the outer diameter of fastened plates D_o is 128 mm. Radius of the loading point r of external force W is varied. W is increased from zero to the amount by which the interface is completely out of contact. The drawing procedure of the bolted joint diagram is as follows:

1) When a certain amount of external force is applied, bolt force is increased along the straight line o-a-b with gradient of k_b, shown in Fig. 4.32. Hence, a new balanced point can be plotted on the straight line, using the amount of bolt force increase ΔF_b. Then, drawing a straight line vertically downward from the point, the corresponding state of fastened plates is located as much as F_c below the initial tightening state. Fig. 4.34 illustrates the procedure how to draw a new balanced point of the bolt-nut connection and fastened plates.
2) By increasing the external force stepwise, as shown in the figure, a pair of ΔF_b and F_c is obtained for each external force. Connecting those points sequentially, the real bolted joint diagram is completed, which represents the nonlinear behavior of fastened plates considering the effect of interface separation.

Fig. 4.35 depicts the bolted joint diagram thus obtained by FE analysis. The loading point radius r of external force W is changed from 12.925 to 64 mm. The former position is just outside the outer surface of the nut and the latter one corresponds to the outer end of the plate surface. When the external force W is applied annularly, the interface completely loses contact when the amount of W is a little over initial bolt force F_b. Since a bolt-nut connection of M16 is clamped with bolt stress of

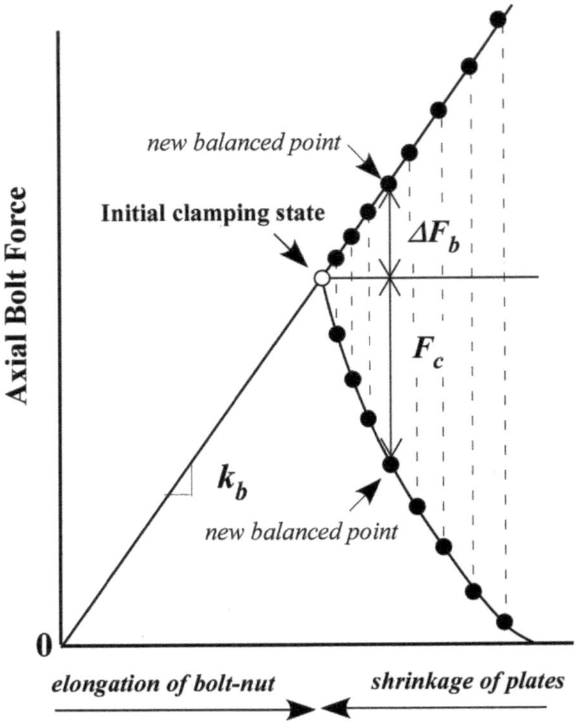

Fig. 4.34 Drawing procedure of bolted joint diagram using numerical results.

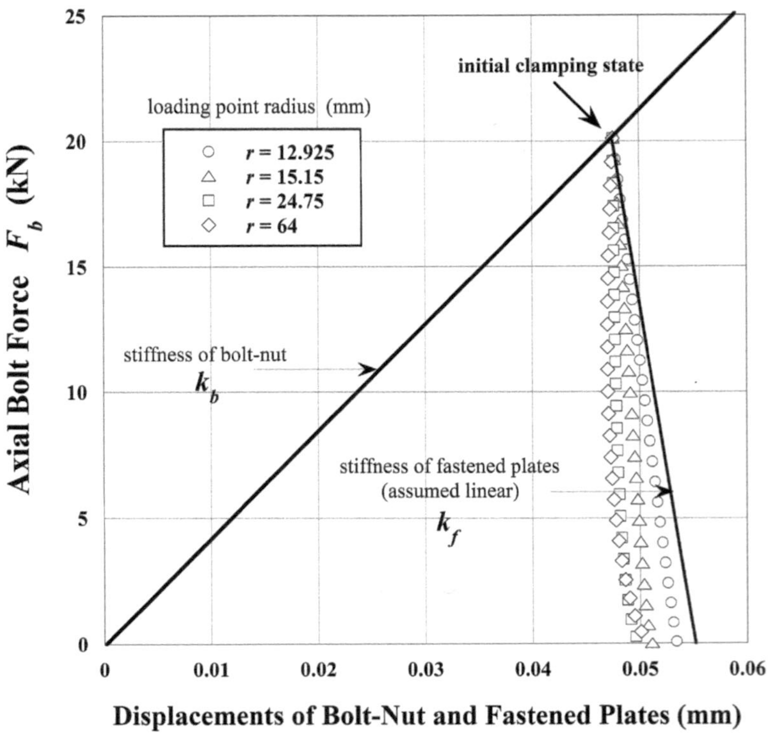

Fig. 4.35 Bolted joint diagram drawn by FE analysis.

100 MPa, it follows that the bolt force F_b is 20.1 kN. Then, the external force W is varied from 0 to 27 kN. The solid line inclined downward right in the figure, indicated by an arrow, represents the spring constant of fastened plates k_f calculated by the conventional method. Spring constant k_f evaluated by FE analysis shows nonlinear behavior against the amount of W. Regarding the effect of loading position r, the relationship between the compression force and displacement of the fastened plates approaches a linear one with smaller r. In other words, the difference from the conventional diagram gets smaller. In contrast, the nonlinearity appears more distinctly as r is increased. That is because the fastened plates deform like a plate under bending moment applied from the periphery, and then the interface separation is gradually progressed.

Next, the focus is placed on the amount of external force W. In the range of small W, the gradient is steep, which indicates high stiffness of fastened plates and small load factor ϕ_u. Accordingly, when W is small, bolt force increase ΔF_b is smaller than predicted by the conventional diagram. Additionally, when W is very small, bolt force only slightly increases and the line representing the plate stiffness changes into the one downward left and the load factor ϕ_u becomes a negative value. It is presumed that this curious phenomenon, though not an important issue from the practical point of view, is caused by the following mechanism. When fastened plates with circular shape are subjected to bending moment from the periphery, rotational deformations occur around the nut-bearing surface so as to reduce the bolt force. A similar phenomenon is observed in pipe flange connections, where the leakage of contained fluids due to bolt force reduction is the primary problem. Then, the load factor ϕ_u sometimes becomes a negative value when the amount of external force W is small [87]. However, as W is increased, the gradient representing the plate stiffness returns to be downward right, as in the ordinary case. In Fig. 4.35, when W is further increased and the compression force acting on the fastened plates is decreased, the gradient representing the plate stiffness shows a sudden decrease, although it is difficult to recognize in the figure. As a result, the bolt force increase ΔF_b exhibits a sharp increase just before the complete separation of the plate interface.

The following is a summary of the above explanation. When external force W reaches a certain value, the separation starts at the plate interface, but the amount of bolt force increase ΔF_b is small in the range of small W. However, as W is further increased and the interface almost loses contact, stress amplitude σ_a exerted on the bolt shows a sudden increase. This phenomenon that the bolt force starts to increase sharply just before the complete separation of plate interface is demonstrated in more detail using helical thread models in Section 4.7. If even larger external force is applied and complete separation occurs, the external force is exerted directly on the bolt, thereby leading to load factor of unity, i.e., $\phi_u = 1$. If the external force is further increased after complete separation, fatigue failures can occur with very high possibility.

4.5.4 Axial bolt force vs external force diagram

Besides the bolted joint diagram, a diagram relating axial bolt force to external force is used to evaluate the fatigue strength of bolted joints. Fig. 4.36 schematically shows the

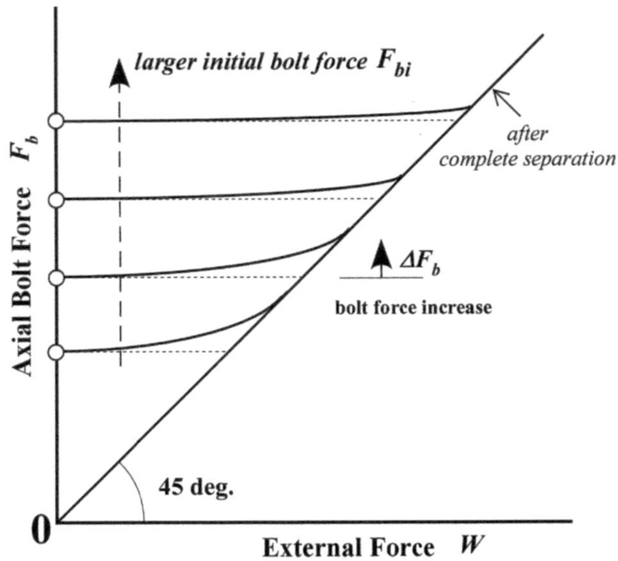

Fig. 4.36 Relationship between bolt force and external force.

relationship between bolt force and external force, referring to the previous studies. The external force is assumed to be applied to the joint as annular pulsating tensile load. The diagram suggests the following mechanical properties of the fatigue strength of bolted joints:

1) There is a nonlinear relationship between bolt force and external force, and the gradient of the curve increases as external force is increased. That is, load factor ϕ_u behaves nonlinearly for varying external force W due in large part to the interface separation. When the loading point of external force is close to the bolt axis or the fastened plates are regarded as thin cylinder, the relationship between ϕ_u and W approaches linear, as explained in the previous section.
2) After the plate interface is completely separated, the external force exerts directly on the bolt. Then, the bolt force F_b in Fig. 4.36 starts to increase along the straight line with a gradient of 45 degrees.
3) As initial bolt force F_{bi} is increased, the interface separation is unlikely to occur. Then, since the bolt force increase ΔF_b is decreased for the same amount of external force W, the load factor ϕ_u gets smaller.

In conclusion, Fig. 4.36 provides an important guideline that as long as the static strength does not matter, raising the initial bolt force F_{bi} is effective to enhance the fatigue strength of bolted joints. In this section and Section 4.5.3, axisymmetric loading was targeted. In actual jointed components, however, external forces are usually exerted in a nonaxisymmetric manner. In some cases, a complete separation does not occur owing to its joint geometry, even if external force is greatly increased. Then, the gradient of the straight line, shown in Fig. 4.36, becomes larger than 45 degrees, which leads to further hazardous situations. In Section 4.6, using FE models of jointed components subjected to nonaxisymmetric loading, a practical diagram for actual bolted joints is presented, which relates external force to bolt force variation.

4.5.5 Estimation method of fatigue strength and stress amplitude of threaded fasteners

A variety of methods have been proposed so far for evaluating the fatigue strength of bolted joints. Some of the methods are introduced in the following.

1) VDI 2230 Blatt1 (2003)

This is well known as a systematic evaluation method of fatigue strength and stress amplitude of screw threads. The evaluation procedure is constructed using the theory of solid mechanics and elasticity. The whole procedure is also translated into Japanese [29]. Although this guideline is for a bolted joint clamped with a single bolt, it can deal with the external force applied eccentrically and take account of the effect of interface separation. This method is considered practical because important values necessary in the joint design process, such as the strength of joint materials, are given in various forms. However, it requires considerable labor to understand the procedure, since a lot of symbols are defined to utilize theoretical formulas of mechanics. In some cases, the statement given there is different from the numerical results by FEM. There is a noteworthy example regarding the effects of interface separation and loading point of external force on the relationship between bolt force variation and external force. In the case of grip length being small, to some extent, calculated results by the VDI procedure become quite different from those by FEM. With regard to VDI procedure, the fatigue strength of bolted joints subjected to eccentric external load is studied [101] and the problems involved in the procedure are examined by FEM [38].

2) Estimation method proposed by Yamamoto [109]

Evaluation formulas for notch factor have been published by the Japan Society of Mechanical Engineers (JSME), the target of which is a round bar with a single annular V-shaped notch subjected to alternating tension and compression loads. Yamamoto modified the formulas so as to be applied to bolted joints. The coefficients appearing in those formulas have been obtained using the data for fatigue limit, introduced in Section 4.4.3. The estimation accuracy for fatigue limit is considered fairly high, because the method is established by combining reliable evaluation formulas for notch factor with reliable experimental data given in Figs. 4.29 and 4.30. It should also be noted that there are some restrictions for the applicable range of this method, since in the foregoing fatigue tests, external loads are applied to bolt head- and nut-bearing surfaces so that the test bolts are subjected to pulsating tensile load.

3) Estimation method proposed by Otaki [8]

Considering that high stress concentration occurs around thread roots, stress distributions along the thread root are calculated using complex stress functions, in which thread geometry is replaced by a series of mountainous curves having the same radius as the thread root. Then, stress concentration factors at thread root are systematically evaluated for metric coarse and fine screw threads, and Whitworth screw threads, in which nominal diameter, thread pitch, and thread root radius are changed in a wide range. Furthermore, stress concentration factors obtained for two-dimensional objects are converted into the factors for three-dimensional ones using Neuber's theory.

Regarding the fatigue strength, Otaki proposed a calculation method that evaluates the notch factor at thread root by use of three-dimensional stress concentration factors. Furthermore, Otaki verified the validity of the proposed method by comparing the fatigue limits of bolted joints obtained in this way with large quantities of experimental data for fatigue limit given in [109]. It is expected that the proposed method has high accuracy, because the fatigue limits of screw threads are estimated by combining the stress concentration factors for thread root, which are derived theoretically, with commonly used formulas in the field of fatigue strength. On the other hand, it involves the same problem concerning the action mode of external loads as Yamamoto's method.

4) Estimation method considering the effect of mean stress

In the previous two methods by Yamamoto and Otaki, it is assumed that fatigue limit is not influenced by the mean stress of the target bolt. Generally, fatigue limit slightly decreases as mean stress is raised in fatigue tests of smoothed test specimens. However, it is said that the effect is not substantial. Yoshimoto assumes that the difference of the fatigue limit between the results obtained by Otaki and Yamamoto is due to mean stress. Hence, using a constant life fatigue diagram or Goodman lines, the ordinate and abscissa of which represent stress amplitude and mean stress, Yoshimoto proposed a theory that accounts for the difference between the results shown by the foregoing two researchers [114].

5) Estimation method using spring and beam models [115]

By combining somewhat the advanced theory of solid mechanics and elementary techniques of numerical analysis, a simple method is proposed to evaluate the stress amplitude exerted on bolts. The method can consider the effect of interface separation and deal with eccentric loadings [116]. It also demonstrates how to draw a bolted joint diagram [117]. Stress amplitudes acting on the bolt cylinder, calculated by the proposed method, are in fairly good agreement with the experimental ones.

Most methods proposed so far for evaluating fatigue strength of screw threads are for bolted joints with axisymmetric geometry and loading. However, actual joints have complex geometries and are subjected to various forms of external loads besides tension and compression, i.e., bending, shear, and twisting. In addition, they are often subjected to multiple types of external loads simultaneously. It is said that stress amplitude is the most dominant factor affecting the fatigue strength of metals. Therefore, numerical methods represented by FEM have come to be used extensively. The target joint is modeled by FEM and the boundary and loading conditions are set corresponding to the actual running condition. The stress amplitudes thus obtained can predictably be used as the guideline for the fatigue strength design. Of particular importance is the stress amplitude at thread root where fatigue cracks are likely to occur. Additionally, the use of FEM makes it possible to evaluate the stress amplitudes for jointed components with identical shape subjected to different forms of external loads. This is a great advantage of the numerical analysis.

4.6 Separation phenomena of plate Interface and stress amplitude

4.6.1 Stress amplitude in bolted joints under eccentric external load

In most cases, external forces in actual bolted joints are applied three-dimensionally. Fig. 4.37A schematically illustrates a bolted joint subjected to an eccentric load. Assuming the jointed portion as being entirely rigid, the magnitudes of forces acting on each point can be calculated using principle of leverage, as shown in Fig. 4.37B. Supposing that the left end of the plate surface, the bolt position, and the location at which external force W is applied are the fulcrum and the points of load and effort, respectively, the force acting on the bolt is calculated by multiplying W by L/L_a, i.e.,WL/L_a. As stated in Section 4.5.3, when external force is applied symmetrically, the additional force applied to the bolt is up to W, even if the plate interface completely loses contact. In contrast, since L/L_a is larger than 1, the bolt force can be greatly increased when eccentric external force is applied. Fig. 4.37C shows a more practical model, in which a bolted joint subjected to an eccentric load is replaced by a cantilever supported by a single spring at the bolt position. In this case, the force acting

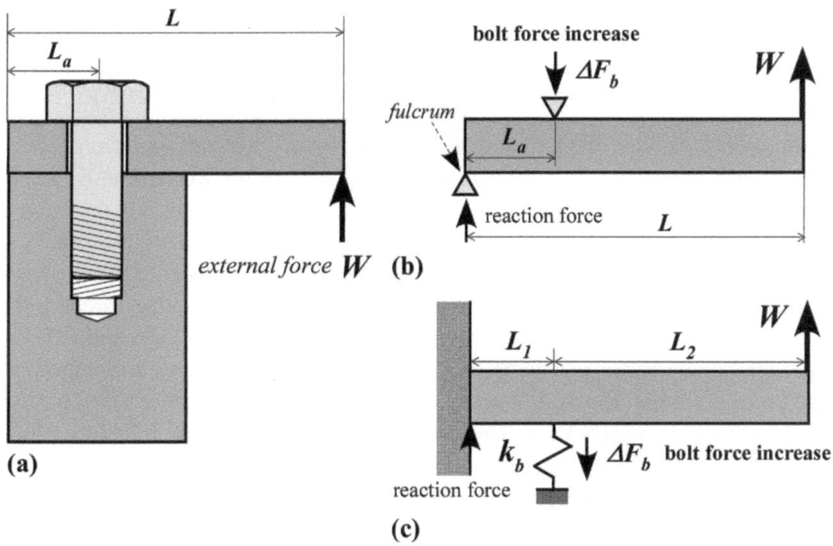

Fig. 4.37 Bolted joint subjected to eccentric external load. (A) Application of external force, (B) introduction of principle of leverage, and (C) cantilever model supported by a single spring.

on the spring k_b that corresponds to bolt force increase ΔF_b is calculated by the following equation, in terms of Young's modulus E and second moment of area I:

$$\Delta F_b = \frac{WL_1^2(2L_1 + 3L_2)}{2L_1^3 + \dfrac{6EI}{k_b}} \tag{4.17}$$

Fig. 4.38 shows how plate interface separation progresses as external force W is increased from the initial tightening state:

1) When the tightening operation is completed, contact pressure due to bolt force F_b is produced at the plate interface, as shown in Fig. 4.38A.
2) When external force W is applied, the separation initiates at the right end of the interface, as shown in Fig. 4.38B.
3) As external force is increased, the separation reaches the bolt hole and spreads to the left side of the bolt.
4) Further increase of external force makes the separation progress toward the left end of the interface, as shown in Fig. 4.38C.

Assuming that the left end is the fulcrum, bolt position is the point of load, and the loading point of W is the point of effort in the complete separation state of the interface, the mechanical behavior explained in Fig. 4.38 is nearly equal to the situation of nail pulling. If the separation spreads to the left end of the interface, bolt force increase ΔF_b shows a drastic increase and reaches WL/L_a, which is L/L_a times larger than in the case of axisymmetric loading, shown in Figs. 4.32 and 4.36. This is an extremely dangerous situation from the viewpoint of fatigue strength. In the case of Fig. 4.38C, the fulcrum exists somewhere at the interface. Accordingly, if bolts were carelessly arranged so that the interface separation is easily progressed, very large stress amplitudes can be applied to the bolts because of a large value of L/L_a in Fig. 4.37A.

4.6.2 Verification of interface separation phenomena by FEM

Using three-dimensional FEM, the relationship between bolt force variation and interface separation is examined for bolted joints subjected to nonaxisymmetric loading. The shape and dimensions of the objective bolted joint are shown in Fig. 4.39 together with the FE model [118]. Nominal diameter of the bolt is M16 with coarse thread. The dimensions of FE model are coincident with the test specimen used in the experiment [101], except that the thread pitch used there is 1.5 mm of fine screw threads. In FE analyses, Young's modulus and Poisson's ratio are set to be 200 GPa and 0.3. Coefficient of friction on contact surfaces is 0.15 and initial bolt stress σ_{bi} is changed to 50, 100, 200, and 300 MPa. Fig. 4.40A represents the relationship between bolt force and external force for $\sigma_{bi} = 100$ MPa. The ordinate and abscissa show bolt force F_b and external force W. In the figure, it is shown how the interface separation progresses. This figure provides two pieces of useful information:

1) The relationship between bolt force and external force is nonlinear, as in the case of axisymmetric loading.

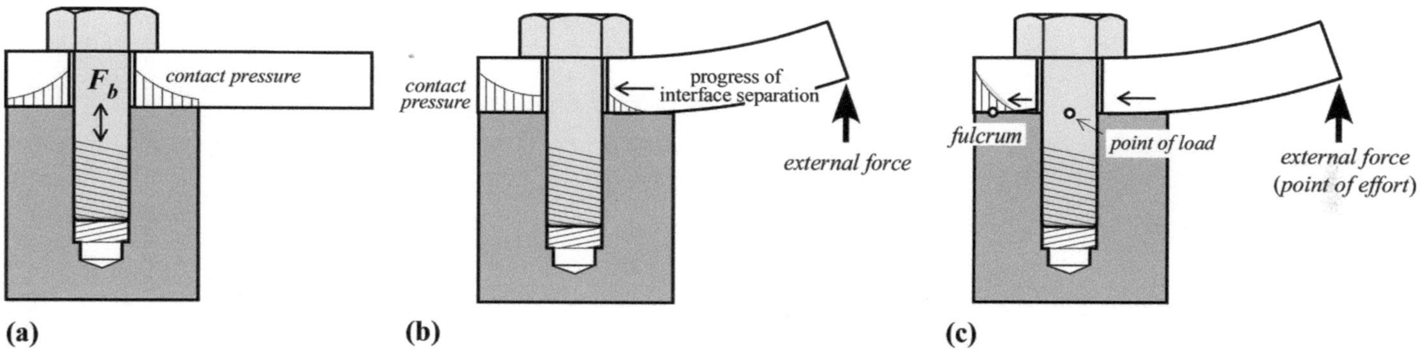

Fig. 4.38 Progress of interface separation with increasing external force. (A) Initial tightening state, (B) separation of right side interface, and (C) separation of left side interface.

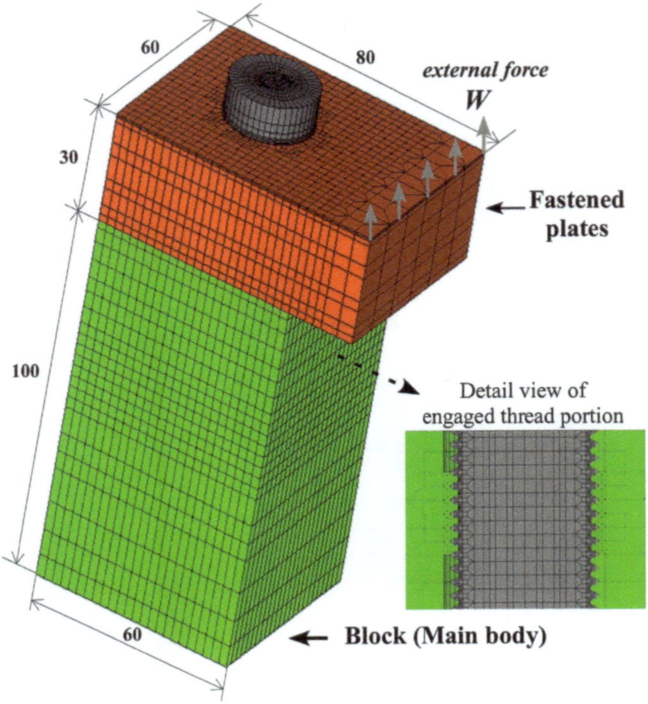

Fig. 4.39 FE model of bolted joint subjected to nonaxisymmetric loading.

2) When the separation progresses beyond the bolt hole, the relationship becomes almost linear. The gradient of the straight line is larger than 45 degrees and is nearly equal to the ratio of L to L_a, i.e., 2.67. This result can be predicted by a simple rigid model, shown in Fig. 4.37B.

Fig. 4.40B shows the results for varying initial bolt stress σ_{bi}. The ordinate and abscissa are shown by dividing F_b and W by initial bolt force F_{bi}, in which F_b/F_{bi} can be replaced by σ_b/σ_{bi}. This figure provides three pieces of useful information:

1) F_b/F_{bi} or σ_b/σ_{bi} can be related to W/F_{bi} with a single curve. The magnitude of W/F_{bi}, at which the separation spreads to the left end of the bolt hole is almost constant, except for the case of $\sigma_{bi}=50\,\text{MPa}$.
2) When external force W is constant, stress amplitude exerted on the bolt can be suppressed by raising the initial bolt force F_{bi} as much as possible within the allowable range of the static strength of joint materials. For example, if F_{bi} is doubled, W/F_{bi} is reduced from 0.3 to 0.15, hence, F_b/F_{bi} is remarkably decreased, as depicted in the figure.
3) When the magnitude of external force W is raised for some reason and the initial bolt force is correspondingly increased at the same rate in order to suppress the stress amplitude, the bolt force is also increased at nearly the same rate, which leads to a highly dangerous situation from the viewpoint of fatigue strength. This phenomenon can be easily understood from Fig. 4.40B because the horizontal value W/F_{bi} is kept unchanged. As a result, this numerical calculation provides a useful guideline on the prevention of fatigue failure. Namely, in order to avoid the increase of stress amplitude when external force is raised, initial bolt force must be raised at a higher rate than the rate of increase of the external force.

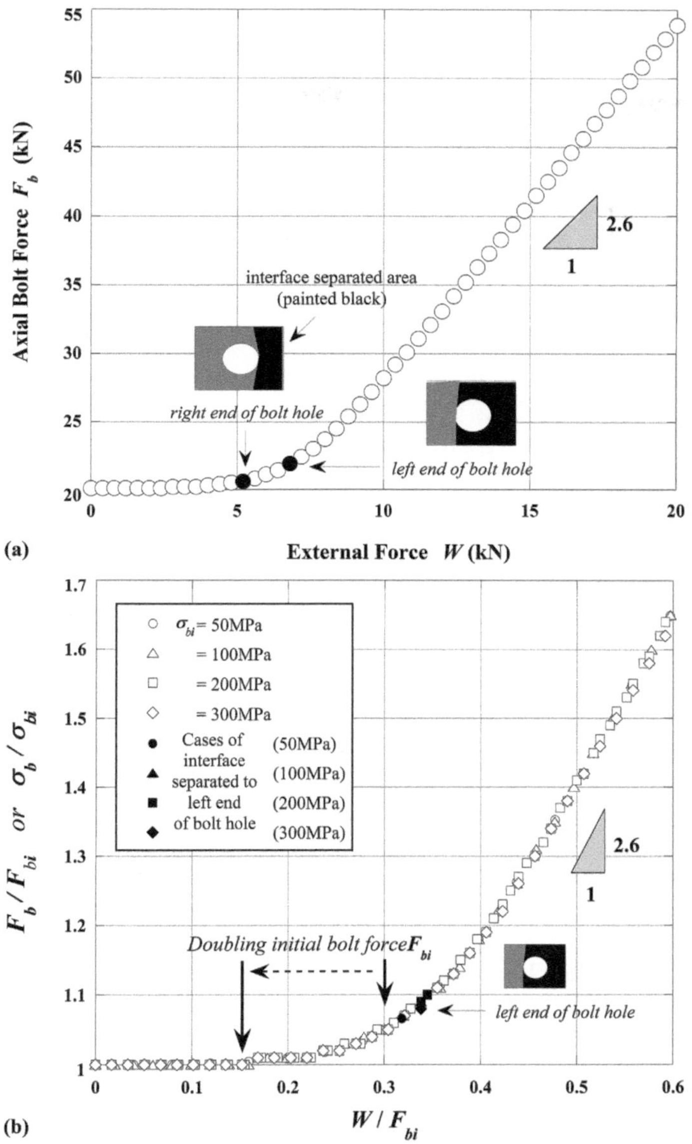

Fig. 4.40 Relationship between bolt force and external force obtained by FE analysis. (A) Case of σ_{bi}=100MPa and (B) effect of σ_{bi}.

4.7 Stress amplitude along thread root

4.7.1 Finite element analysis using helical thread models

As shown in fig. 2.1, fatigue failures of bolt-nut connections are observed in most cases around the first bolt thread root, where the maximum stress concentration occurs, or at the bolt thread roots two or three pitches away from the nut-bearing surface. This implies a close relationship between stress concentration and fatigue failure. In contrast, fatigue failures of body-side engaged threads frequently occur at the female thread root engaged with the far end bolt thread, as already explained in Section 4.4.2. Stress concentration at this position is comparatively small, which is shown in Fig. 4.26B. Therefore, in order to clarify the fatigue failure mechanism of threaded fasteners, it is necessary to evaluate the distributions of stress amplitude along the thread root. In the following sections, the stress amplitude distributions along the thread root in engaged threads are analyzed by helical thread models for bolt-nut connections and body-side engaged threads. To clearly show the effect of helical thread geometry, the shape of the joint is assumed to be axisymmetric, except for the engaged thread portion. External force is applied as annular pulsating tensile load.

4.7.2 Stress amplitude and fatigue failure of bolt-nut connections

In this section, the fundamental characteristics of stress amplitude along thread root of coarse screw threads are shown first [119]. Next, the characteristics of stress amplitude are compared between coarse and fine screw threads [94]. In Fig. 4.41, the shape, dimensions and loading conditions of the analytical model are shown for a bolt-nut connection of M16 with coarse thread. Thread pitch P is 2 mm. Fastened plates are composed of two identical hollow cylinders. Its thickness and outer diameter are 24 and 128 mm, and the inner diameter is 17.5 mm, corresponding to bolt hole diameter d_h of second grade. The nut is simply modeled as a round-shaped nut whose outer diameter is equal to mean diameter B. Initial bolt stress σ_b is 100 MPa and coefficient of friction is 0.15. Bolt force F_b is 20.1 kN, corresponding to $\sigma_b = 100$ MPa. The magnitude of external force W is changed as the ratio of W to F_b, such that W/F_b is 1/3, 2/3, 1, 4/3, and 5/3. The external force is applied on the plate surface. The distance of the loading point from the bolt axis r is changed to 16, 32, and 64 mm. The position of $r = 16$ mm is located just outside the outer surface of the nut, since mean diameter B is 25.375 mm.

Fig. 4.42 shows the FE model. Stress amplitude is defined by dividing the difference of the axial stress at thread root between under external load and in the initial tightening state by two. Numerical results are shown in Fig. 4.43A and B for the cases of $r = 16$ and 64 mm, respectively. The ordinate and abscissa represent the stress amplitude and the distance from the nut-bearing surface. The number of engaged threads is five, since the nut thickness is set to be 10 mm. γ_c in the figure denotes the contact ratio, which is the ratio of the contact lengths at the interface in the radial direction under external force and in the initial state. Namely, γ_c is 1 in the initial state and is zero when the interface completely loses contact. In the range of external force W shown in Fig. 4.43, the interface is partly in contact, and no distinct peak is observed in the stress amplitude distribution when W/F_b is 1/3 and 2/3. Then, as W/F_b reaches 1, γ_c becomes fairly small and two sharp peaks appear at the first bolt thread root and the root of thread runout. This indicates that as external force exceeds a certain level, the progress of the interface separation is accelerated and the stress amplitude is rapidly increased.

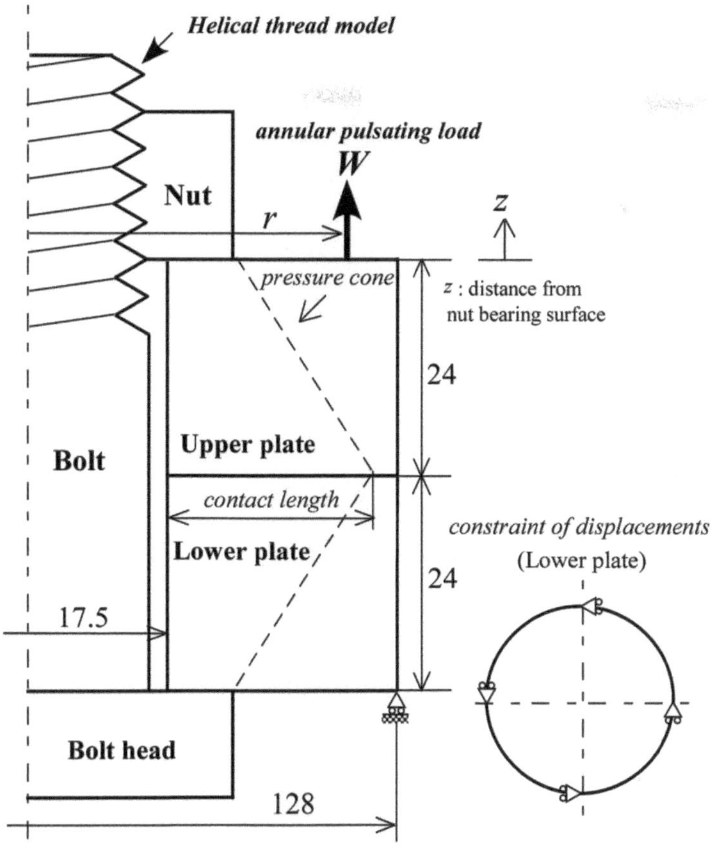

Fig. 4.41 Shape, dimensions, and loading conditions of analytical model.

This phenomenon is consistent with the numerical results by axisymmetric FE analysis shown in Fig. 4.35; that is, load factor is not constant for varying external force and it shows a rapid increase just before the complete interface separation. When external force W is further increased, the interface of fastened plates completely loses contact and the stress amplitude starts to increase sharply, showing a similar pattern to the distribution of thread root stress, as shown in Fig. 4.18. In Fig. 4.44, the numerical results for $r=16$ mm are shown, where the interface completely loses contact. It is recognized that the stress amplitudes at the first bolt thread root and the root of thread runout rapidly approach the critical level, by which fatigue failures occur after the complete separation.

Stress amplitudes produced in fine screw threads are analyzed by using the FE model shown in Fig. 4.21, which was used for evaluating stress concentration. In Fig. 4.45, the effect of thread pitch P on the stress amplitude distribution along thread root is examined for M16, in which P is changed to 2, 1.5, and 1 mm [94]. Initial bolt stress σ_{bi} is 100 MPa and the radius of the loading point r is 64 mm. The magnitude of external force W is equated to bolt force F_b, i.e., $W/F_b=1$. Focusing on the maximum

Fig. 4.42 FE model of bolted joint with helical thread geometry.

stress amplitude, there is almost no difference between the cases of $P = 2$ and 1.5 mm. However, in the case of ultrafine pitch screw threads with $P = 1$ mm, the stress amplitude is reduced distinctly. The fatigue strengths of coarse and fine screw threads with identical nominal diameter can be predicted to a certain extent, by the relationship between fatigue limit σ_w and the ratio of thread pitch to nominal diameter, P/d, depicted in Fig. 4.30.

The corresponding values of P/d are indicated by three arrows on the abscissa in Fig. 4.30. According to Fig. 4.30, fatigue limit σ_w is first reduced with decreasing P and then σ_w starts to increase. Therefore, it is considered that $P = 1.5$ mm belongs to the region in which the minimum value of σ_w is generated. $P = 1$ mm belongs to ultrafine threads, where the decrease in thread pitch leads to a large reduction in fatigue limit. In other words, a slight difference in thread pitch can cause a large variation of fatigue limit in the region of ultrafine threads. It should be pointed out here that loading conditions are different between the numerical analysis and the experiments. Besides, it is not necessarily appropriate to discuss the magnitude of fatigue limit only by the stress amplitude. However, it can be said that the numerical results in Fig. 4.45 are consistent, to some extent, with the experimental ones given in Fig. 4.30.

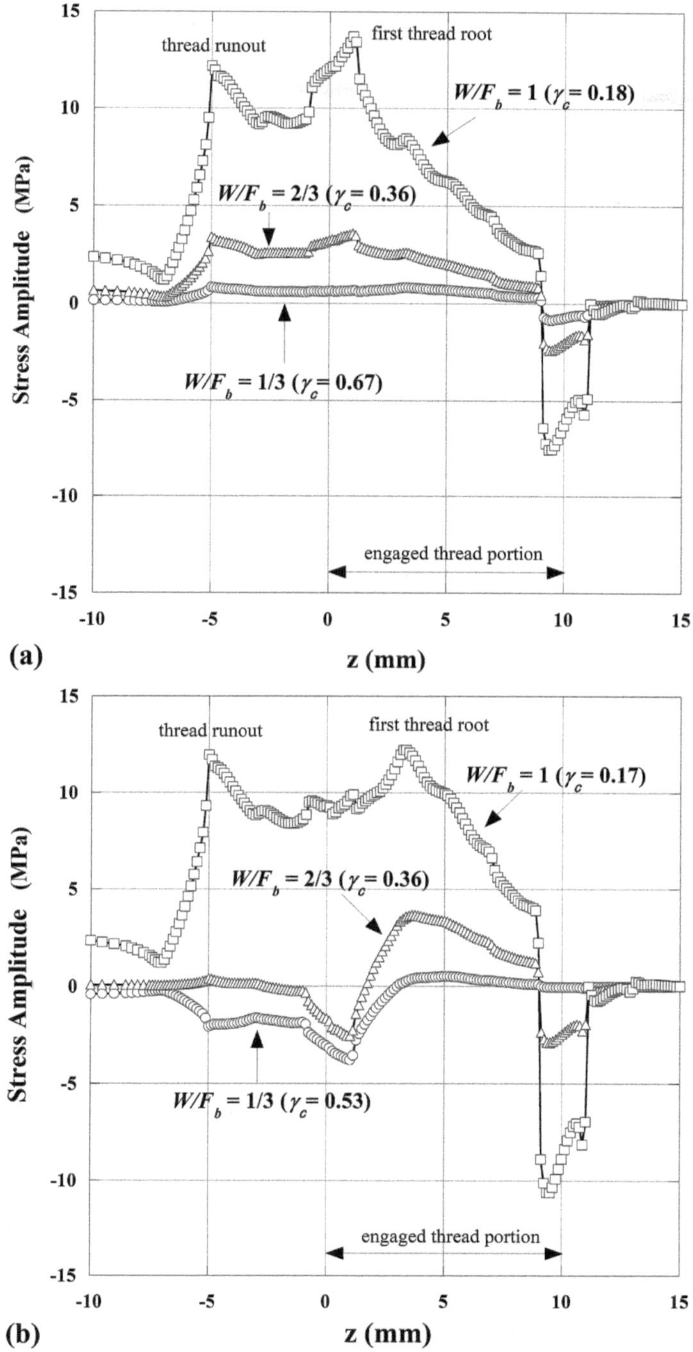

Fig. 4.43 Stress amplitude distributions along bolt thread root in a case of partial contact. (A) $r = 16$ mm and (B) $r = 64$ mm.

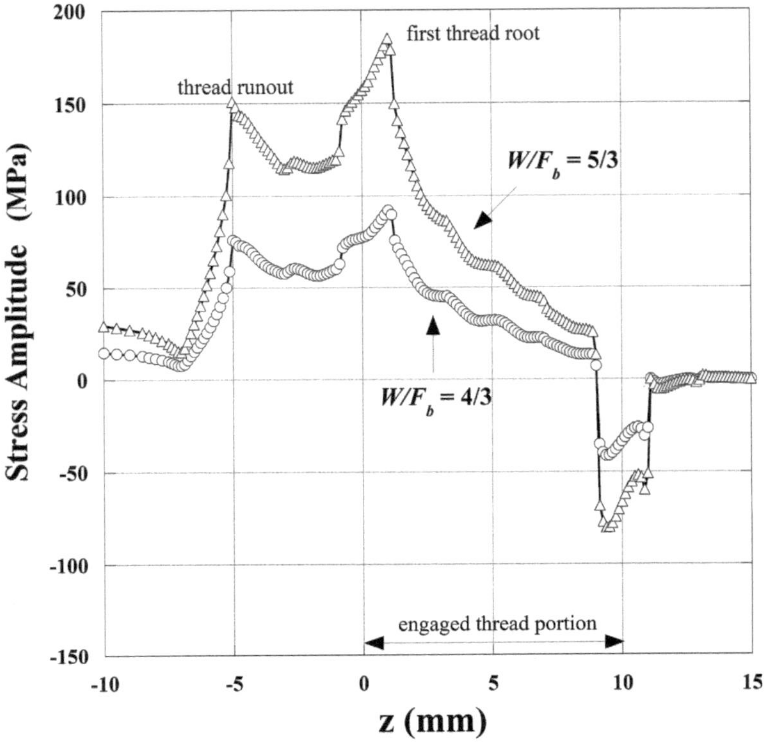

Fig. 4.44 Stress amplitude distributions along bolt thread root in a case of complete separation.

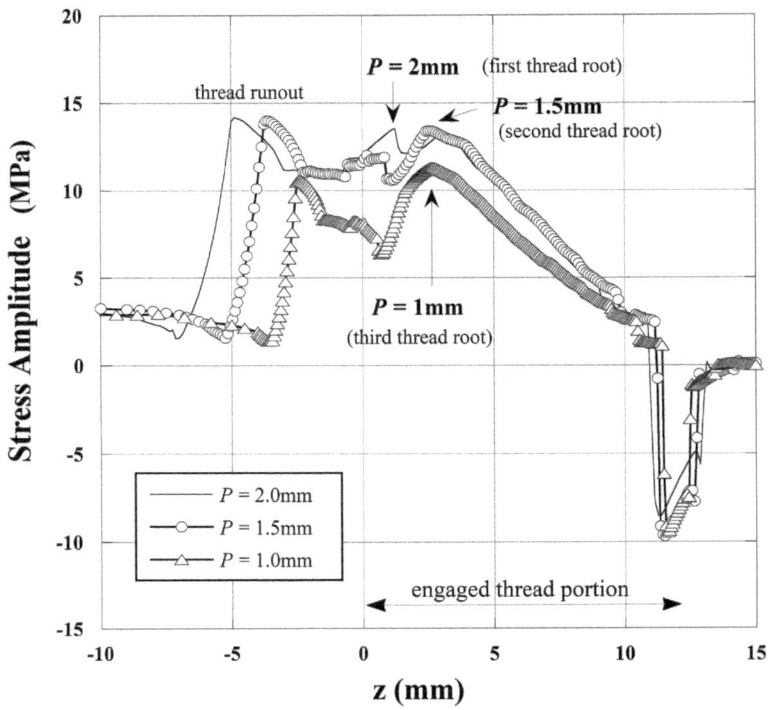

Fig. 4.45 Effect of thread pitch on stress amplitude distributions along bolt thread root.

4.7.3 Stress amplitude and fatigue failure of engaged threads in main body side [20]

Stress concentrations in body-side engaged threads were discussed in Section 4.3.3. They show some different behaviors from those in bolt-nut connections, due to high stiffness of female threads. As for fatigue failures, they sometimes occur at different locations from bolt-nut connections, as explained in Section 4.4.2. For example, the large end of the connecting rod of a diagonally split type is the most fragile part, due to fatigue failure, in medium speed diesel engines. In those cases, fatigue cracks occur not only in the crank pin bolts, which tighten the cap to the large end, but also in the female threads machined in the large end portion, whose structure is shown in Fig. 4.46. The mechanism of such types of fatigue failure is examined from the view-point of stress amplitude. The FE model shown in Fig. 4.25 is used for numerical analysis again. Initial bolt stress σ_{bi} is changed to 100, 200, and 300 MPa, among which the case of $\sigma_{bi}=100$ MPa is the standard condition. Bolt forces F_b corresponding to the above initial bolt stresses are 20.1, 40.2, and 60.3 kN, respectively. Other analytical conditions are the same as the numerical analysis of bolt-nut connections.

Fig. 4.47A and B shows the stress amplitude distributions along the bolt thread root and body-side female thread root. Loading position r of the external force W is 16 mm. In the bolt side, the maximum stress amplitude of about 18 MPa is generated at the third thread root. On the other hand, in the body-side female threads, the stress amplitude occurring at the far end root exceeds 30 MPa. This result is consistent with the accidents occurring in actual machines and reported in the previous paper. That is, in body-side engaged threads, fatigue failures are likely to occur not only at the bolt thread root but also at the far end thread root of female threads [108]. Fig. 4.47C shows the stress amplitude distributions along the female thread root of bolt-nut connections under the same analytical conditions. The maximum stress amplitude in this case is

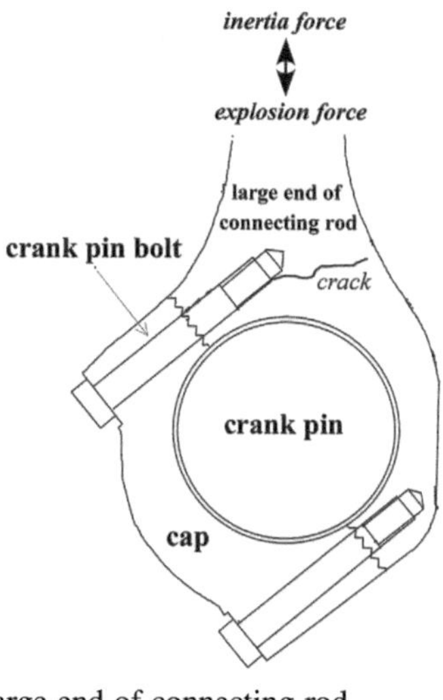

Fig. 4.46 Fatigue failure of large end of connecting rod.

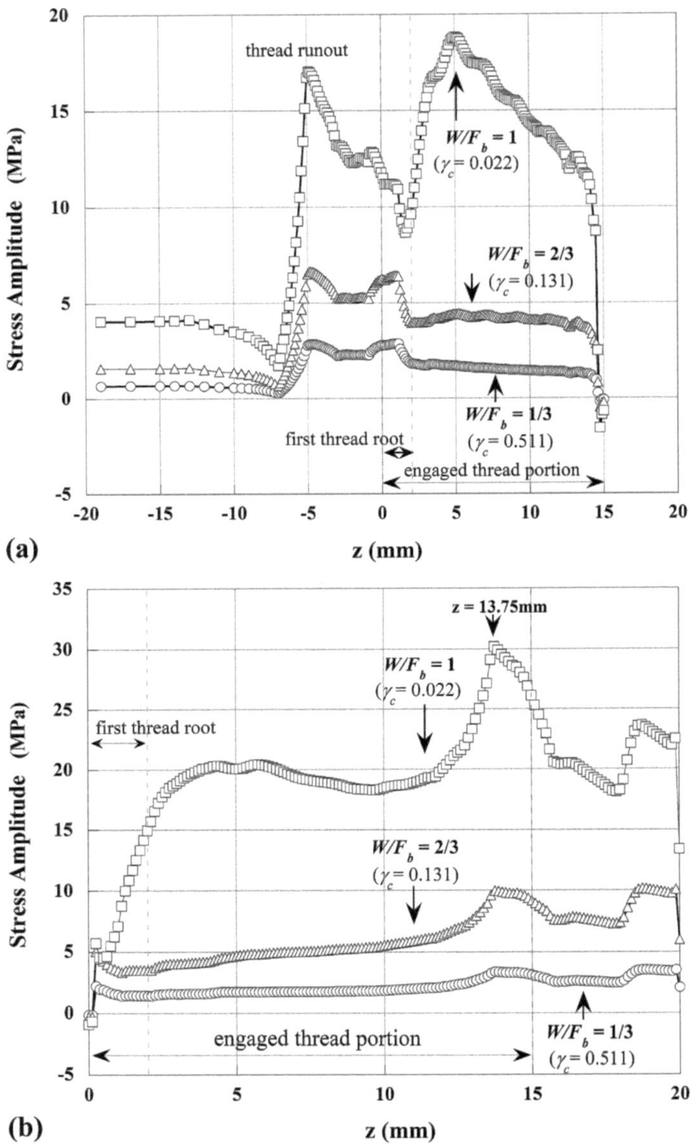

Fig. 4.47 Stress amplitude distributions along thread root in a case of partial contact. (A) Male thread, (B) female thread, and

considerably smaller than that of the body-side female thread. Accordingly, it is concluded that the fatigue failure occurring at the far end female thread root is a phenomenon inherent to the body-side engaged threads. Meanwhile, the maximum stress amplitude in Fig. 4.47B is about 30 MPa, which is not so high as to cause fatigue failure. That is because the initial bolt stress σ_{bi} is set to be low, i.e., $\sigma_{bi} = 100$ MPa and the

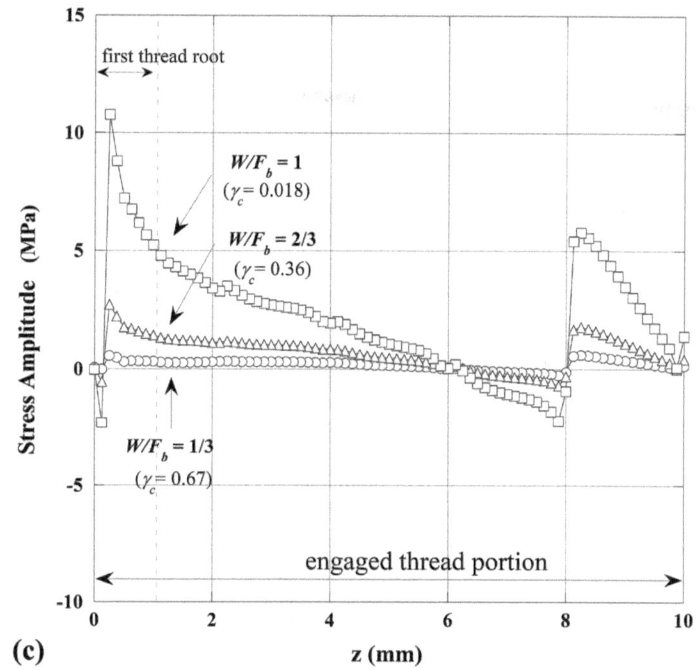

Fig. 4.47, Cont'd (C) nut side (bolt-nut connection).

corresponding bolt force F_b is 20.1 kN. In the case treated here, when the external force W slightly larger than 20.1 kN is applied, the interface becomes completely separated, then the stress amplitude starts to increase sharply. Therefore, if the bolt force and the external force are further increased at the same rate, the maximum stress amplitude can easily reach a critical value, leading to a fatigue failure.

Fig. 4.48A and B shows the stress amplitude distributions along the bolt thread root and body-side female thread root after the occurrence of complete separation at the plate interface, due to larger external force. The magnitude of W/F_b is changed to 4/3 and 2. In this case, the stress amplitudes in the body-side female threads show a significant increase. However, the maximum stress amplitude occurring in bolt threads is even higher than that in female threads. Shown in Fig. 4.49A and B are the relationship between the maximum stress amplitude and the external force W, in which W is normalized with respect to bolt force F_b. γ_c denotes the contact ratio previously defined. In the male thread side, the maximum stress amplitude occurs at the point symbolized by 2.5P until W/F_b exceeds 1 and reaches complete separation. Meanwhile, the location 2.5P represents the third bolt thread root and is two and half pitches away from nut-bearing surface. After the complete separation, the maximum stress amplitude occurs at the first bolt thread root. For comparison, stress amplitudes in a bolt-nut connection are also given in the figure by a dashed line. The figures in the parentheses represent the contact ratio γ_c. In the female thread side, the maximum stress amplitude is produced at the far end thread root until W/F_b exceeds about

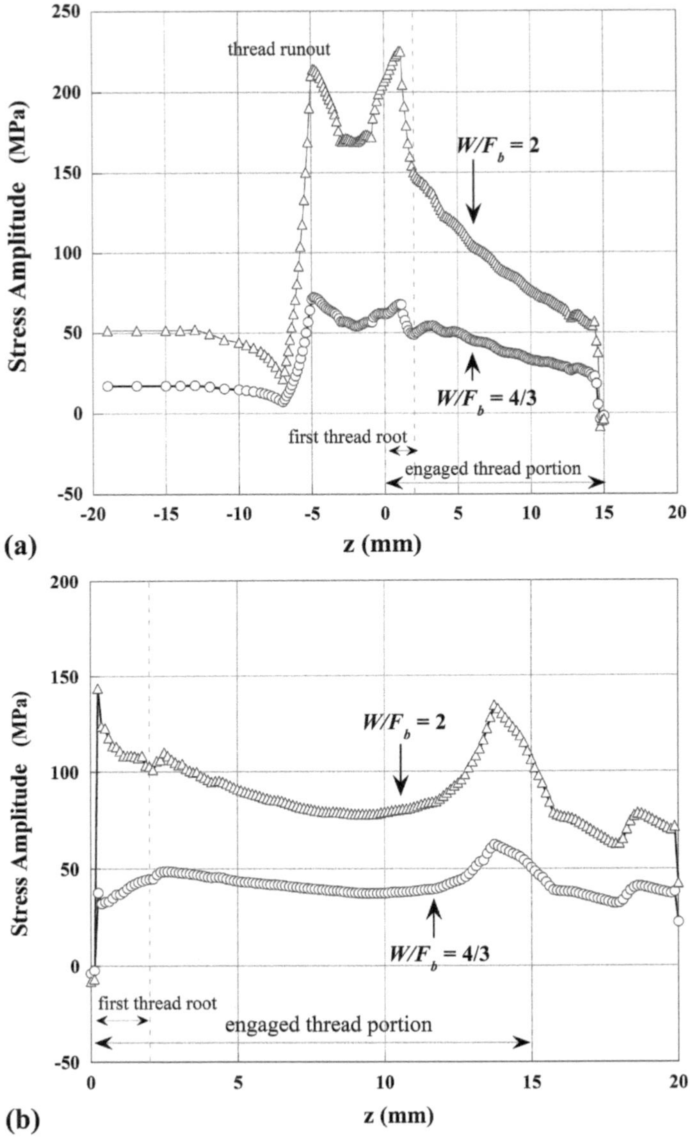

Fig. 4.48 Stress amplitude distributions along thread root in a case of complete separation. (A) Male thread and (B) female thread.

1.5 after the complete separation. This result supports the position at which fatigue failures are likely to occur in the body-side female threads of actual jointed portions, as explained in Figs. 2.1 and 4.46.

As the interface separation progresses, the stress amplitude at thread root is increased. In order to suppress the amplitude to a low level, it is effective to raise the initial bolt force. Therefore, from the practical point of view, it is necessary to

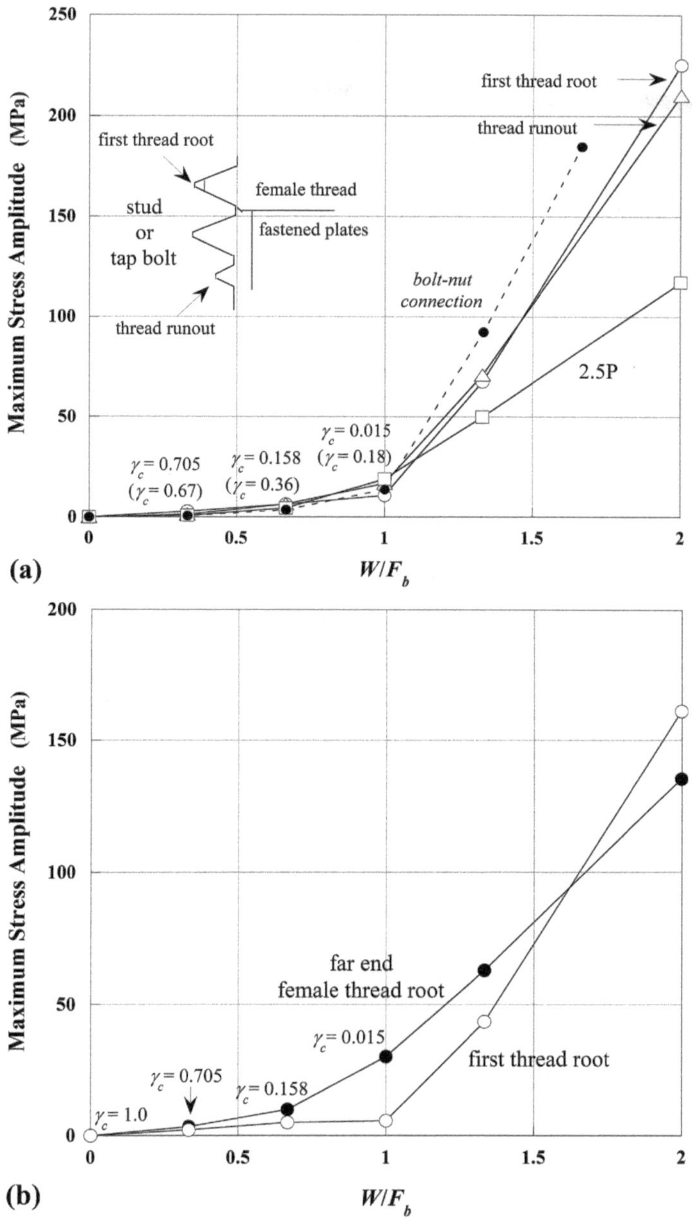

Fig. 4.49 Effect of bolt force and external force on the maximum stress amplitude. (A) Male thread and (B) female thread.

clarify how much bolt force increase is required against the predicted increase of external force. Fig. 4.50 shows the effects of bolt force F_b and external force W on the maximum stress amplitude occurring in body-side female threads. Initial bolt stress σ_{bi} is varied from 100 to 500 MPa with increments of 100 MPa. It is found from the figure that the maximum stress amplitude is almost proportional to initial bolt

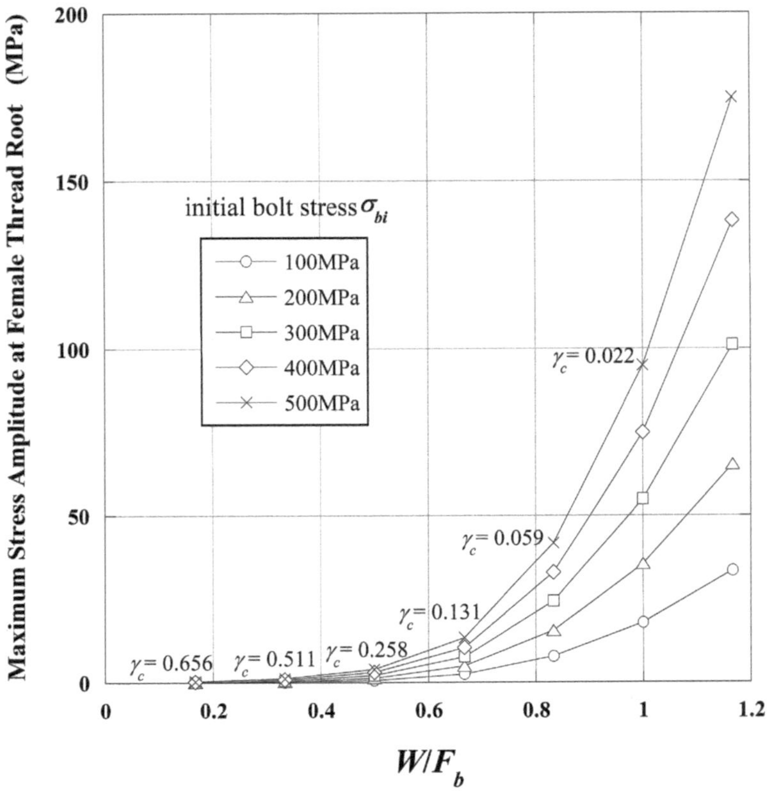

Fig. 4.50 Effect of bolt force and external force on the maximum stress amplitude in body-side female threads.

stress when the ratio of external force to bolt force, W/F_b, is constant. From the practical point of view, it can be rephrased as follows:

1) Under constant external force, the maximum stress amplitude can be reduced by raising initial bolt force.
2) When bolt force is raised at the same rate as the increase of external force, the maximum stress amplitude is also increased proportionally, thus leading to a highly risky state.

The second phrase is particularly important when designing bolted joints. Anyway, the aforementioned results indicate the same phenomena as those in the case of bolted joints subjected to nonaxisymmetric loading, as shown in Fig. 4.40B in Section 4.6.2. In conclusion, the primary countermeasure for lowering the stress amplitude is to set the initial bolt force as high as possible.

4.7.4 Stress amplitude and plastic deformation [20]

Plastic deformations are produced around thread roots with comparative ease under low bolt force, because high stress concentration is generated there. The effect of plastic deformations on the stress amplitude at the thread root is examined in this section. Using the same FE models and analytical conditions as those in the previous section, a constitutive equation proposed by Swift is applied to express the stress–strain relationship in the plastic region:

$$\sigma = 839.48\left(0.0002 + \varepsilon^{p}\right)^{0.0743} \tag{4.18}$$

The above equation was obtained by tensile tests of bolt materials [48]. σ and ε^{p} denote true stress and plastic strain, respectively. Yield stress σ_Y is found to be 446 MPa from the equation. Fig. 4.51 shows the stress amplitude distributions along body-side female thread root when bolt stress σ_b is 400 MPa, which is about 90% of σ_Y. External force is applied three times as annular pulsating tensile load. Numerical results by elastic analysis are also given in the figure. Comparing to the results by elastic analysis, a considerable difference is observed in the first cycle, but the difference becomes small after the second cycle. Although not shown in the figure, similar results are obtained for $\sigma_b = 300$ MPa. Therefore, for the purpose of evaluating the stress amplitude at thread root, it is considered that elastic analysis can provide results with practically sufficient accuracy. Illustrated in Fig. 4.52 is the relationship between the

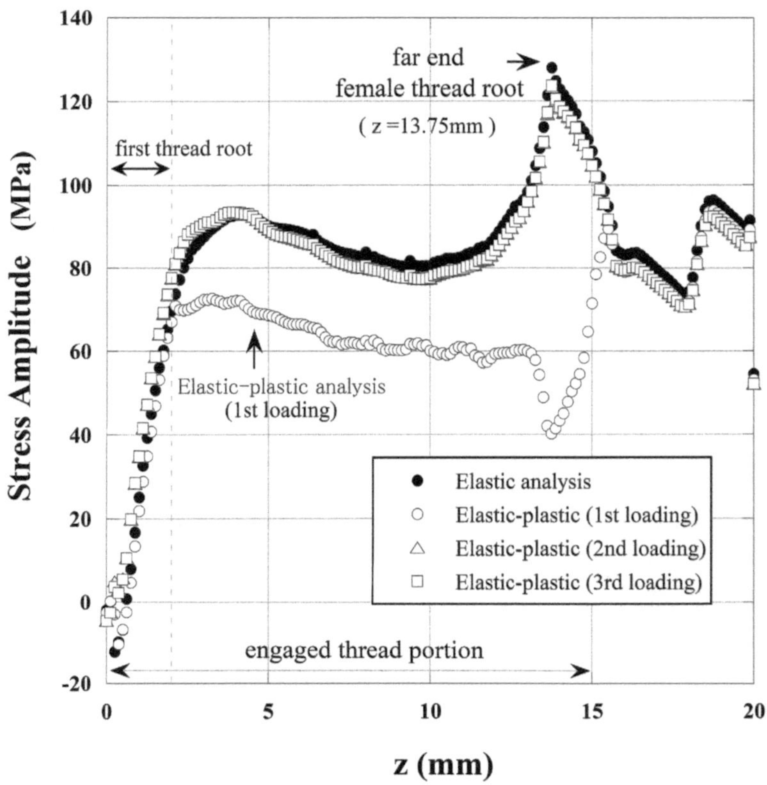

Fig. 4.51 Effect of plastic deformation on stress amplitude.

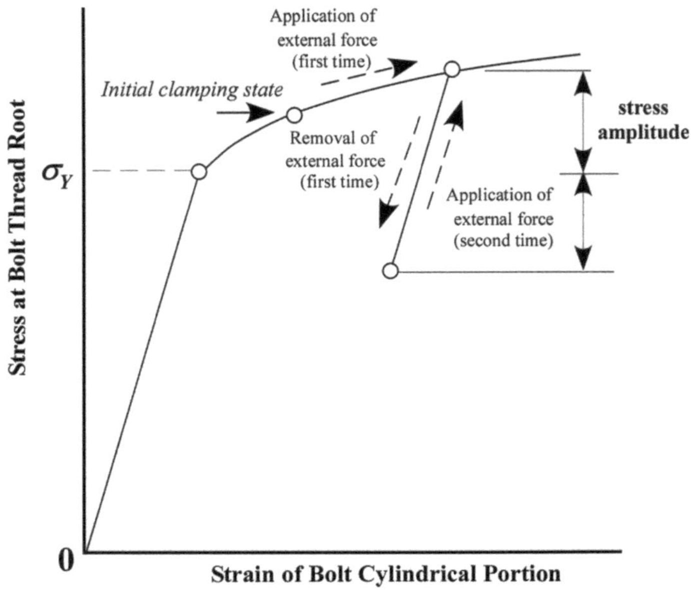

Fig. 4.52 Stress–strain behavior in bolts subjected to plastic deformations.

stress at the bolt thread root and the strain of the bolt cylindrical portion. It explains the phenomena shown in Fig. 4.51. In the first loading cycle, plastic deformations are progressed, which raises the yield stress after the second loading cycle. However, after the second loading cycle, the stress amplitude becomes almost the same value as the one obtained by elastic analysis, since the subsequent unloading and reloading processes move on the same straight line repeatedly.

4.8 Improvement measures of fatigue strength of threaded fasteners

Fatigue strength of threaded fasteners is influenced by a variety of factors, such as joint materials, nominal diameter, thread pitch, joint geometry, tightening and loading conditions, magnitude of external force, stress gradient at thread root, etc. Therefore, several countermeasures can be considered to raise the fatigue strength. Meanwhile, the most dominant factor affecting the fatigue strength is the stress amplitude exerted on threaded portions. Hence, in order to raise the fatigue strength, it is important to suppress the bolt force variation when subjected to external force. More specifically, the stress amplitude at thread root, at which cracks are likely to initiate, should be lowered. Some practical methods to lower the stress amplitude are listed below.

1) Design the bolt stress as high as possible within the allowable range of static strength

Aiming at preventing interface separation or suppressing the extent of separation as much as possible, initial bolt stress should be higher as long as the static strength does not matter. It is shown in Fig. 4.50 that when the ratio of external force to bolt force is kept constant, the maximum stress amplitude occurring at bolt thread root increases almost proportional to bolt stress. Accordingly, when external force is increased for some reason, bolt force must be raised to an extent that exceeds the increasing rate of external force.

2) Use of low stiffness bolts

As is evident from the bolted joint diagram, bolt force variation becomes small when a bolt is deformable and fastened plates are difficult to deform under external loads. In other words, it is desirable for the spring constant of the bolt-nut connection k_b to be small and that of fastened plates k_f to be large. Slender bolts or hollow bolts, as shown in Fig. 4.53A and B, are sometimes used to realize the former countermeasure. In slender bolts, the cylindrical portion diameter is smaller than nominal diameter. A crank pin bolt used for clamping the large end of internal combustion engines, shown in Fig. 4.53C, is a kind of slender bolt. Axial stiffness of the bolt is decreased in proportion to the cross-sectional area of the bolt. The product of Young's modulus E times second moment of area I is termed bending stiffness. The amount of I is significantly reduced just by slightly reducing the diameter. Hence, slender bolts are effective to lower the stress amplitude of the joint under bending loads.

In Fig. 4.54A, the effectiveness of slender bolts is examined by evaluating the stress amplitude distribution along bolt thread root using helical thread FE models. The analytical object is a bolt-nut connection of M16 with coarse thread, in which the diameter of cylindrical portion is reduced to be equal to thread root diameter. Other analytical conditions are the same as those in Section 4.7.2. It is found that stress amplitudes in slender bolts around the first bolt thread root are lower than those in ordinary bolts, depicted by solid lines. However, it the complete separation occurs at the interface due to the increase of external force W, slender bolts have no effect on the reduction of stress amplitude around the first bolt thread root, which is shown in Fig. 4.54B. As for hollow bolts, shown in Fig. 4.53B, their axial stiffness is reduced in proportion to the cross-sectional area. However, it can hardly be expected to enhance the fatigue

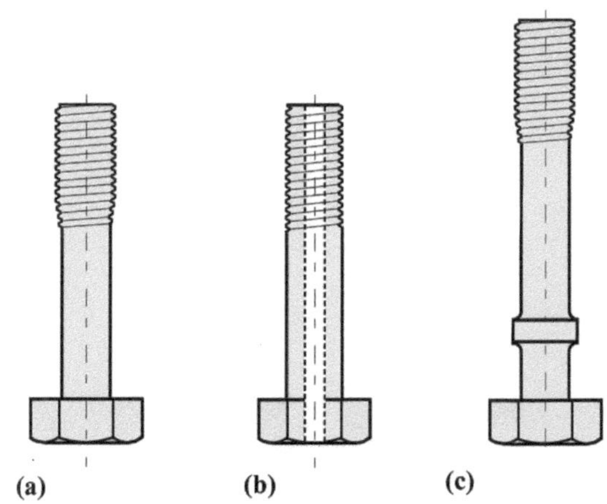

Fig. 4.53 Special bolts aiming at lowering stress amplitude.
(A) Slender bolt, (B) hollow bolt, and (C) crank pin bolt.

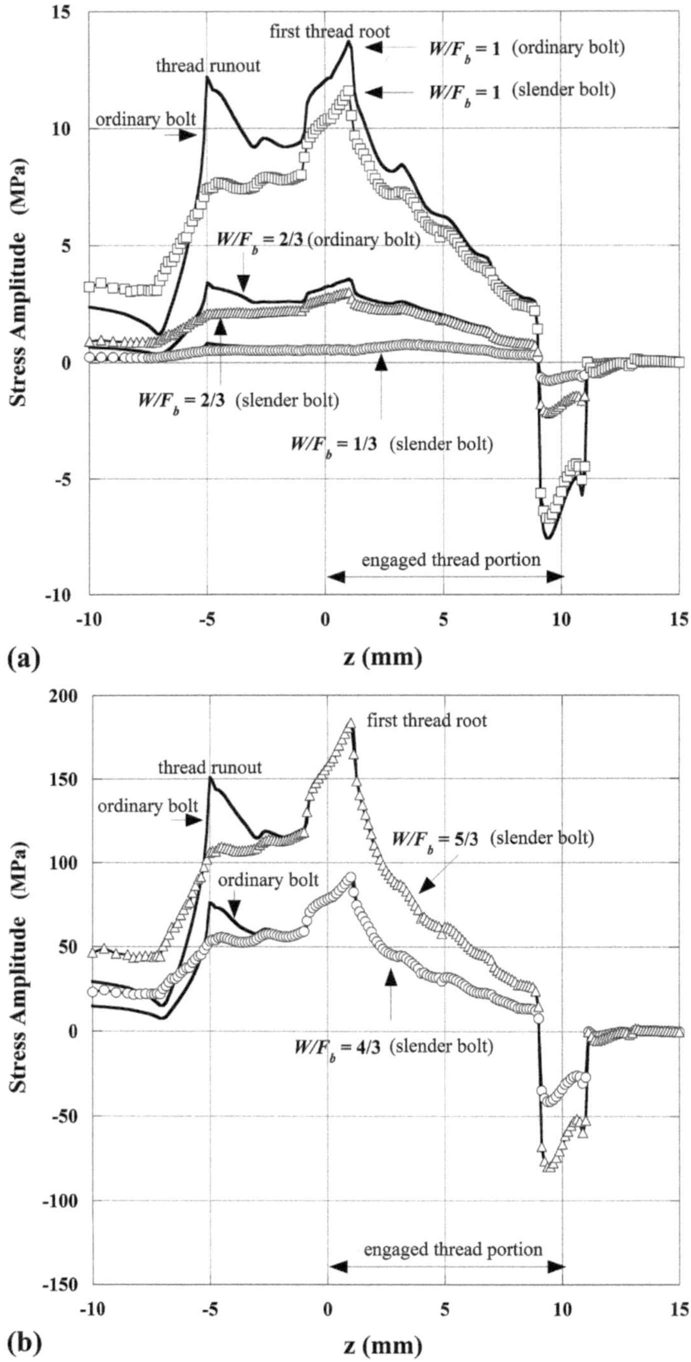

Fig. 4.54 Effect of slender bolts in lowering stress amplitude. (A) Partial contact and (B) complete separation.

strength under bending loads, because the reduction rate in bending stiffness, represented by second moment of area, is very small. The logic is easily understood using an elementary theory of solid mechanics.

Bolt stiffness can also be reduced by manufacturing the target bolt using materials with low Young's modulus. Referring to table 1.4, Young's modulus of titanium alloy is just over 50% of that of carbon steel, which implies that if bolts made of titanium alloy with identical geometry are used instead of carbon steel bolts, a similar effect to the use of a bolt that is twice as long is expected in lowering the bolt stiffness. In Fig. 4.55, stress amplitudes along bolt thread root are compared for varying bolt materials, i.e., pure titanium, titanium alloy, and carbon steel, where fastened plates are made of carbon steel in all cases [25]. In the numerical analyses, the FE model shown in Fig. 4.42 is used and the boundary and loading conditions are the same as in Section 4.7.2. External forces are applied annularly at the distance of 16 mm from the bolt axis. The plate interface is partly in contact. Using titanium or titanium alloy bolts, the maximum stress amplitude occurring at the first bolt thread root is reduced to about two-thirds of that of carbon steel bolts. Although involving a problem of high cost, titanium and titanium alloy bolts have low Young's modulus and low coefficient of linear expansion, thereby being unlikely to loosen under thermal load as well. It is considered, therefore, that titanium and titanium alloy bolts possess great potential for application to high value-added joints.

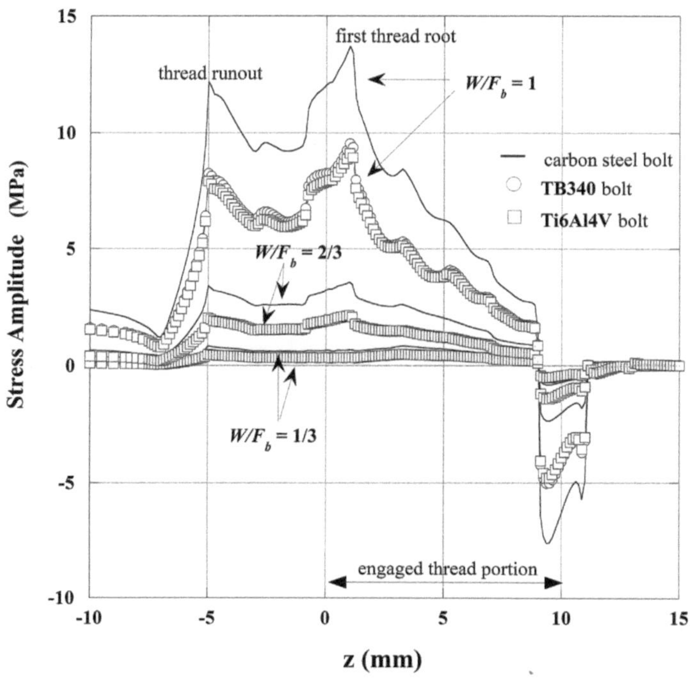

Fig. 4.55 Effect of titanium bolts in lowering stress amplitude.

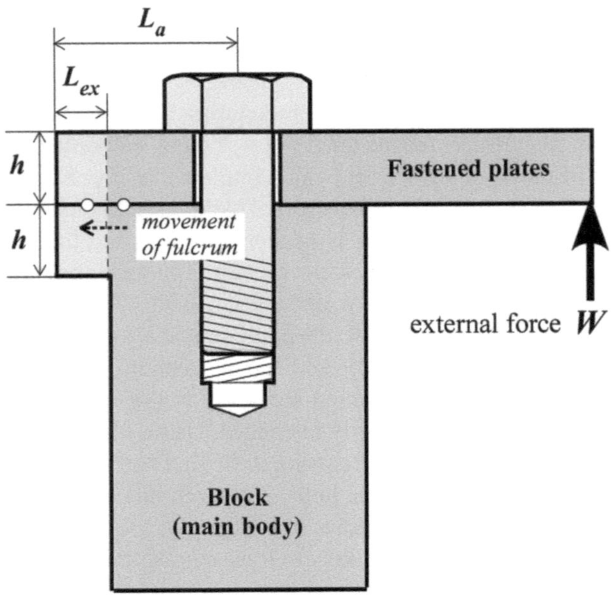

Fig. 4.56 Bolt arrangement aiming at lowering bolt force variation.

3) Design the bolt locations so as to lower bolt force variation

Bolts should be aligned so as to lower the bolt force variation against the external force to be applied. For example, when subjected to eccentric external loads, as shown in Fig. 4.37A, bolt force variation due to bending moment can be reduced by locating the bolt as close to the point of effort as possible, which is evident from the explanation in Section 4.6.1. This method is effective when the width of fastened plates is larger above a certain extent against the bolt diameter.

4) Modification of joint geometry so as to lower bolt force variation

If large bending moment is exerted on the objective bolt due to external force, the interface separation is likely to occur and is easily progressed. Hence, it is sometimes effective to modify the joint geometry for lowering bolt force variation. A concrete example is shown in Fig. 4.56. Protrusions of length L_{ex} are added at the end of jointed portion for the FE model illustrated in Fig. 4.39, by which the distance from the bolt axis to the left end of the fastened plates is increased.

In Fig. 4.57A and B, the relationship between the maximum stress occurring at bolt thread root and external force W is depicted. The plate thickness h is 30 and 20 mm, respectively. The maximum stress is produced at the thread root of the right side of the bolt, in which the tensile stress is increased due to the action of bending moment induced by the external force. In either case, the increase rate of the thread root stress is decreased owing to the added protrusions. Specifically, the gradient of the curve that relates the maximum stress to W is reduced in the region where the interface separation is considerably progressed. In the case of $h = 20$ mm, the effect of reducing the maximum stress is saturated at smaller L_{ex} compared to the case of $h = 30$ mm, and the

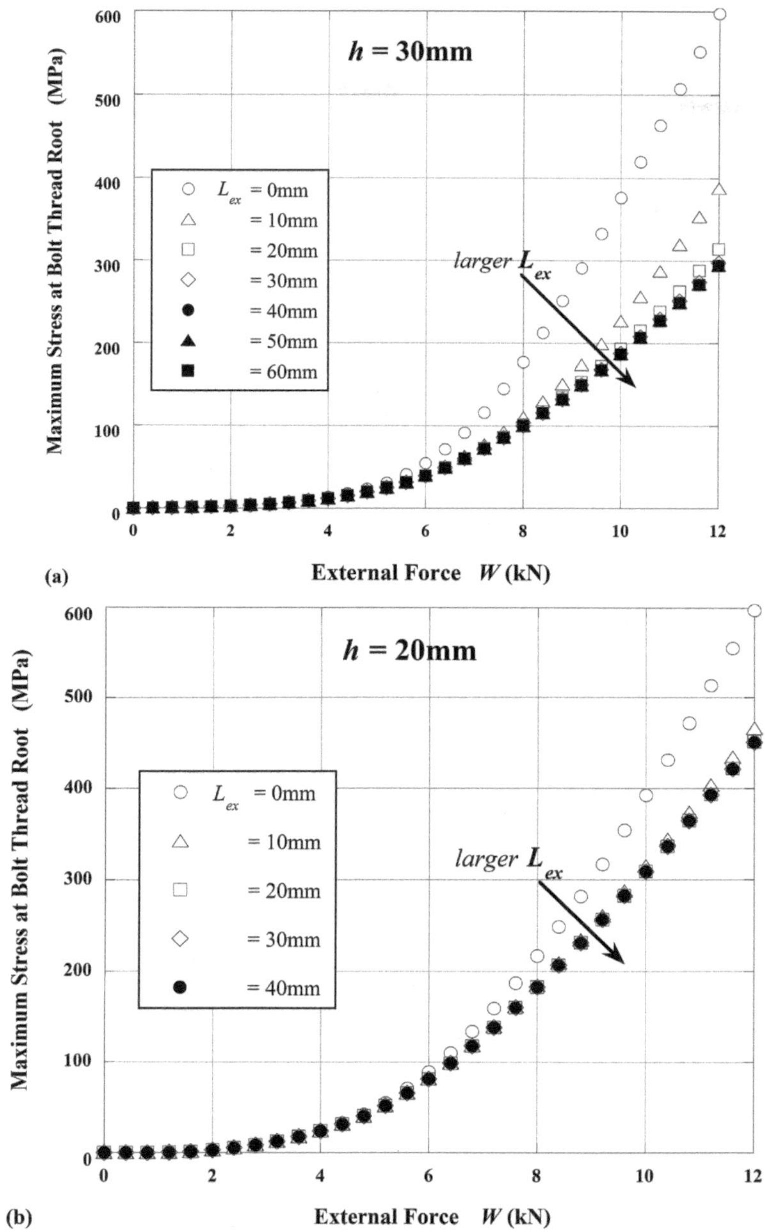

Fig. 4.57 Effect of added protrusions in lowering stress amplitude. (A) Case of h=30 mm and (B) h=20 mm.

degree of reduction effect is also lower. This mechanism can be explained, in connection with the bending moment exerted on bolt, by the plate stiffness and the distance between the bolt axis and fulcrum. The second moment of area I of beam or plate is proportional to the cube of plate thickness. Accordingly, since the bending stiffness of plate in proportional to the cube of h when identical materials are used, the bending deformation of $h=30$ mm is considerably smaller than that of $h=20$ mm. This implies that the case of $h=30$ mm is more suitably simulated by the simple model illustrated in Fig. 4.37B, where the jointed portion is wholly assumed to be a rigid body. As a result, the distance between the bolt axis and fulcrum becomes shorter in the case of $h=20$ mm due to larger deformations, as predicted from the deformation patterns shown in Fig. 4.38B and C. Hence, it leads to the increase in the magnification ratio of external force L/L_a. To summarize the above, the fulcrum should be removed from the bolt axis as far as possible to suppress the bolt force variation when subjected to external force.

5) Considerations in the effectiveness of long bolts

Since the spring constant of bolt-nut connection k_b is almost inversely proportional to bolt length, it is effective to use long bolts. Comparing the case of Fig. 4.58A to the shape of the jointed components on the left side in Fig. 4.58B, the use of a longer bolt with identical nominal diameter decreases k_b, but the spring constant of fastened plates k_f also decreases correspondingly due to the increase in plate thickness. Therefore, load factor of the jointed components, subjected to annular pulsating tensile load, cannot be lowered. This is evident because k_b and k_f are decreased at the same rate in the bolted joint diagram. Meanwhile, actual joints are often subjected to bending moment due to eccentric load, as shown in Fig. 4.58B. To cope with those cases, a calculation method is proposed by VDI 2230 [29] for evaluating the stress amplitude in bolts, whose target is a joint subjected to eccentric load. However, the proposed method by VDI does not necessarily provide sufficient accuracy for varying grip length. Hence, in order to examine the effect of long bolts, it is recommended to evaluate the stress amplitude occurred in bolts with the help of FEM by modeling the joint, including the loading point.

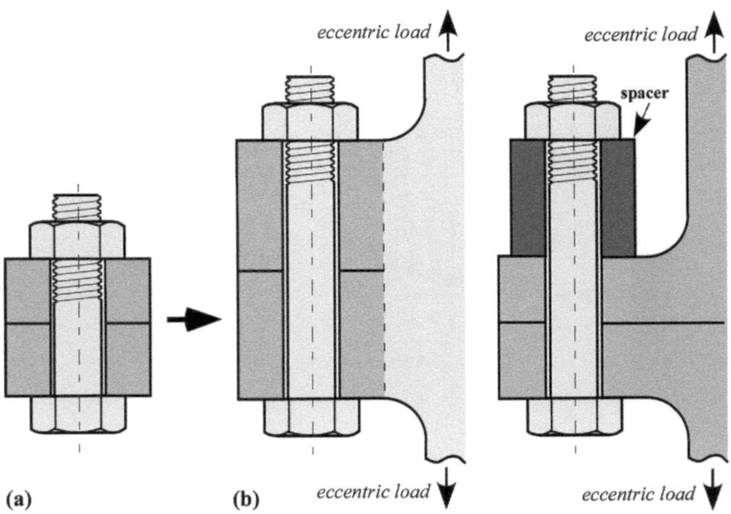

Fig. 4.58 Effect of long bolts in lowering stress amplitude. (A) Short bolt and (B) long bolt with identical nominal diameter.

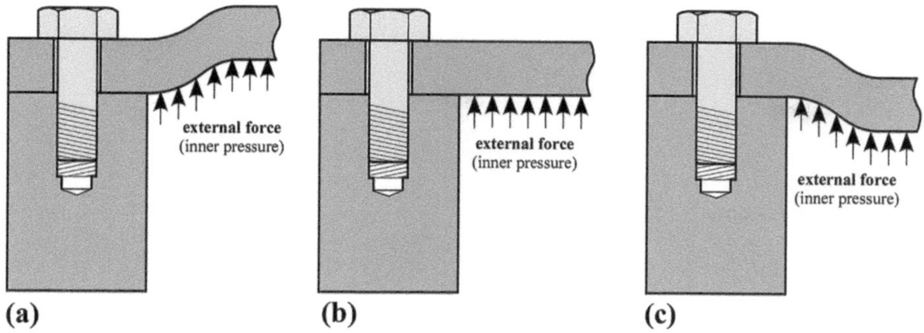

Fig. 4.59 Relationship between cover shape of pressure vessel and load factor. (A) Convex cover shape, (B) flat cover shape, and (C) concave cover shape.

For the bolted joints used in pressure vessels, shown in Fig. 4.59, the relationship between joint geometry and load factor has been reported by Junker. He concluded that the load factor decreases in the order of convex, flat, and concave cover shapes. The idea is also followed by VDI 2230. However, no distinct difference is observed among the three cases, according to the numerical results by FEM, in which the joint geometry is precisely modeled, with constant grip length and bolt locations being unchanged in all cases. This phenomenon can be explained by the fact that, under the same external force, the magnitudes of axial tension and bending moment acting on the target bolt are not influenced by the shape of cover. Hence, when designing a joint like the one illustrated in Fig. 4.59, it is of primary importance to avoid or suppress the interface separation for lowering the stress amplitude. As a concrete measure, bolt force should be designed as high as possible within the range that other problems won't occur, and the strategies listed from 2) to 4) are selectively introduced in accordance with the situation.

– Learning by numerical example 4.1: Bending stiffness of slender bolt and hollow bolt

For M16 bolts with coarse thread, the bending stiffness of slender bolts and hollow bolts is compared to that of ordinary solid bolts. The bolt body diameter of slender bolts is equal to the thread root diameter. The inner diameter of hollow bolts is half the nominal diameter. Assuming equal Young's modulus, bending stiffness can be compared by means of second moment of area I. The second moment of area of a hollow bar, whose inner and outer radii are a and b, is expressed as $I=\pi\ (b^4-a^4)/4$. Substituting $b=8$ mm and $a=0$ mm, the second moment of area I for ordinary solid bolts is calculated. I for slender bolts is calculated by substituting $b=13.546/2$ mm, which is half the thread root diameter, and $a=0$ mm. I for hollow bolts is obtained by substituting $b=8$ mm and $a=4$ mm. The root diameter of slender bolts is calculated by use of the expression shown in numerical example 2.2.

The second moment of area I is calculated to be 3217, 1653, and 3015 mm^4 for solid bolts, slender bolts, and hollow bolts, respectively. The bending stiffness of slender bolts and hollow bolts is 51.4% and 93.7%, respectively, of the stiffness of solid bolts, which substantiates the effectiveness of slender bolts in lowering the stress amplitude.

Bolted joints under thermal load

5

5.1 Fundamentals of thermal and mechanical behaviors of bolted joints

5.1.1 Thermal deformation and thermal stress [120]

When material bodies are heated, they are elongated; when cooled, they are shrunk, in contrast. The amount of elongation or shrinkage δ is calculated as the product of coefficient of linear expansion α_{ex}, temperature change ΔT, and the original length of the object L.

$$\delta = \alpha_{ex}\Delta TL \tag{5.1}$$

The average coefficient of linear expansion of carbon steel is about 12×10^{-6}, as given in table 1.5. For example, as the temperature of a round bar of 1 m length is increased by as much as 100 °C from the ambient temperature, the bar elongates by about 1.2 mm. When the temperature of a material body changes, either thermal expansion or shrinkage occurs, and thermal stresses are generated by constraining the thermal deformation. A thin wire placed on the ground under the blazing sun is elongated due to the temperature increase, but thermal stress is not generated in it because the deformation is not constrained. In the case of a thick bar, thermal stress is generated during the temperature rising process due to the temperature gradient produced in the body. Then, as the whole temperature becomes constant after a sufficiently long time, the bar is regarded as being in the free expansion state and the thermal stress disappears, as in the case of a thin wire.

When railroad rails are elongated beyond an allowable value under the blazing sun, the rails adjacent to each other mutually constrain the deformations, thereby producing large compression forces. In actual machines and structures, thermal stresses due to temperature gradient, occurring within the body, frequently becomes an issue. For instance, when high temperature fluids flow in a pipeline, the inner surface temperature of the pipe is elevated and a negative temperature gradient is generated toward the outer surface. As a result, the portion around the inner surface is subjected to compression stress because its free expansion is constrained. In contrast, tensile stress is generated around the outer surface so as to be balanced with the compression stress. When hot coffee is spilled on a table, some amount of compression stress is generated on the table surface. That is because the free expansion of the heated portion is constrained by the surrounding materials under room temperature.

The most fundamental equation of thermal stress is derived by considering the case of a straight bar that is uniformly heated or cooled and the deformations are

The Mechanics of Threaded Fasteners and Bolted Joints for Engineering and Design.
https://doi.org/10.1016/B978-0-323-95357-3.00001-0

completely constrained at both ends. Omitting the derivation process, only the resulting equation is shown below:

$$\sigma = \alpha_{ex} \Delta T E \tag{5.2}$$

In Eq. (5.2), the product of coefficient of linear expansion α_{ex} and temperature change ΔT is termed thermal strain. When the deformation is completely constrained, the magnitude of thermal stress σ is determined by α_{ex}, ΔT and Young's modulus E regardless of its length L. Supposing that $\alpha_{ex} = 12 \times 10^{-6}$, $E = 200\,\text{GPa}$, and $\Delta T = 100\,°\text{C}$, high compression stress of $\sigma = 240\,\text{MPa}$ is generated. In actual practice, however, the magnitude of thermal stress becomes considerably smaller because the ends of the heated object cannot be completely constrained.

Thermal stresses are also generated when thermal expansions or shrinkages of two bodies are mutually constrained. When bolted joints are subjected to thermal load, a bolt-nut pair and fastened plates constrain the opponent deformation. Accordingly, the thermal stress produced by the foregoing mechanism has a high degree of similarity to the bolt force variation due to thermal load. In Fig. 5.1, the temperatures of two bars are raised as much as ΔT_I and ΔT_{II} from the ambient one, in which axial deformations are mutually constrained. It is assumed that the two bars are not bent even after elongating and $\Delta T_I > \Delta T_{II}$ and $\alpha_{exI} > \alpha_{exII}$. The amounts of free expansion of bar I and bar II are expressed as $\alpha_{exI} \Delta T_I L$ and $\alpha_{exII} \Delta T_{II} L$. Since the elongation of bar I is larger than that of bar II, bar I is compressed and bar II is elongated. Supposing that the stresses produced in bar I and bar II are σ_I (<0) and σ_{II} (>0), the following equilibrium equation holds with respect to displacements, because the two bars are elongated by the same amount:

$$\alpha_{exI} \Delta T_I L + \frac{\sigma_I}{E_I} L = \alpha_{exII} \Delta T_{II} L + \frac{\sigma_{II}}{E_{II}} L \tag{5.3}$$

where E_I and E_{II} represent Young's modulus of each bar.

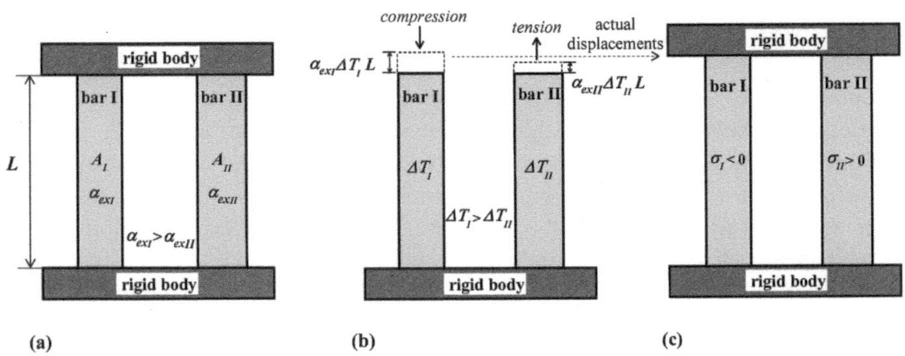

Fig. 5.1 Thermal stress generated in two bars whose deformations are mutually constrained. (A) Initial state, (B) free expansion of two bars, and (C) stresses generated by thermal load.

Denoting the cross-sectional area of each bar by A_I and A_{II}, compression force produced in bar I, $\sigma_I A_I$, is equal in size and opposite in sign to tensile force produced in bar II, $\sigma_{II} A_{II}$:

$$\sigma_I A_I = -\sigma_{II} A_{II} \tag{5.4}$$

Solving Eqs. (5.3) and (5.4) as simultaneous equations, the expressions for σ_I and σ_{II} are derived:

$$\sigma_I = -\frac{\alpha_{exI} E_I \left\{ \Delta T_I - \frac{\alpha_{exII}}{\alpha_{exI}} \Delta T_{II} \right\}}{1 + \frac{A_I E_I}{A_{II} E_{II}}} = -\overline{K} \alpha_{exI} E_I \Delta T_I$$

$$\sigma_{II} = -\frac{A_{II}}{A_I} \sigma_I \tag{5.5}$$

where $\overline{K} = \dfrac{1 - \frac{\alpha_{exII}}{\alpha_{exI}} \frac{\Delta T_{II}}{\Delta T_I}}{1 + \frac{A_I E_I}{A_{II} E_{II}}}$.

The first equation of Eq. (5.5) is given in the form of multiplying $\alpha_{exI} E_I \Delta T_I$ by $\overline{K}$, where the former term represents the thermal stress in bar I constrained at both ends. $\overline{K}$ is termed restraint coefficient.

– Learning by numerical example 5.1: Elongation and thermal stress due to thermal load

Elongations of round bars of 50 mm length, made of carbon steel, stainless steel, and aluminum alloy are compared when the bar temperature is raised by as much as 100 °C from the ambient temperature. Substituting $\Delta T = 100$ °C, $L = 50$ mm, $\alpha_{ex} = 11.8 \times 10^{-6}$, 17.3×10^{-6}, and 23.2×10^{-6} into Eq. (5.1), the amounts of elongation are calculated to be 0.059, 0.087, and 0.116 mm, respectively. Coefficients of linear expansion of each material are the values at room temperature, cited from table 1.5. Thermal stresses produced in the bars, of which both ends are completely constrained, are calculated to be 244, 337, and 169 MPa, respectively, using Young's modulus of 207, 195, and 73 GPa, given in table 1.4. The product of coefficient of linear expansion and Young's modulus, $\alpha_{ex} E$, represents the amount of thermal stress per temperature rise of one degree. Therefore, it is used as an index of the magnitude of thermal stress generated in various materials. Values of $\alpha_{ex} E$ for the three materials are 2.44, 3.37, and 1.69 MPa, which indicates that it is easy to generate large thermal stresses in stainless steels.

5.1.2 Mechanism of bolt force variation due to thermal load

When bolted joints are subjected to thermal load, bolt forces are increased or decreased. The magnitudes of bolt force variation in those cases are sometimes in the critical level to threaten joint safety, or can be at a negligible level. If bolt forces are greatly increased, plastic deformations become a problem in the jointed components, including threaded fasteners. On the contrary, when bolt forces are remarkably decreased, loosening of threaded fasteners matters. Listed in the following are the mechanisms of bolt force variation due to thermal load:

1) Effect of the temperature difference between a bolt-nut pair and fastened plates

The temperatures in a bolted joint show a smooth distribution except at the interfaces, such as nut- and bolt head-bearing surfaces. However, if the temperature difference between a bolt-nut pair and fastened plates exceeds a certain value, bolt force is substantially varied. Approximate values of bolt force variation can be estimated by the mean temperatures of a bolt-nut pair and fastened plates.

2) Effect of the material combination of a bolt-nut pair and fastened plates

Even if the joint temperature is almost entirely uniform, the combination of different materials between a bolt-nut pair and fastened plates produces bolt force variation to a greater or lesser extent. This is because the difference in coefficients of linear expansion α_{ex} between the two bodies generates different thermal expansions, as found from Eq. (5.1). Therefore, even a slight difference can be a problem, because the bolt force variation is proportional to the difference in α_{ex}.

3) Effect of bolt length

The amount of elongation or shrinkage δ due to thermal load is proportional to the original length L of the objective body, as indicated by Eq. (5.1). In actual joints, when longer bolts with identical nominal diameter are used, bolt force variation is increased due to the effect of varying stiffness. The use of long bolts is effective for enhancing the fatigue strength or preventing the loosening of threaded fasteners. In contrast, care should be taken under thermal load because large bolt force variation is expected.

4) Effect of joint temperature variation with time

In many cases, bolt force reaches the maximum in a relatively short time after the start of heating or cooling; then it decreases and approaches a constant value. The magnitude of the maximum bolt force is increased as the thermal conductivity of joint materials becomes small. A typical example is a bolted joint made of stainless steel. The maximum bolt force is also increased with increasing heat flux supplied to the jointed components.

5) Effect of heat flow direction

When jointed portions are subjected to thermal load, the direction of heat flow is perpendicular to or along the bolt axis in many cases. In the former case, a greater or lesser bending moment is generated in addition to bolt force variation, due to the difference in thermal expansion between the upper and lower sides of fastened plates regarding the heat flow direction. When the heat is supplied to the plate surface of nut side or bolt head side, it is an example of the latter case. There is one more importance case: when the heat flows radially into fastened plates toward the bolt axis, the temperature of fastened plates is necessarily higher than the bolt-nut temperature. In this case, if the bolt-nut pair and fastened plates are made of identical materials, the bolt force is increased. However, if the coefficient of linear expansion of the bolt-nut material is larger than that of fastened plates, the bolt force could be reduced as time goes by. For instance, combining a bolt-nut pair made of stainless steel with fastened plates made of carbon steel, the bolt force may be significantly reduced because the coefficient of linear expansion of stainless steel is about 1.5 times larger than that of carbon steel.

6) Effect of stiffness reduction due to temperature dependence of Young's modulus

Young's modulus is usually decreased with increasing temperature, as shown in table 1.4. When bolts, nuts, and fastened plates are all made of identical materials and are heated up to the same temperature, the thermal strains generated in them are also equal, as well as the amounts of elongation. However, bolt force may be decreased due to the reduction in Young's modulus, as predicted by Hooke's law. As an important example, the stiffness of sheet gaskets, widely used in pipe flange connections, is largely reduced at high temperature. Hence, it may cause a significant reduction in bolt forces in the running condition.

7) Effect of the change in environmental conditions around jointed components

When the coefficient of heat transfer on the joint surface is large, it might be the cause of bolt force reduction, because the surface temperature and the temperature gradient inside the joint become changeable. For example, in the case of large machines or structures operating at high temperature and located outside, bolt forces may significantly vary if the surface temperature substantially changes due to rainfall or other reasons.

8) Effect of heat flow through the interface or small clearance gap existing around the joint

Regarding the heat flow through the interface and small clearance gap, its amounts can be quantitatively evaluated by thermal contact coefficient h_c and apparent thermal contact coefficient h_e, explained in section 2.7. Since thermal contact coefficient h_c varies according to thermal conductivity and Vickers hardness of contacting bodies, contact pressure, and surface roughness, the amount of h_c varies depending on the joint materials, surface conditions, and tightening conditions. As a result, when the joint is subjected to thermal load, the contact pressure changes due to the difference in thermal expansion among clamped parts, and hence, the amount of h_c varies with time. Apparent thermal contact coefficient h_e varies with gap size. In bolted joints, h_e becomes an issue at the gap existing around bolt holes and at the clearance flank in engaged threads. The gap size around bolt holes increases in the order of first, second, and third grade of bolt hole diameter d_h. In the case of M16 bolt, d_h is 17, 17.5, and 18.5 mm, respectively; thereby the clearance between the bolt and fastened plates being 0.5, 0.75, and 1.25 mm. This means that the heat becomes easier to transfer as the bolt hole grade is increased. On the other hand, the gap size around the clearance flank is very small, hence, a considerable amount of heat flows through it. Accordingly, the heat flow causes a substantial change in gap size due to the temperature increase around engaged threads.

5.1.3 Simple formula for calculating bolt force variation

In order to evaluate the bolt force variation under thermal load, the introduction of numerical analysis is often required, especially for the transient state where not much time has passed after the start of heating. On the other hand, when the temperature field is in a steady state after a sufficiently long time and the temperature gradients in the bolt-nut pair and fastened plates are not so large, approximate values of bolt force variation can be obtained by utilizing the equations given in Section 5.1.1.

Fig. 5.2 illustrates a bolted joint whose bolt force is changed due to thermal load. Although the temperatures in actual bolted joints usually show complex distribution patterns, some simple formulas are derived for calculating bolt force variation using the mean temperature of each part. Let temperature increases in the bolt-nut pair and fastened plates be ΔT_b and ΔT_f, coefficients of linear expansion α_b and α_f, and Young's moduli E_b and E_f. The amounts of bolt stress variation and compression stress variation in fastened plates are denoted by $\Delta\sigma_b$ and $\Delta\sigma_f$. Bolt stress σ_b is defined by use of bolt cylindrical area A, whose diameter is equal to nominal diameter d. Cross-sectional area of fastened plates is changed according to the joint configuration shown in fig. 1.12. In this section, for simplicity, a thin cylinder is treated, which is uniformly compressed because of its small outer diameter. Equilibrium equations of displacements and forces in the bolt-nut pair and fastened plates can be expressed in the same form as Eqs. (5.3) and (5.4) by denoting grip length as L_f:

$$\begin{aligned} &\alpha_b L_f \Delta T_b + \frac{\Delta\sigma_b}{E_b} L_f = \alpha_f L_f \Delta T_f + \frac{\Delta\sigma_f}{E_f} L_f \\ &\Delta\sigma_b A = -\Delta\sigma_f A_f \end{aligned} \tag{5.6}$$

where A_f represents the cross-sectional area of fastened plates.

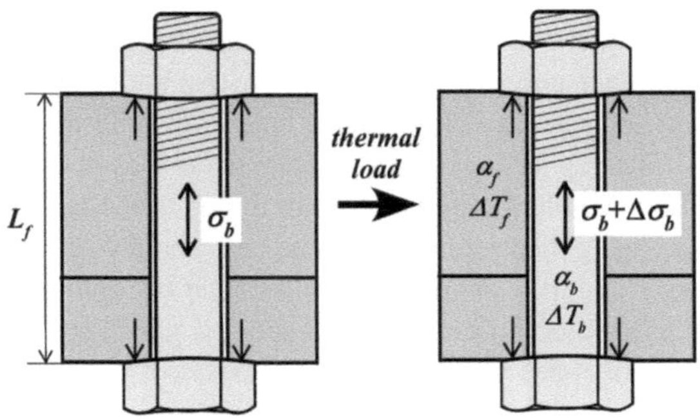

Fig. 5.2 Bolt force variation in a bolted joint subjected to thermal load.

Solving Eq. (5.6) as simultaneous equations, $\Delta\sigma_b$ and $\Delta\sigma_f$ are obtained. Then, AE_b/L_f and A_fE_f/L_f that represent the axial stiffness are replaced by the spring constants of bolt-nut pair and fastened plates, k_b and k_f. Considering that $\Delta\sigma_bA$ is equal to bolt force variation ΔF_b, the following equation is derived;

$$\Delta F_b = -\frac{(\alpha_b\Delta T_b - \alpha_f\Delta T_f)L_f}{\frac{1}{k_b} + \frac{1}{k_f}} \tag{5.7}$$

Various assumptions have been made to derive Eq. (5.7). However, it can provide approximate values of bolt force variation once the mean temperature variations of the bolt-nut pair and fastened plates under thermal load and their material constants are given. When the bolt-nut pair and fastened plates are made of an identical material and their coefficients of linear expansion are also equal, Eq. (5.7) is rewritten as follows by equating α_f to α_b:

$$\Delta F_b = -\frac{\alpha_b(\Delta T_b - \Delta T_f)L_f}{\frac{1}{k_b} + \frac{1}{k_f}} \tag{5.8}$$

When the bolt-nut pair and fastened plates are made of different materials and the difference in temperature between them is small, Eq. (5.7) is rewritten as follows by substituting $\Delta T_b = \Delta T_f = \Delta T$:

$$\Delta F_b = -\frac{(\alpha_b - \alpha_f)\Delta T L_f}{\frac{1}{k_b} + \frac{1}{k_f}} \tag{5.9}$$

In many cases, bolts, nuts, and fastened plates are made of identical or similar materials. In which case, the difference in temperature between the bolt-nut pair and fastened plates is commonly rather small in steady state. Referring to Eqs. (5.7)–(5.9), the mechanical behavior of bolted joints under thermal load is summarized as follows, regarding bolt force variation:

(1) Bolt force variation ΔF_b is proportional to the difference in coefficients of linear expansion between a bolt-nut pair and fastened plates.
(2) Since ΔF_b is in proportion to ΔT, which is the temperature variation from the initial temperature, bolt force variation is increased as the temperature rise or drop gets larger.
(3) As for the effect of grip length L_f, ΔF_b is apparently increased in proportion to L_f. However, the rate of variation is actually lowered because spring constants of a bolt-nut pair and fastened plates, k_b and k_f, are decreased due to the increase of L_f.

As an example of case (1), combining a bot-nut pair made of carbon steel and fastened plates made of austenitic stainless steel, bolt force F_b is increased, $\Delta F_b > 0$, due to $\alpha_b < \alpha_f$ when the joint temperature is raised. On the contrary, F_b is decreased, $\Delta F_b < 0$, when cooled. In the case of a bolt-nut pair and fastened plates being made of different materials, the bolt force variation is proportional to the difference in coefficients of linear expansion between them, according to Eq. (5.9). This relationship suggests the difficulty in constructing a joint with different materials. Hence, when combining different materials in a bolted joint, it is highly recommended to use accurate values of the coefficient of linear expansion. This situation is quite different from the case in which there exists a difference in Young's modulus in connection with Hooke's law.

For instance, comparing the cases of Young's modulus being 200 and 206 GPa, the difference in stress is about 3% in elastic analysis. In contrast, only a slight difference between α_b and α_f can produce large bolt force variation ΔF_b, because it is proportional to the product of the difference $(\alpha_b - \alpha_f)$ times temperature increase ΔT. It is therefore concluded that more attention should be paid in determining the values of α_b and α_f than those of Young's modulus. Regarding the effect of grip length explained in (3), both fatigue strength and antiloosening performance are enhanced with larger grip length. On the other hand, it is also important to remember that larger grip length produces larger bolt force variation under thermal load.

- Learning by numerical example 5.2: Effect of difference in coefficient of linear expansion on bolt stress variation

For a bolt-nut connection of M16 with coarse thread, bolt stress variation $\Delta\sigma_b$ is calculated for the cases of the ratio of grip length to nominal diameter, L_f/d, being changed to 2, 4, 6, 8, and 10. Other conditions are the same as those in numerical example 2.1.

A bolt-nut pair and fastened plates are made of carbon steel and stainless steel, respectively. Temperature increase from the ambient ΔT is supposed to be 100 °C. Material constants and thermal properties at room temperature are used referring to tables 1.4 and 1.5. Young's moduli are 207 and 195 GPa for carbon steel and stainless steel, respectively, and coefficients of linear expansion are 11.8×10^{-6} and 17.3×10^{-6}. Substituting $\Delta T = 100$ °C into Eq. (5.9) and using the foregoing values, bolt force variation ΔF_b is calculated and then $\Delta\sigma_b$ is obtained. The resulting values of $\Delta\sigma_b$ are found to be 45.9, 55.0, 58.9, 61.0, and 62.4 MPa in ascending order of L_f/d. If the material combination is switched, bolt stress is naturally reduced by the same amount. The rate of bolt stress variation is decreased as grip length L_f increases. The mechanism can be explained by considering the relationship between grip length and the two spring constants, k_b and k_f. In addition, for bolted joints with similar form, the amount of $\Delta\sigma_b$ is hardly influenced by nominal diameter of bolts.

- Learning by numerical example 5.3: Effect of temperature difference among jointed components on bolt stress variation

A bolt-nut pair and fastened plates are all made of carbon steel. Mean temperature of the bolt-nut pair is supposed to be 10 °C lower than that of the fastened plates. Under the same conditions as those in numerical example 5.2, bolt stress variation $\Delta\sigma_b$ is calculated by substituting -10 °C into the term representing the temperature difference in Eq. (5.8). Young's modulus is set to be 207 GPa. The resulting values of $\Delta\sigma_b$ are found to be 10.0, 12.0, 12.8, 13.3, and 13.6 MPa in ascending order of L_f/d. $\Delta\sigma_b$ is increased with increasing grip length, but the rate of increase is gradually decreased.

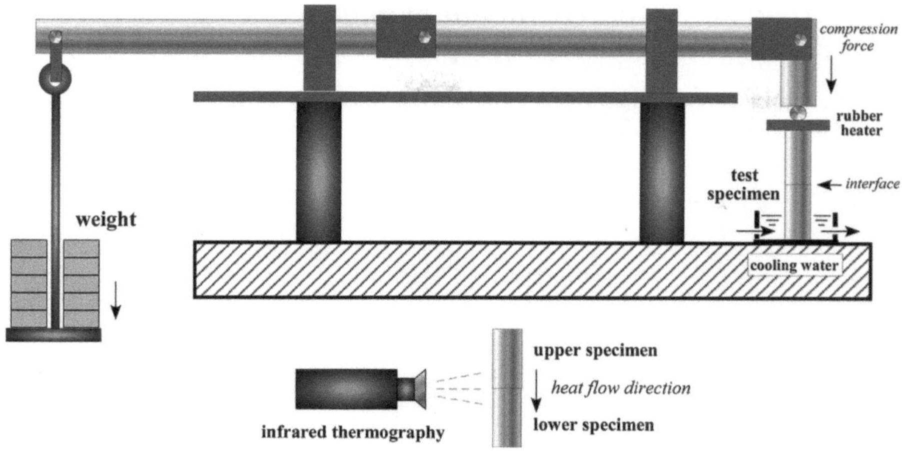

Fig. 5.3 Experimental setup for measuring thermal contact coefficient.

5.2 Evaluation method of amount of heat transferred through contact surface

5.2.1 Measuring method of thermal contact coefficient

Thermal contact coefficient h_c (W/m^{2}K) represents the amount of heat flux q (W/m^2) flowing through the contact surface per unit temperature difference, which has the same dimension as coefficient of heat transfer:

$$h_c = \frac{q}{\Delta T} \tag{5.10}$$

To evaluate h_c, it is necessary to measure heat flux q and temperature difference ΔT between mating surfaces. Fig. 5.3 shows a measuring device for thermal contact coefficient. Utilizing a lever mechanism, two round-bar test specimens are pressed at constant pressure, and the upper end surface of the upper specimen is heated by a rubber heater. Temperature distributions around the interface between the two specimens are measured by infrared thermography. ΔT is obtained by extrapolating the measured surface temperatures of both specimens to the interface. Using the temperature gradient evaluated by the temperature distributions, heat flux q is calculated with the help of Fourier's law. The calculation procedure is explained in detail in [121]. In the case of the two specimens being made of identical materials, evaluated is the thermal contact coefficient at the interface of identical materials [122]. In the case of the combination of two different materials, thermal contact coefficient at the interface of dissimilar materials [121] can be evaluated in the same manner. Mating surfaces of the two specimens are machined so as to achieve common values of arithmetic mean roughness R_a, 0.8, 1.6, 3.2, and 6.4 μm. Contact pressure at the interface is raised to about 50 MPa.

5.2.2 Thermal contact coefficient at interface composed of identical materials [122]

By conducting experiments for commonly used materials for threaded fasteners and jointed portions, i.e., carbon steel (S45C, SS400), stainless steel (SUS304), aluminum alloy (A2024), and pure aluminum (A1050), the following equation has been derived for evaluating thermal contact coefficient at an interface composed of identical materials:

$$h_c = 10^5\left[c_1\lambda\frac{\left(\frac{p_n}{Hv}\right)^{2/3}}{Rat^{m_h}} + \frac{c_2}{Rat^{n_h}}\right] \tag{5.11}$$

$$Rat = Ra_1 + Ra_2$$
$$c_1 = 0.06\in[0.055, 0.065]$$
$$c_2 = 0.09\in[0.085, 0.095]$$
$$m_h = 0.8\in[0.8, 0.9]$$
$$n_h = 0.7\in[0.7, 0.8]$$

In Eq. (5.11), λ, p_n, and Hv represent thermal conductivity (W/mK), contact pressure (MPa), and Vickers hardness, respectively. Ra_t is the sum of arithmetic mean roughness of mating surfaces, Ra_1 and Ra_2. c_1, c_2, m_h, and n_h are related constants. Eq. (5.11) indicates that thermal contact coefficient at the interface composed of identical materials h_c is proportional to two-thirds power of the ratio of contact pressure to Vickers hardness, p_n/Hv, and it is directly proportional to thermal conductivity λ. It also shows that h_c is decreased as surface roughness increases, which makes the heat less likely to be transmitted. Eq. (5.11) is also applicable to an interface composed of pure titanium (TB340), titanium alloy (Ti6Al4V), and heat-resistant nickel alloy (Inconel 600) [25].

Since thread surface, nut-bearing surface, and bolt head-bearing surface in a bolted joint are subjected to high contact pressure, thermal contact coefficients there become high. As for the interface existing in fastened plates, contact pressure decreases radially outwards with distance from the bolt axis, according to the concept of the pressure cone. There is one important point concerning h_c: the distribution patterns of contact pressure at the interface in bolt-nut connections are quite different from those at the interface adjacent to body-side engaged threads. When subjected to thermal load, bolt force is changed and the contact pressure is varied correspondingly. Hence, when evaluating thermal and mechanical behaviors of bolted joints by FEM, it is possible to take account of the effect of contact pressure varying with heating or cooling time using Eq. (5.11), because h_c can be expressed as a function of contact pressure p_n.

5.2.3 Thermal contact coefficient at interface composed of dissimilar materials [121]

In bolted joints, different materials are sometimes used for threaded fasteners and fastened plates. To cope with such cases, research on thermal contact resistance at interfaces composed of dissimilar materials was performed intensively from the late 1950s to the 1960s. It has been reported that the amount of thermal contact resistance varies depending on the heat flow direction even for the same combination of contacting materials [123]. However, few studies have been reported that propose an equation for quantitatively evaluating the thermal contact resistance. Accordingly, for an interface composed of commonly used engineering materials, i.e., carbon steel, stainless steel, and aluminum alloy, thermal contact coefficient h_c at the interface composed of dissimilar materials is measured by the experimental setup shown in Fig. 5.3. Using the experimental results, the following empirical equation is proposed in the same form as Eq. (5.11):

$$h_c = 10^5 \left[c_1 \lambda_1 \frac{\left(\frac{p_n}{Hv_1} \right)^{2/3}}{Rat^{m_h}} + c_2 \lambda_2 \frac{\left(\frac{p_n}{Hv_2} \right)^{2/3}}{Rat^{m_h}} + \frac{c_3}{Rat^{n_h}} \right] \tag{5.12}$$

$$\begin{aligned} Rat &= Ra_1 + Ra_2 \\ c_1 &= 0.01 \in [0.008, 0.017] \\ c_2 &= 0.025 \in [0.013, 0.036] \\ c_3 &= 0.15 \in [0.1, 0.2] \\ m_h &= 1.0 \in [0.7, 1.1] \\ n_h &= 0.8 \end{aligned}$$

The symbols are commonly used with Eq. (5.11). Two different materials are identified by adding subscripts “1” and “2” to thermal conductivity λ and Vickers hardness Hv. The first term in the bracket is related to the material at higher temperature and the second one is at lower temperature.

Fig. 5.4 shows the measurement results for the combination of carbon steel and stainless steel. It is found that thermal contact coefficient h_c for the heat flow from carbon steel to stainless steel is lower compared to the case of the heat flow in the reverse direction. This situation can be generalized as the heat flow from higher thermal conductivity materials to lower ones. Therefore, it can be concluded that thermal contact coefficient at an interface composed of dissimilar materials becomes higher when the material on the low temperature side has higher thermal conductivity. Eq. (5.12) is also applicable to cases where pure titanium or titanium alloy is combined with carbon steel or stainless steel [25].

5.3 Evaluation method of amount of heat transferred through a small gap [124]

As explained in section 2.7, a small gap exists between the bolt cylindrical portion and the bolt hole surface of fastened plates, ranging from about 0.2 to 5 mm. The gap size around the clearance flank of engaged threads, at which thread surfaces of male and female threads are not in contact, is even smaller. In any case, those mating surfaces cannot be treated as an insulated boundary and a certain amount of heat is transmitted, although they are separated by a thin air layer. Fig. 5.5 illustrates the measuring

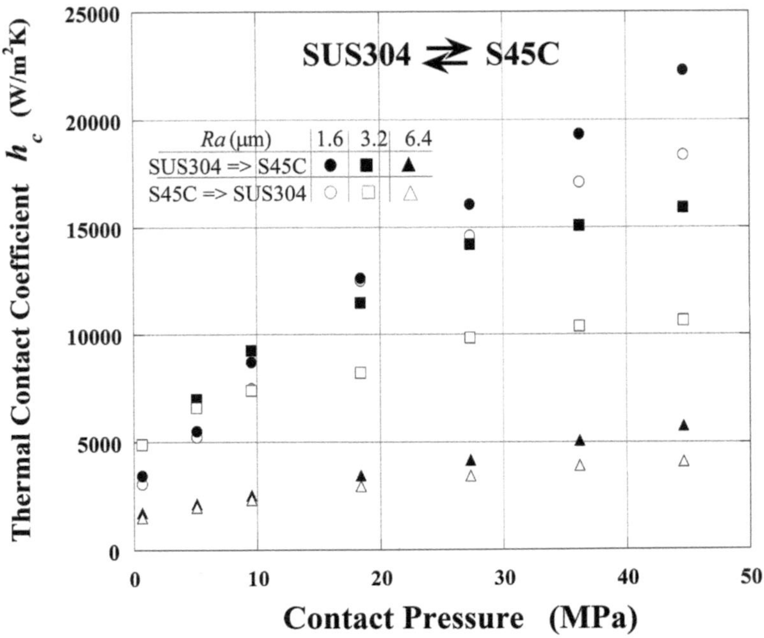

Fig. 5.4 Thermal contact coefficient at an interface composed of dissimilar materials.

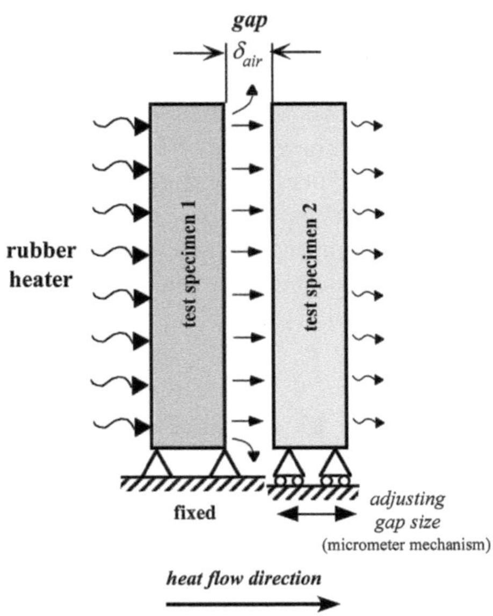

Fig. 5.5 Experimental setup for measuring apparent thermal contact coefficient.

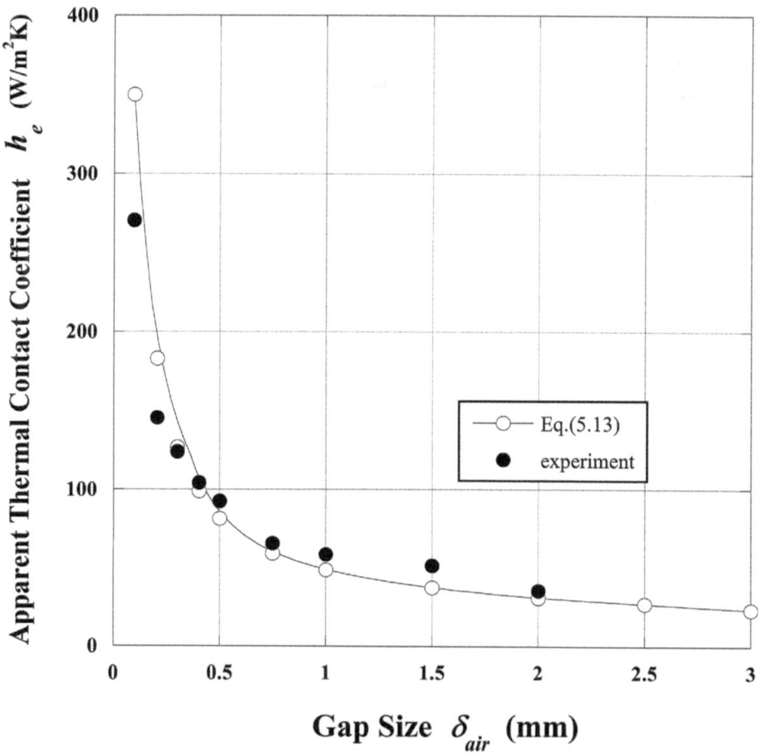

Fig. 5.6 Relationship between apparent thermal contact coefficient and gap size.

method of the heat flow through a small gap. The gap size is finely adjusted utilizing a micrometer mechanism. A rubber heater is mounted on the outer surface of the left test specimen, and the measurement is carried out at various levels of gap size. Apparent thermal contact coefficient h_e is calculated as the amount of heat flux per unit temperature difference between the two plates. It is obtained by the balance of the heat flowing through the gap and the heat dissipated into the environment.

Fig. 5.6 shows the experimental results: h_e increases as the gap size is decreased. This tendency becomes remarkable when the gap size approaches as small as 1 mm. In the region of the gap size being less than 0.5 mm, h_e shows a steep increase. The figure also shows that some amount of heat still flows through the gap at about 3 mm. Assuming that the heat flow consists of heat conduction through the air layer, convection, and radiation, h_e is expressed by the following equation:

$$h_e = \frac{\lambda_{air}}{\delta_{air}} + h_{cv} + h_r \tag{5.13}$$

where λ_{air} and δ_{air} represent thermal conductivity of air and the gap size, respectively.

The coefficient of convection heat transfer h_{cv} is measured to be 25 W/m^{2}K. The coefficient of radiation heat transfer h_r [124] can be calculated using the emissivity of the object, etc. It is found from Fig. 5.6 that the proposed equation, Eq. (5.13), approximates the experimental results fairly well.

5.4 Thermal contact coefficient and apparent thermal contact coefficient in bolted joints

Fig. 5.7 schematically shows the heat flow in a bolted joint subjected to thermal load. Thermal contact coefficient h_c and apparent thermal contact coefficient h_e become an issue at the interface listed below:

(1) Thermal contact coefficient h_c: nut-bearing surface, bolt head-bearing surface, plate interface, contact surface of pressure flank.
(2) Apparent thermal contact coefficient h_e: around bolt hole, around clearance flank, plate interface outside pressure cone.

Heat flow through a plate interface under contact pressure is evaluated by h_c. Meanwhile, at an interface that is lightly in contact, the amount of heat flow is evaluated by h_c or h_e, depending on whether the gap exists or not. In the case of gap size being regarded as zero, zero contact pressure is substituted into the empirical equation of h_c, since h_e becomes infinite according to Eq. (5.13). As shown in Eqs. (5.11) and (5.12), thermal contact coefficient h_c increases with increasing contact pressure. Mean contact pressures on nut- and bolt head-bearing surfaces range from about 60% to 80% of the bolt stress, referring to the area ratio of bolt cylindrical portion to nut-bearing surface, depicted in fig. 3.32. Therefore, h_c becomes very large, i.e., small thermal contact resistance. However, the effect on the total heat flow is limited because of its small area.

In the engaged thread portion, thread contact pressure increases toward the nut-bearing surface from the top face of the nut, but the contact area is also small.

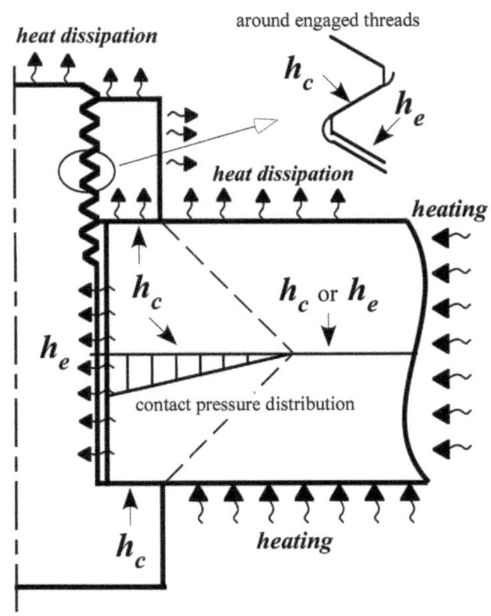

Fig. 5.7 Heat flow in a bolted joint subjected to thermal load.

In contrast, contact pressure at the plate interface is rather low because of its large contact area. In that case, thermal contact resistance there is increased with distance from the bolt axis, due to the specific contact pressure distribution illustrated in Fig. 5.7, thereby significantly affecting the temperature distribution. The point to notice is that a certain amount of heat flows even outside the pressure cone. This occurs between plates that are lightly in contact or where small gaps exist due to the warpage of plates. Furthermore, since h_c is proportional to thermal conductivity of contacting materials, it has a large effect on temperature distributions in bolted joints made of materials with low thermal conductivity like stainless steel. Another important factor is surface roughness. As surface roughness is increased, h_c decreases and the effect on temperature distributions reveals itself remarkably.

Apparent thermal contact coefficient h_e sharply increases with decreasing gap size, as shown in Fig. 5.6. Bolt hole diameter d_h of first, second, and third grades changes approximately from 0.2 to 5 mm for bolts with nominal diameter ranging from M6 to M64, as stated in section 2.7. This implies that a fair amount of heat flows through the gap around the bolt hole when small bolts are used with a small bolt hole. As for clearance flank in engaged threads, the gap size is very small and is varied when subjected to thermal load. If small clearance exists at the plate interface outside the pressure cone, due to plate deformations, the introduction of h_e may be required, depending on the heat flow direction.

In the tightening operation using a bolt heater, explained in section 3.5, the amounts of heat flow through nut- and bolt head-bearing surfaces can be evaluated by thermal contact coefficient h_c to estimate the total heat flow from the bolt to fastened plates. In this case, h_c becomes very large due to high contact pressure, but its effect is limited because the total amount of heat is small owing to the small contact area. The amount of heat flow from bolt body to bolt hole surface of fastened plates is calculated by apparent thermal contact coefficient h_e. Although the value of h_e is considerably smaller than that of h_c, the total amount of heat flow becomes fairly large, because the related surface areas are large. The tightening process by use of a bolt heater can be analyzed in detail by incorporating h_c and h_e into FE formulation [81]. On the other hand, if the target is a bolted joint with large grip length and the objective is to estimate the approximate value of bolt force to be generated, it can be achieved by a simplified analysis, as explained in section 3.5.2 [82], in which only the heat flow through a small gap is considered using apparent thermal contact coefficient h_e.

5.5 Analysis of thermal and mechanical behaviors of bolted joints by FEM

5.5.1 Evaluation of thermal and mechanical behaviors using an axisymmetric FE model

In order to evaluate the behavior of a bolted joint subjected to thermal load in detail, it is necessary to perform three-dimensional analysis. On the other hand, if heat flow is almost symmetric around the bolt axis, it can be dealt with by axisymmetric analysis.

Moreover, when aiming at the collection of fundamental data necessary in the design process, such as bolt force variation and heat flow in the jointed portion, axisymmetric analysis could be an effective means.

In this section, the target is a bolted joint of cylindrical shape clamped with a single bolt, which is subjected to axisymmetric heat flow inwards from the outer surface. How the difference in material properties of the joint materials to be used and surface roughness affect the bolt force variation with time and the joint temperature distribution is estimated. Nominal diameter of a bolt-nut pair is M16. Fastened plates are supposed to be a single hollow cylinder. The grip length is 48 mm, the outer diameter is 96 mm, and the inner diameter is 17.5 mm, corresponding to a bolt hole diameter of second grade. Initial bolt stress is 100 MPa and the total heat of 100 W is supplied, as a uniform heat flux, from the side surface of the hollow cylinder. The other surfaces exposed to the ambient temperature are assumed to be a convection boundary with coefficient of heat transfer of 5 W/m^{2}K. The amounts of arithmetic mean roughness Ra on thread surface and nut- and bolt head-bearing surfaces are set to be equal, and Ra is changed to 1.6, 3.2, and 6.4 μm, where 3.2 μm is the standard condition. The effect of thermal contact resistance is considered by means of thermal contact coefficient h_c, given in Eq. (5.11), for thread surface and nut- and bolt head-bearing surfaces. Regarding the clearance flank in engaged threads and the small gap between bolt cylindrical portion and bolt hole surface of fastened plates, apparent thermal contact coefficient, given in Eq. (5.13), is used instead of h_c.

Fig. 5.8 shows the temperature distributions in bolted joints made of carbon steel S45C, stainless steel SUS304, and aluminum alloy A2024 after heating for 60 min. Three-digit numbers in the figure represent the temperatures for each isothermal line. In the numerical analysis, material constants at room temperature are used, given in tables 1.4 and 1.5. The differences between the maximum and minimum temperatures are 8 °C in aluminum alloy, 12 °C in carbon steel, and 33 °C in stainless steel. The effect of the difference in thermal conductivity emerges markedly. In Fig. 5.9, the effect of the joint material on bolt stress variation with time is depicted by dividing bolt stress σ_b at each time by initial bolt stress σ_{bi} of 100 MPa. Comparing to the temperature distributions in Fig. 5.8, it is found that the joint materials with lower thermal

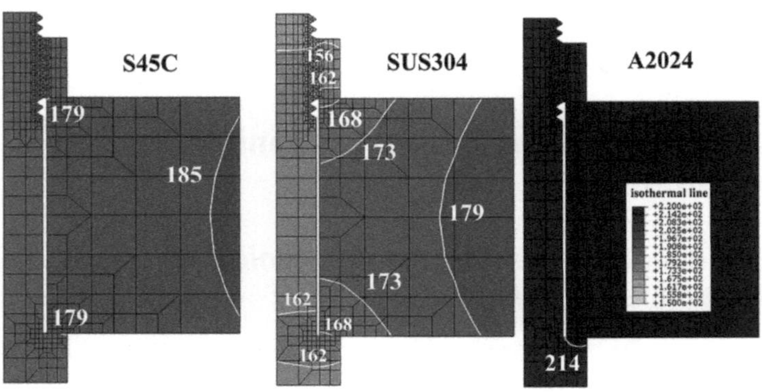

Fig. 5.8 Comparison of temperature distributions in bolted joints made of different materials.

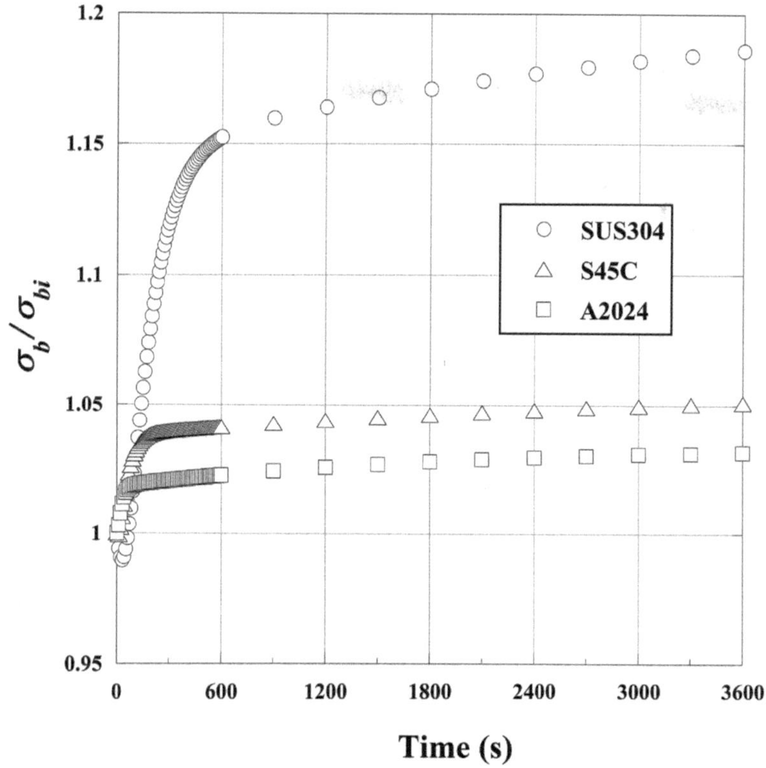

Fig. 5.9 Effect of joint material on bolt stress variation with time.

conductivity show larger bolt stress variation, because low thermal conductivity produces large temperature difference between the bolt-nut pair and fastened plates. In Fig. 5.10, the effect of surface roughness Ra is shown for joints made of carbon steel. As Ra is increased, thermal contact coefficient h_c at the interface is decreased. Thus, the temperature difference between the bolt-nut pair and fastened plates increases and the bolt stress is somewhat increased.

Fig. 5.11 shows the effect of the difference in coefficient of linear expansion between the bolt-nut pair and fastened plates. By way of comparison, the results for carbon steel, shown in Fig. 5.9, are also depicted in the figure. It shows that only a slight reduction in the coefficient of linear expansion of the bolt-nut pair, from 11.8×10^{-6} to 11×10^{-6} and 10×10^{-6}, causes a considerable increase in bolt stress. In contrast, if the coefficient of the fastened plates is relatively small, bolt stress is significantly reduced. This figure makes an important point that when a bolted joint composed of different materials having fairly different coefficients of linear expansion—such as the combination of carbon steel and stainless steel—is subjected to thermal load, it may cause a significant loss in bolt force or a denting on nut- and bolt head-bearing surfaces. Even if the bolt-nut pair and fastened plates are both made

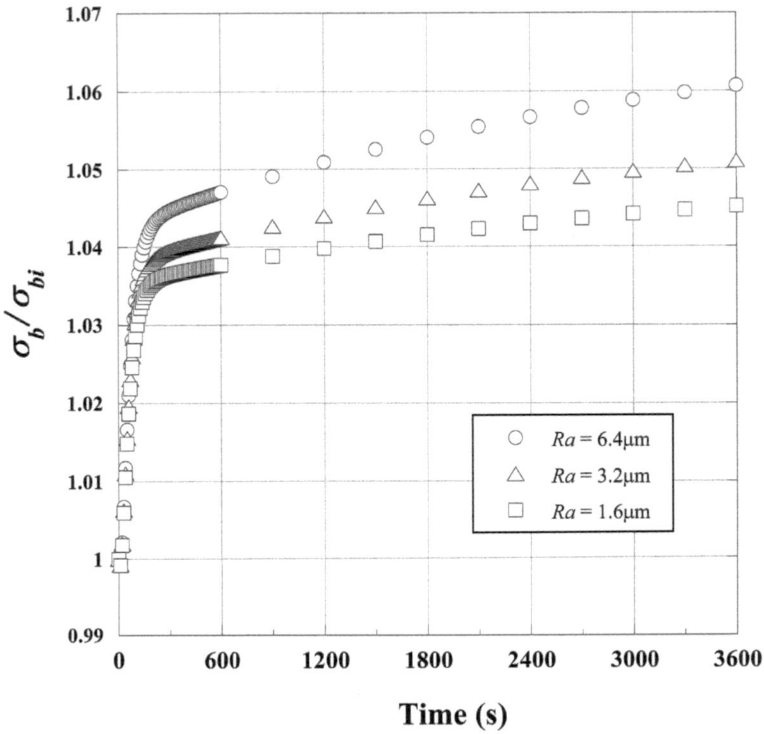

Fig. 5.10 Effect of surface roughness on bolt stress variation with time.

of carbon steel-based materials, their coefficients of linear expansion are not exactly the same. Therefore, as shown in Fig. 5.11, a small difference in coefficient of linear expansion will inevitably exist and this produces a certain amount of bolt force variation, even if the temperature of the jointed portion is uniformly elevated. This phenomenon can be easily predicted from Eqs. (5.7) and (5.9) in Section 5.1.3 and a calculation example treated in numerical example 5.2.

Meanwhile, threaded fasteners used in electronic components and their related products are often subjected to thermal load. In these cases, the material of threaded fasteners is sometimes different from that of fastened plates due to various reasons, and this may significantly change the bolt force in the running condition, owing to the difference in coefficients of linear expansion, even if the temperature increase of the target joint is not so large. Fig. 5.11 would be of assistance to avoid such an unfavorable situation. Approximate values of bolt force variation due to thermal load can be estimated using Eqs. (5.7)–(5.9).

In connection with the foregoing discussion, bolt force reduction due to thermal load can be mitigated by making positive use of the difference in coefficient of linear expansion between a bolt-nut pair and fastened plates. For example, titanium alloys are a good candidate because of high strength, though being expensive and easily

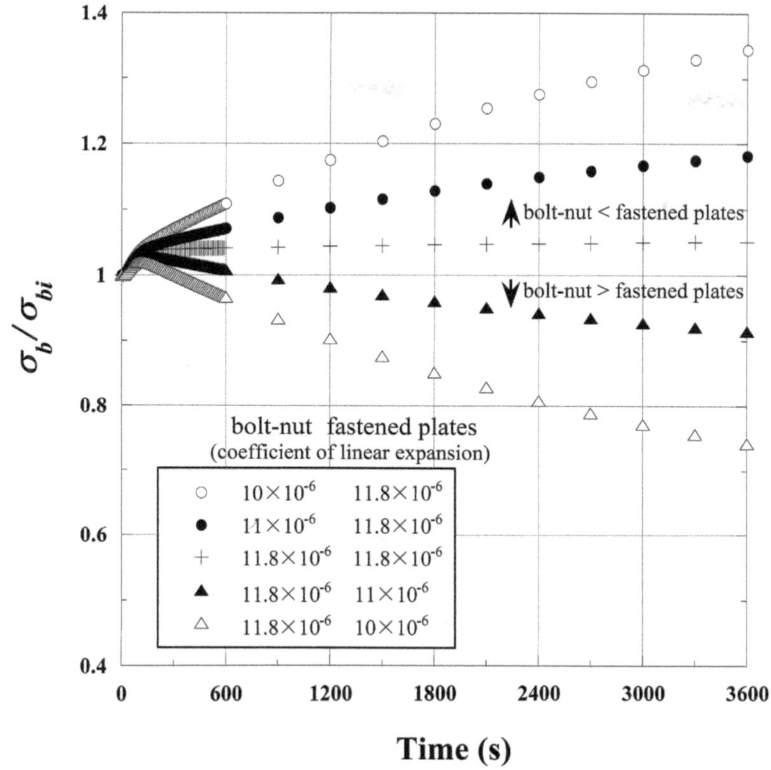

Fig. 5.11 Effect of the difference in coefficient of linear expansion on bolt stress variation with time.

seized. Titanium alloy's coefficient of linear expansion is considerably lower than that of carbon steel, as given in table 1.5. Therefore, the bolt force reduction can be suppressed by employing a bolt-nut pair made of titanium alloys as a substitute for carbon steel bolts and nuts in the target joint, where loosening is likely to occur due to thermal load. It has been reported in [25] that for fastened plates made of carbon steel or stainless steel, the use of a bolt-nut pair made of titanium or titanium alloys can lower the bolt force reduction in the cases of axisymmetric heat flow treated here and three-dimensional heat flow treated in the next section.

5.5.2 Evaluation of thermal and mechanical behaviors using a three-dimensional FE model

Actual bolted joints are rarely subjected to axisymmetric thermal load. In many cases, heat flow direction is perpendicular to the bolt axis, and consequently a certain amount of bending moment is produced in the bolt. Therefore, it is necessary to evaluate the variation of bending moment with time, in addition to bolt stress variation. Moreover,

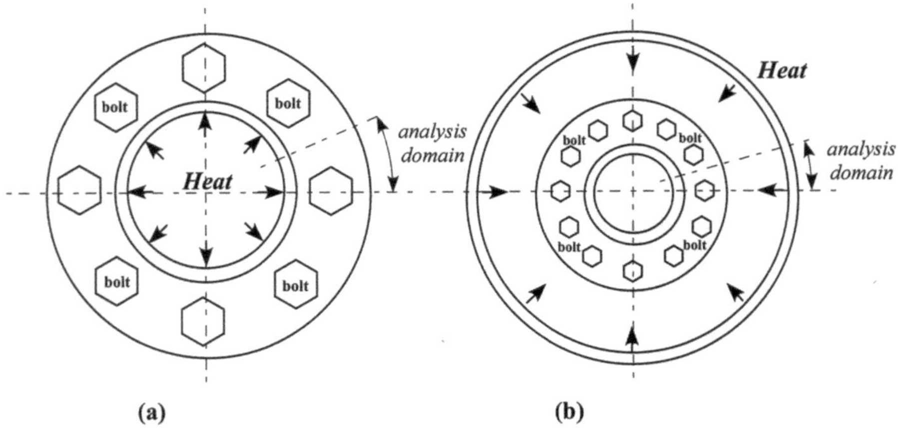

Fig. 5.12 Heat flow pattern in actual bolted joints under thermal load. (A) Pipe flange and (B) disk brake.

it is useful from the viewpoint of machine design to clarify the rate of heat flow through each part of a bolted joint.

Fig. 5.12 illustrates an example of a bolted joint under thermal load [125]. It shows the heat flow patterns around bolted joints that clamp pipe flanges and disk brakes. In the case of pipe flange connections, the heat supplied from the inner surface by contained fluids flows radially outwards. In contrast, in the case of disk brakes, the heat generated on the brake pad surface flows radially inwards. Considering the geometric symmetry, temperature and stress fields can be evaluated by setting the fan-shaped area as the analysis domain. Fig. 5.13 shows a rectangular parallelepiped model for evaluating the thermal behavior of a bolted joint, in which heat flow direction is assumed to be perpendicular to the bolt axis, aiming at covering commonly encountered cases. When the number of clamping bolts is increased and the sector angle decreases, the models in Fig. 5.12 reduce to this rectangular parallelepiped one.

The heat flowing into the domain from the left surface, which is uniformly heated, is dissipated by convection from the right surface. The upper and lower surfaces are assumed to be insulated in order to coincide with the fan-shaped domain, shown in Fig. 5.12. In the figure, Q_{total} denotes the total amount of heat that flows through the bolt cross section including the fastened plates. Q_{th}, Q_{shk}, Q_{hd}, and Q_f represent the amounts of heat flowing through engaged threads, bolt cylindrical portion, bolt head, and fastened plates, respectively. Assuming that the surfaces are insulated boundaries except for the heat supplied and heat dissipation surfaces, the ratios of Q_{th}, Q_{shk}, Q_{hd}, and Q_f to Q_{total} add up to unity.

$$\frac{Q_{th} + Q_{shk} + Q_{hd} + Q_f}{Q_{total}} = 1 \tag{5.14}$$

Fig. 5.14 shows an example of an FE model. Considering the symmetry of temperature and stress fields with respect to heat flow direction, one half of the objective bolted joint is modeled. The target bolt is M16. As for the block size, the ratio of grip length L_f

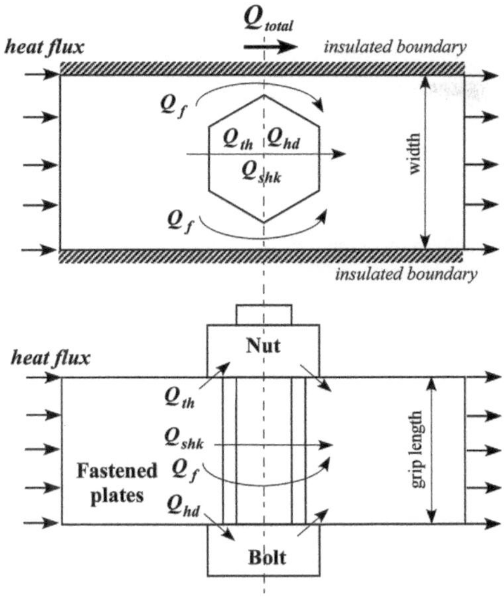

Fig. 5.13 Heat flow proportion in a bolted joint under thermal load.

to nominal diameter d, L_f/d, is changed to 4, 5, and 6. The block width is twice the nominal diameter. The block and a bolt-nut pair are supposed to be made of identical materials, and three kinds of materials are selected, i.e., carbon steel S45C, stainless steel SUS304, and aluminum alloy A2024. When performing numerical analysis, material constants given in [125] are used, and the values are slightly different from those given in tables 1.4 and 1.5. Initial bolt stress is 100 MPa and the coefficient of friction is set to be 0.15 on all contact surfaces. Surface roughness Ra is 3.2 μm. $L_f/d = 4$ is set to be the standard condition. The intensity of heat flux is 11.74 W/m^{2}K, corresponding to the case of applying total heat of 24 W to the standard model. Coefficient of heat transfer at heat dissipation surface is set to be 25 W/m^{2}K and the ambient temperature is 20 °C.

Fig. 5.15A and B represent the variation of the heat flow proportion through each part for joint materials of carbon steel S45C and stainless steel SUS304. Just after the start of heating, most of the total heat Q_{total} flows through fastened plates, designated by Q_f. Then, the heat flow proportions through engaged threads Q_{th}, bolt cylindrical portion Q_{shk}, and bolt head Q_{hd} gradually increase. After a certain amount of time, each heat flow proportion reaches a constant value. No significant difference is observed in the heat flow proportions between carbon steel and stainless steel, but the latter case takes a fairly long time until reaching a steady state, due to its low thermal conductivity. Fig. 5.16 shows the bolt stress variation with time for three kinds of joint materials. The ordinate represents the bolt stress σ_b at each time divided by the initial bolt stress σ_{bi}. Unlike the case of axisymmetric heating, the bolt stress σ_b reaches a peak in a relatively short time after the start of heating. Then, it decreases and becomes constant. Since the heat flow direction in general structures or machines is perpendicular to the bolt axis in many cases, it is considered that the bolt stress sometimes varies with time in the manner shown in Fig. 5.16. Due to the difference

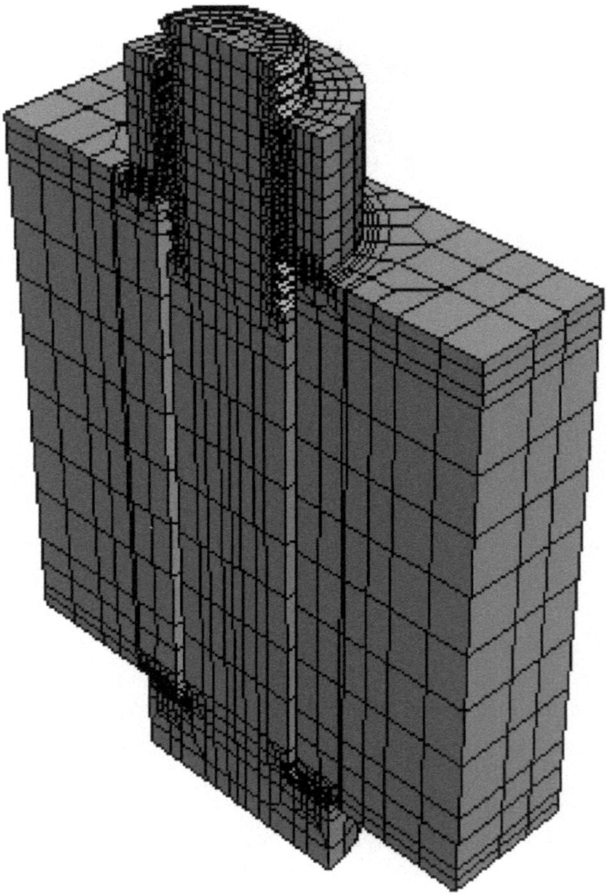

Fig. 5.14 Example of an FE model for evaluating thermal behavior of a bolted joint.

in temperature between the bolt-nut pair and fastened plates, the joint materials with lower thermal conductivity produce higher peaks of bolt stress σ_b.

When the heat flows perpendicular to the bolt axis, the bolt is deformed in a convex shape toward the upstream side, due to the difference in thermal expansion between the upstream and downstream sides. The magnitude of bending stress generated by this mechanism is important from the viewpoint of joint safety when subjected to thermal load. Fig. 5.17 shows the variation of bending stress with time for the standard condition defined above. The parameter in the figure is the ratio of grip length L_f to nominal diameter d. For stainless steel SUS304, numerical results for $L_f/d=6$ are also shown. The ordinate represents the bending stress σ_{bnd} at each time divided by the initial bolt stress σ_{bi}. Although not shown in Fig. 5.16, bolt stress σ_b slightly decreases as grip length L_f is decreased. On the contrary, the bending stress σ_{bnd} shows a considerable increase as grip length is decreased. Furthermore, the absolute value of bending stress is larger than the bolt stress variation shown in Fig. 5.16. In particular,

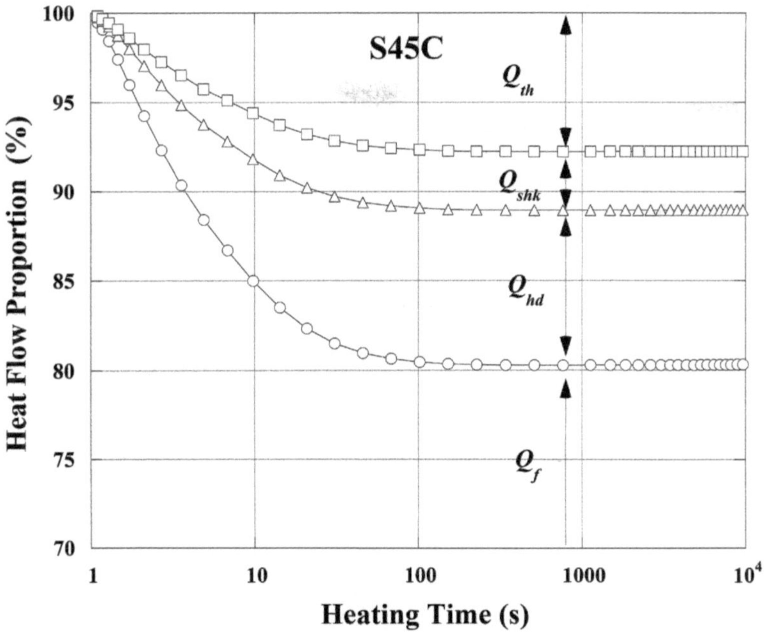

(a)

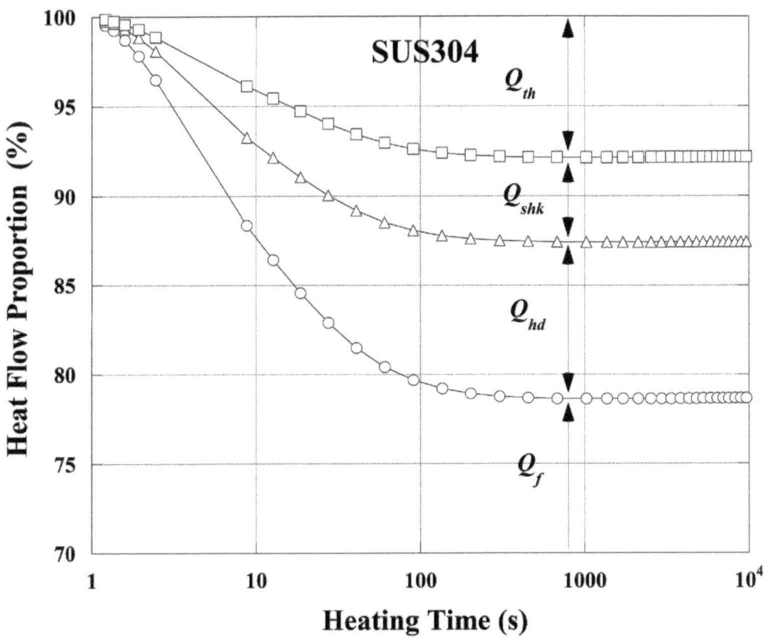

(b)

Fig. 5.15 Variations of heat flow proportion through each part of a bolted joint.***

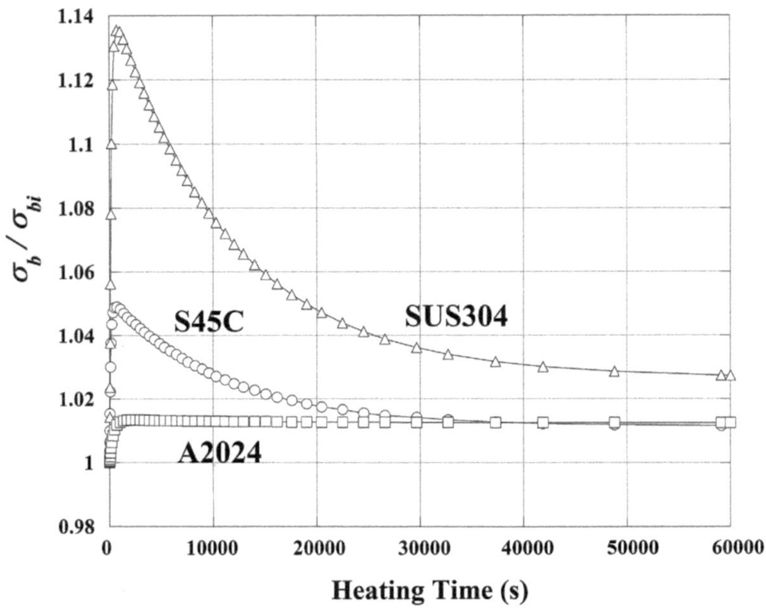

Fig. 5.16 Effect of joint material on bolt stress variation with time.

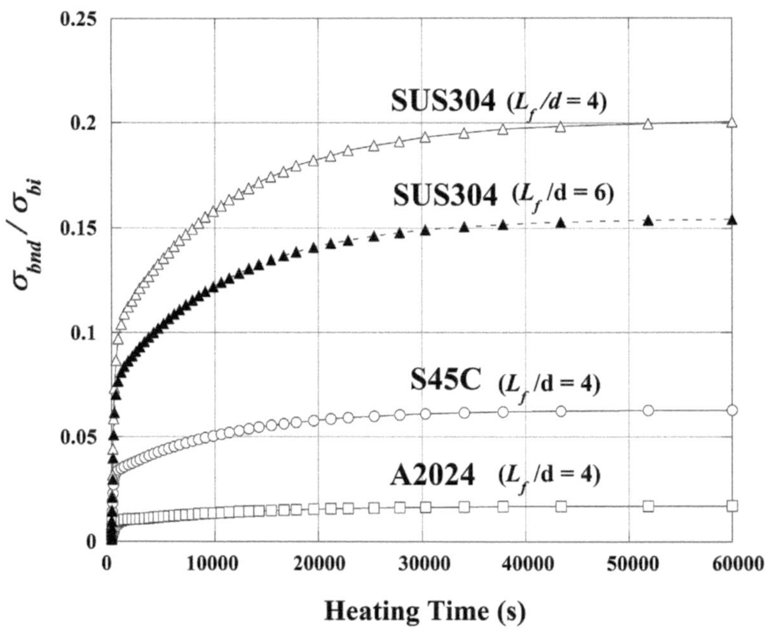

Fig. 5.17 Effect of joint material on bending stress variation with time.

large bending stress is produced in the joint made of stainless steel with small grip length. That is due to a large temperature difference between the upstream and downstream sides. Accordingly, it is concluded that attention should be paid when materials with low thermal conductivity are used in a joint with small grip length under thermal load, because large bending stress is generated in addition to the bolt stress variation.

5.6 Seizure of threaded fasteners

5.6.1 Conditions prone to seizure

Seizure of threaded fasteners sometimes becomes a problem in joints composed of materials with low thermal conductivity, such as stainless steel, titanium, etc. Regarding the seizure mechanism, numerical calculations have been conducted under some assumptions in order to evaluate the temperature distribution under the bolt head for the tightening operation by applying torque to the bolt head [126]. However, these dealt with the cases performed by very high tightening speed. Therefore, the research unfortunately did not provide the information to clarify the mechanism of the seizure sometimes occurring in a hand-tightening operation. Major causes of seizure when tightening by the torque control method, are listed in the following:

(1) The target joint is composed of materials with low thermal conductivity such as stainless steel and titanium, including threaded fasteners.
(2) Coefficients of friction on thread surface and nut- and bolt head-bearing surfaces are high, on which sliding friction occurs in the tightening process.
(3) In connection with the preceding condition, thread surface and nut-bearing surface have not been sufficiently lubricated.
(4) When the processing accuracy of threaded fasteners and fastened plates is low, thread surface or nut-bearing surface tends to be in partial contact, hence, the actual contact surface becomes very small.
(5) The target bolt is tightened with high bolt stress under any of the above conditions.
(6) Tightening speed is high.
(7) Grip length is large due to the use of long bolts.
(8) The seizure is more likely to occur in fine screw threads than in coarse screw threads.

The amount of temperature increase on nut-bearing surface or bolt head-bearing-surface in the tightening process can be estimated by the product of contact pressure, coefficient of friction, and sliding speed. In the case of tightening operation from the nut side, letting contact pressure be p_{nu}, coefficient of friction μ_{nu}, and sliding speed v, $\mu_{nu}p_{nu}v$ represents the amount of heat generated per unit time and unit area on the nut-bearing surface. By use of $\mu_{nu}p_{nu}v$, the preceding phenomena from (1) to (8) are examined. Regarding (1), assuming that the amount of heat generation is constant, the surface temperature is likely to be increased in the materials with low thermal conductivity because of a poor dissipation of heat, thereby leading to the seizure. The conditions of (2) and (3) are related to coefficient of friction, (4) and (5) are to contact pressure p_{nu}, (6) is to sliding speed v. The conditions of (7) and (8) are associated with the mechanical consideration that the required tightening energy is almost in proportion to grip length and is nearly in inverse proportion to thread pitch, explained in section 3.8. By designing the joint so as to mitigate the situations listed above, the occurrence of seizure can be considerably reduced. However, the problem in the actual tightening operation is that the seizure sometimes occurs in unexpected situations.

5.6.2 Hypothesis proposed on seizure occurrence

Surface roughness is considered to be deeply related to seizure. Fig. 5.18 shows some examples of seizure occurrence on nut- and bolt head-bearing surfaces and plate surface, which are all made of stainless steel. Tightening operation was done by hand-tightening without lubrication. In the experiments, surface roughness *Ra* is changed from the original value, being smaller than 0.5 μm, which corresponds to a smooth surface, to the one larger than 2 μm obtained by use of sandpaper. As a result, it has been confirmed that seizure is likely to occur on the surface with very small surface roughness. Since coefficient of friction tends to be increased with increasing surface roughness [71], it is considered that the seizure occurrence cannot be explained only by the amount of coefficient of friction.

In the numerical analyses conducted in [126], a perfect contact condition is assumed on the bolt head-bearing surface. However, this apparent contact area A_{ap} is much larger than real contact area A_{re}, on which a lot of microprojections existing on the mating surfaces are in contact. The relationship between A_{ap} and A_{re} is

Fig. 5.18 Example of seizure occurring on nut- and bolt head-bearing surfaces and plate surface. (A) Bolt head-bearing surface and plate surface, (B) nut-bearing surface, and (C) plate surface.

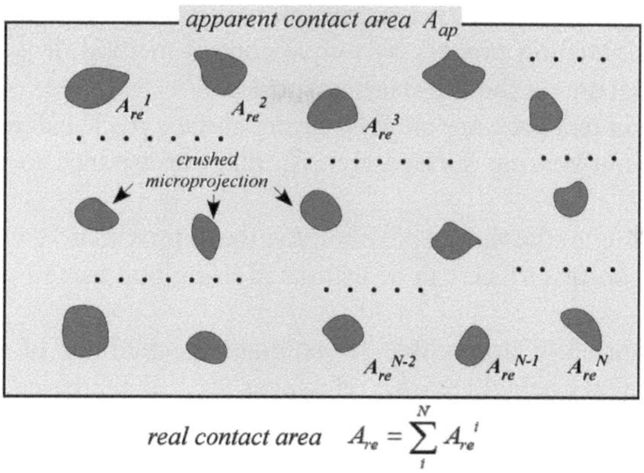

Fig. 5.19 Apparent contact area A_{ap} and real contact area A_{re}.

schematically shown in Fig. 5.19. Area ratio of real to apparent contact areas A_{re}/A_{ap} significantly changes depending on contact pressure. In the field of tribology, when evaluating the temperature on a sliding surface, on which two contacting bodies pressed together are moving at a certain relative speed, an estimation method utilizing exact solutions of heat conduction theory is proposed by assuming that a single microprojection slides on the surface of a semiinfinite body [127,128]. Fig. 5.20 illustrates the heat conduction model. Mean temperature increase ΔT_m on the surface of a microprojection can be calculated in the following procedure. Let coefficient of friction be μ, plastic flow stress of the material p_m, and sliding speed v. Then, $\mu p_m v$ represents the amount of heat generated on the surface per unit time and unit area. Assuming that the shape of the microprojection in contact with the mating flat surface is a circle of radius a_p, ΔT_m is expressed in the following form for the case of relatively low sliding speed:

$$\Delta T_m = \frac{0.849\mu p_m v a_p}{2\lambda} \tag{5.15}$$

where λ denotes thermal conductivity of the materials in contact, and p_m can be equated to three times the yield stress σ_Y [129].

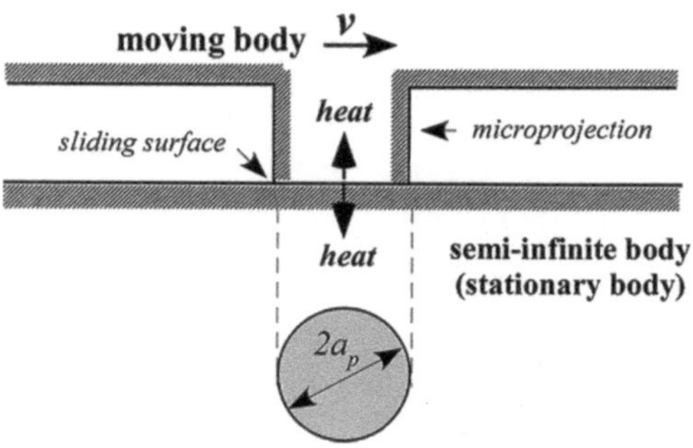

Fig. 5.20 Heat conduction model for evaluating sliding surface temperature.

The above equation provides important information on the mechanism of seizure occurring in the tightening process by torque control method. It is explained below how the seizure occurs on the nut-bearing surface:

Step 1: Mean contact pressure on nut-bearing surface p_{nu} is calculated by dividing bolt force F_b by nut-bearing surface area A_n that corresponds to apparent contact area A_{ap}.

Step 2: According to the theory of tribology, the approximate value of real contact area A_{re} on nut-bearing surface can be estimated using bolt force F_b and plastic flow stress p_m.

Step 3: It is virtually impossible to estimate the number of microprojections N composing the real contact area A_{re}. However, as N is increased, the contact area per one microprojection necessarily decreases. That is, with decreasing N, the contact area increases conversely.

Fig. 5.21 shows the numerical results of the mean temperature on the sliding surface of a microprojection, calculated by Eq. (5.15). The abscissa represents the sliding speed. The parameter in the figure is the radius of microprojection a_p. The temperature of the sliding surface shows a sharp increase as sliding speed v is increased. The effect of a_p is very large as well. This means that as the number of microprojections

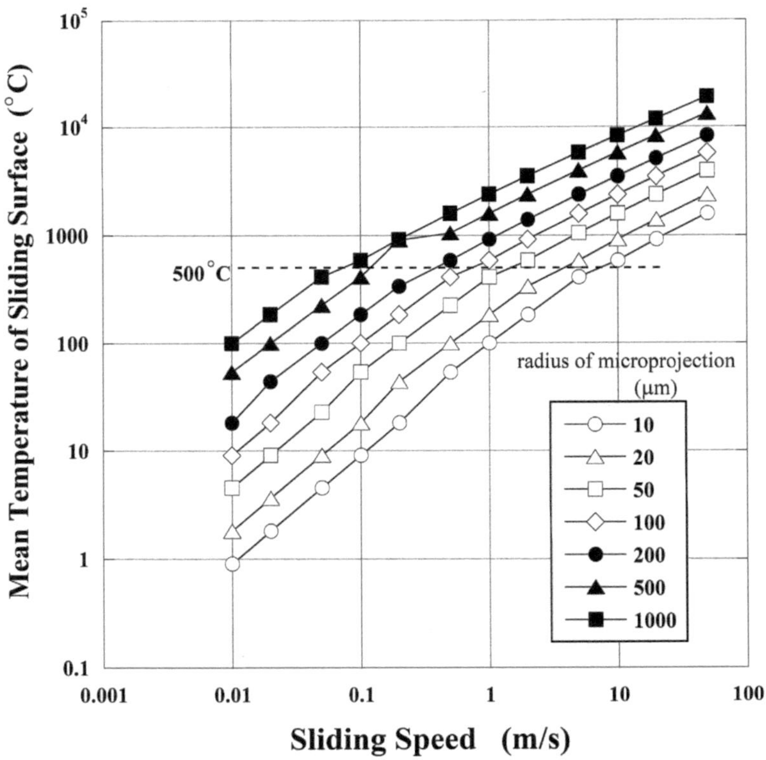

Fig. 5.21 Sliding surface temperature calculated by heat conduction theory.

N composing real contact area A_{re} gets smaller under constant bolt force F_b, the radius of microprojection a_p necessarily increases. As a result, the sliding surface temperature is raised according to Eq. (5.15). From the above consideration, the seizure phenomena shown in Fig. 5.18, which indicate that in stainless steels the seizure is more likely to occur with decreasing surface roughness, can be explained using heat conduction theory. For a constant bolt force F_b, assuming that the number of microprojections in contact N decreases with decreasing surface roughness, mean radius per unit microprojection a_p is increased, which leads to the temperature increase of the sliding surface. It is worth noting here that there are numerous microprojections on each sliding surface. If, unfortunately, multiple microprojections are located very closely, the surface temperature on each microprojection is further raised due to the interaction. Accordingly, the seizure between the mating surfaces can occur abruptly. Although being a hypothesis, it may be the reason why the seizure occurs or does not occur under the same tightening conditions.

There are some field reports that seizure sometimes or often occurs in threaded fasteners with large nominal diameter and fine screw threads. When large bolts are used and nominal diameter d exceeds 100 mm, thread pitch P is not increased correspondingly, hence, lead angle β becomes very small. As a result, the required tightening energy considerably increases. In addition, large screw threads inevitably involve some geometric errors due to the processing difficulty, which possibly leads to the condition of partial contact on the sliding surfaces, i.e., thread surface and nut-bearing surface. Furthermore, in the field, it is difficult to apply the lubricant onto the sliding surface uniformly, comparing to small screw threads.

In conclusion, it is strongly desirable to further clarify the mechanism of seizure occurrence by comprehensive experiments, FEM, and rigorous analyses, in which the properties of joint materials are taken into account and tribology and heat conduction theories are ingeniously introduced.

Loosening of threaded fasteners

6

6.1 Rotation loosening and nonrotation loosening

Loosening of screw threads is divided into two categories: loosening occurring with or without rotation of threaded fasteners such as bolts and nuts. In this book, the former is termed rotation loosening and the latter nonrotation loosening. In either case, bolt force is reduced and it may lead to fatigue failure. In other words, even if rotation loosening can be prevented, fatigue failure could occur when bolt force is significantly reduced due to nonrotation loosening. Nowadays, reduction in the occurrence of rotation loosening has, to a large extent, been achieved using special threaded fasteners, and can be almost completely prevented, depending on the types of target structures or machines and the related service conditions. In contrast, for nonrotation loosening, it inevitably occurs to some extent, which is represented by the bolt force reduction due to embedment. It is often caused by the progress of plastic deformation of microprojections existing at the interface. Furthermore, rotation or nonrotation loosening occurs when the joint is subjected to thermal load, due to the difference in thermal expansion between threaded fasteners and fastened plates, the change in mechanical and thermal properties of the joint materials, etc.

Rotation loosening generally occurs when threaded fasteners are subjected to repeated external load. On the other hand, nonrotation loosening is generated, in addition to the embedment mentioned above, by creep and relaxation due to thermal load and depression of the nut- or bolt head-bearing surface owing to high bolt force, etc., all of which can occur without being subjected to repeated loading. These phenomena progress with time and are commonly promoted when external loads are applied repeatedly.

6.2 Why bolted joints are easy to loosen

The possibility of loosening occurrence is significantly changed depending on joint geometry, tightening condition, loading condition of external force, environment surrounding the joint, etc. Therefore, it sometimes occurs that an antiloosening device effective for a certain type of joint does not work well for another type of joint. One of the root causes of the phenomenon is that threaded fasteners are used in a wide variety of conditions compared to other machine elements like gears, bearings, springs, etc. For instance, the role of ordinary gears is to transmit torque and rotation. Hence, the magnitude of load acting on the gear can be estimated with fairly high accuracy by the values of torque and rotation speed. In the case of bearings, they are classified into radial and thrust bearings according to the loading direction. They are also categorized into rolling and sliding bearings with respect to friction

The Mechanics of Threaded Fasteners and Bolted Joints for Engineering and Design.
https://doi.org/10.1016/B978-0-323-95357-3.00005-8

conditions. Some types of rolling bearings can support both radial and thrust loads simultaneously, but the loading form does not change significantly even if the magnitude of the applied load is varied.

In contrast, threaded fasteners, typified by bolts and nuts, are subjected to external forces in a wide variety of forms. For example, threaded portions of dental implants, which have come to be widely used recently, are subjected to all kinds of external forces such as compression, bending, twisting, and sometimes tension, depending on the foods to be chewed. Furthermore, for threaded fasteners used in large machines and precision instruments, the loading form significantly varies according to the location of use.

From the above considerations, it is presumed that there is a limit to the evaluation of antiloosening performance by a single testing method, since threaded fasteners are used for joints with various configurations and are subjected to various types of external forces. For instance, the NAS3350 vibration test is a severe testing method in which an impact load is applied to a bolted joint repeatedly in the direction perpendicular to the bolt axis. Hence, it has been widely employed in Japan to evaluate the antiloosening performance of threaded fasteners. However, there are cases where thread loosening occurs in a relatively short time, although the target joint could have endured the NAS vibration test. The reason can be explained by the diversity of the usage of threaded fasteners stated above. In this way, it is extremely difficult to completely prevent loosening. Accordingly, several examples are introduced in the following in order to design bolted joints that are hard to loosen, by enumerating the ways such joints are easy to loosen.

(1) Bolted joint with small grip length subjected to mechanical external force

Bolted joints consisting of thin plates, clamped by short bolts, are very easy to loosen. Since thin plates have very high stiffness, bolt elongations are small. Therefore, when the embedment on contact surfaces progresses and/or the plate warpage is flattened due to repeated external forces, only a slight change in the amount of bolt elongation significantly reduces the bolt force.

(2) Bolted joint with large grip length subjected to thermal load

In bolted joints subjected to thermal load, the amount of thermal expansion is proportionally increased as grip length is increased for the same temperature rise, as explained in section 5.1.3. Hence, a joint with larger grip length is easier to loosen. Since this phenomenon is the exact opposite of (1), particular attention should be paid to the loosening of threaded fasteners when the target joint is subjected to thermal and mechanical loads simultaneously.

(3) Bolted joint consisting of many plates

As the number of fastened plates increases, loosening due to embedment becomes easier to progress, because the number of contact surfaces increases correspondingly. In addition, the embedment is easily progressed with larger surface roughness at the interface, which leads to bolt force reduction.

(4) Bolted joint clamped with low bolt force

The amount of bolt force reduction due to embedment, estimated under a certain assumption, is independent of initial bolt force, which is explained in Section 6.4.3. Hence, by setting the initial bolt force high, the residual bolt force is still high even if bolt force reduction occurs. Therefore, the problems caused by loosening can be avoided. When the joint is subjected to repeated external load, the slip occurring on contact surfaces becomes the major cause of loosening. In order to mitigate this unfavorable phenomenon, it is effective to set the initial bolt force high so as not to generate slip. In contrast, for bolt force reduction due to creep, it has been pointed out that high initial bolt force merely accelerates the progress of thread loosening and no effect is found on the mitigation of bolt force reduction [130].

6.3 Bolt force reduction due to rotation loosening

6.3.1 Mechanism of rotation loosening

Rotation loosening occurs due to the relative slip, caused by repeated external load, on thread surface and/or nut-bearing surface. The mechanism of the generation of rotation loosening is classified into four types, illustrated in Fig. 6.1.

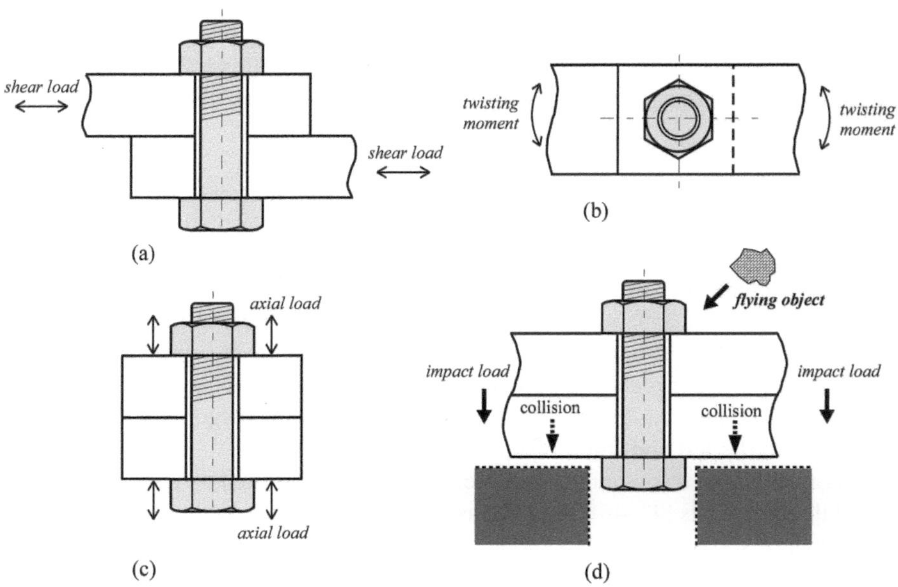

Fig. 6.1 Various mechanisms of the occurrence of rotation loosening. (A) Perpendicular to bolt axis, (B) around bolt axis, (C) along bolt axis, and (D) impact load.

1) Repeated shear load applied perpendicular to bolt axis

This case corresponds to the situation of the external force being applied in the direction perpendicular to the bolt axis or in a similar form. The loosening of this type frequently becomes a problem in various industrial fields.

2) Repeated twisting load applied around bolt axis

This case becomes an issue in the following situations: screw threads used for attaching spectacle temples in daily life, female threads processed in the center of rotation shaft to attach a part in industrial fields, etc. In any case, twisting torque is repeatedly applied around the bolt axis.

3) Repeated axial load applied in the direction of bolt axis

This case corresponds to the situation of the external force being applied along the bolt axis or in a similar form. There are fewer cases that loosening occurs by this mechanism, comparing to 1) and 2).

4) Impact load

Impact load becomes an issue in the following cases: external load is applied to the joint at very high speed, e.g., due to the falling of an object, the collision of small stones repeatedly against the jointed portion, etc.

In relation to the above phenomena, rotation loosening often occurs by the following mechanism. When tightening threaded fasteners by torque control method, the bolt cylindrical portion is twisted elastically, to some extent, due to the helical thread geometry. Accordingly, the relative slips occurring on thread surface and nut-bearing surface can cause rotation loosening. The mechanism of rotation loosening and its evaluation method are explained in detail in [130–132]. In this book, therefore, the mechanism of rotation loosening is only briefly introduced, as follows.

(1) Repeated shear load applied perpendicular to bolt axis

According to the theory proposed by Yamamoto, when shear load is repeatedly applied to a bolted joint, the bolt's cylindrical portion is elastically twisted. Then, in the service condition, the relative slips on thread surface and nut-bearing surface occurring while the twisting moment is released are considered to cause rotation loosening [131]. Fig. 6.2 schematically shows how rotation loosening occurs due to the relative slip on the nut-bearing surface, in which one cycle of repeated shear load is applied to the upper plate of a bolted joint, composed of two plates, in the direction perpendicular to bolt axis [132]. The figure shows that the slip is generated by the bolt inclination to the left and right, repeatedly.

This mechanism of loosening phenomenon has been simulated by FE analysis using helical thread models [44–46]. In [131], assuming that rotation loosening occurs when the amount of slip on the contact surface exceeds a certain limit, antiloosening performance is compared by manufacturing a dedicated testing machine and measuring the limit value. Test objects are threaded fasteners equipped with antiloosening functionality and various antiloosening mechanisms and devices, which have been proposed for the prevention of loosening. Using the experimental results, antiloosening performance

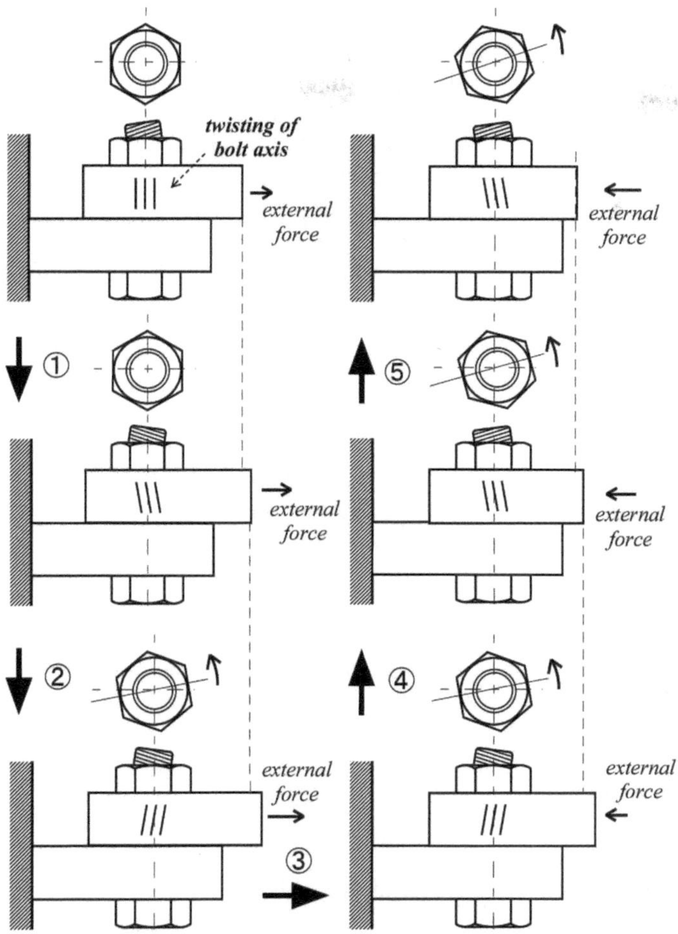

Fig. 6.2 Yamamoto's theory explaining the mechanism of rotation loosening by repeated shear load.

has been judged, where the performance of an ordinary hexagon nut is selected as the evaluation standard. It is given in ascending order as follows:

> *Toothed washer < spring washer < conical spring washer < ordinary hexagon nut < double-nut locking, hexagon nut with flange, prevailing torque type hexagon nut with nylon insert < anaerobic adhesive.*

It has sometimes been reported in academic conferences that washers have no effect on loosening prevention or that the use of washers may even promote the loosening. However, the use of a flat washer can enhance antiloosening performance, because it allows an increase of the initial bolt force. Detailed discussions are given in the subsequent sections. The use of anaerobic adhesives naturally prevents slip occurrence. It has also been reported, based on the experimental results, that double-nut locking, hexagon nut with flange, and prevailing torque type hexagon nut have almost the same degree of antiloosening performance. In addition, a unique research report has been published that tries to explain the loosening mechanism by introducing the concept of loosening rotation torque [130].

(2) Repeated twisting load applied around bolt axis [130,133]

For the case of a joint being subjected to repeated twisting load around the bolt axis, it is proposed that the start condition of rotation loosening, due to the accumulation of nut rotation in the loosening direction, can be expressed by the following equation:

$$\frac{1}{2}F_b d_2 \tan\left(\rho'_{th} - \beta\right) < T_2 < \frac{1}{2}F_b d_2 \tan\left(\rho'_{th} + \beta\right) = T_1 \tag{6.1}$$

The above equation indicates that nut-bearing torque T_2, explained in Chapter 3, is an intermediate value between thread torque T_1 in the tightening process and loosening thread torque given in eq. (3.19). Furthermore, in defining a critical twisting angle that won't produce the slip on thread surface and nut-bearing surface, it is recommended to use the angle as the evaluation index for rotation loosening. To prevent this type of loosening, it is preferable to use a bolt-nut pair with smaller thread pitch P, and the equivalent friction diameter of nut-bearing surface d_{nu} should be larger. Regarding coefficients of friction, the coefficient on thread surface μ_{th} should be smaller. On the contrary, the coefficient on nut-bearing surface μ_{nu} is preferably larger. It is also recommended to set the initial bolt force higher and design the grip length longer and the diameter of bolt body smaller.

(3) Repeated axial load applied in the direction of bolt axis

It has been confirmed by Goodier et al. through experiments that loosening occurs when tensile load is applied to a bolted joint in the axial direction repeatedly so that the bolt force is increased and decreased [134]. The loosening mechanism is clarified by Sato et al. [135]. In [78], it is concluded that loosening phenomenon due to the repeated axial load need not be considered from the design point of view, because the maximum bolt force variation that won't cause loosening is very large; hence, fatigue failure is more likely to become an issue.

(4) Impact load and inertia force

When external force is applied to a bolted joint at very high speed, i.e., as an impact load, rotation loosening can occur due to the propagation of a stress wave. Supposing the situation on a construction site that small stones impact the jointed portion of an excavator, this type of rotation loosening is reproduced by a small ball being brought into collision against a nut [136]. Regarding the case of a bolted joint subjected to impact load in the axial direction, the loosening mechanism is studied experimentally by dropping the jointed portion freely, which is explained by the fact that the loosening is caused by the propagation of a stress wave [137,138]. In the case of the impact load being applied in the direction perpendicular to the bolt axis, the loosening mechanism is the same as that in (1), but the progress of loosening is further promoted [139]. When external force does not increase so steeply as impact load, it can be treated as inertia force.

Rotation loosening of the screw threads by which the blade of an electric fan is attached to the motor shaft is an example of loosening due to inertia force. In this joint, left-hand screw threads are used so that inertial torque acts in the direction to further tighten the threads when the blade starts rotating. As for the nuts used for attaching tires to the axle shafts of a large vehicle, in the past, right-hand screw threads were used for right-hand wheels and left-hand screw threads were for left-hand wheels; in recent years, however, right-hand screw threads have been used for the wheels on both sides. It has been pointed out in [131] that the usage of screw threads in this way following the old practice is related to the cause of wheel falling-off accidents of large vehicles, in which many of the accidents occur in the left-hand wheels. However, in most accidents of this type, wheel bolts or inner nuts are ruptured due to metal fatigue, and the threaded fasteners are kept in the initial state without rotation. Therefore, rotation loosening is considered to be a secondary cause. The wheel falling-off accidents in Japan, due to the fatigue failure of threaded fasteners, are discussed in detail in section 8.2.

6.3.2 Bolt force reduction due to return rotation of nut

If a nut rotates in the loosening direction, bolt force reduction occurs correspondingly. Once a nut starts to loosen, the loosening is progressed with time in many cases, due to the repetition of external force. Meanwhile, the degree of risk, which is caused by a slight rotation of nut in the loosening direction, greatly varies depending on the joint geometry and materials. In the case of the fastened plates consisting of thin plates with small grip length, only a small return rotation of the nut causes a large reduction in bolt force, because the rotation angle required to produce the target bolt force is small. This means that thin plates clamped by bolt-nut pairs are in a state that is easy to completely loosen. The process of bolt force reduction caused by the return rotation of nut is very complicated from the mechanical point of view. Therefore, in this section the difference in the degree of risk is evaluated with respect to the difference in joint geometry and materials. To achieve this, the equation that relates nut rotation angle to bolt force is used, which has been derived for the tightening process by elastic angle control method, discussed in section 3.3. In eq. (3.32), replacing rotation angle ϕ by return rotation angle $\Delta\phi$, bolt force F_b by bolt force reduction ΔF_b, the following equation is obtained.

$$\Delta\phi = \frac{2\pi}{P}\Delta F_b\left(\frac{1}{k_{th}} + \frac{1}{k_s} + \frac{1}{k_{cyl}} + \frac{1}{k_{hd}} + \frac{1}{k_f}\right) \tag{6.2}$$

In Eq. (6.2), assuming constant $\Delta\phi$, bolt force reduction ΔF_b increases with larger thread pitch P and larger spring constants, which represent the stiffness of each part of the bolted joint. Therefore, bolt force reduction ΔF_b becomes larger in coarse screw threads than in fine ones for the same nominal diameter. Regarding the joint stiffness, as grip length L_f is decreased, bolt force is significantly reduced for constant $\Delta\phi$, because the amounts of k_s, k_{cyl}, and k_f are increased. As for joint materials, bolt force reduction ΔF_b is decreased in bolted joints made of aluminum alloy or titanium with low Young's modulus, since the spring constant of each part is proportional to Young's modulus.

Learning by numerical example 6.1: Bolt force reduction due to return rotation of nut

The amounts of bolt stress reduction $\Delta\sigma_b$ occurring per one degree of nut return rotation are compared for bolt-nut connections of M16 with coarse thread, in which the ratio of grip length L_f to nominal diameter d, L_f/d, is changed to 1, 2, 4, 6, 8, and 10. Other calculation conditions are the same as those in numerical example 3.8. Calculations are performed by Eq. (6.2) without considering the effect of interface stiffness. Bolt stress reductions thus calculated are 21.2, 14.2, 8.5, 6.1, 4.7, and 3.9 MPa, which indicates that only a slight return rotation produces a large stress reduction in thin plates.

6.3.3 Prevention methods of rotation loosening

A wide variety of special threaded fasteners or mechanisms have been proposed aiming at the prevention of rotation loosening. For the antiloosening strategies that have widely been employed so far, threaded fasteners or mechanisms commonly used are ranked in Section 6.3.1 according to the antiloosening performance, based on the experimental results for the case of bolted joints being subjected to repeated shear loading [131]. Prevention methods of rotation loosening are classified as the method using special threaded fasteners, etc., and the method that devises tightening condition or joint geometry. Regarding the former category, a variety of methods are proposed, such as threaded fasteners with antiloosening functionality, addition of special parts that enhance antiloosening performance around the joint, etc.

Many antiloosening strategies intend to raise the effect of frictional work on thread surface and nut-bearing surface; antiloosening parts based on geometrical ingenuity, for instance, by changing thread profile, are also available. Since it is impossible to introduce all types of antiloosening methods, some representative ones are briefly presented in this section. In [130], how loosening can be prevented is discussed from a wide perspective. In [140], using plain illustrations, several practical antiloosening parts are introduced. In [141], antiloosening methods are classified into eight types and detailed explanations are given for each method, in which it is also shown how each method corresponds to JIS. Please refer to those references for further information.

(1) Use of special threaded fasteners

- **1)** Special nut with altered thread profile or inserting some inclusions, causing interference between male and female threads to be generated in the tightening process, thereby increasing the resistance to return rotation of nut. These types of nuts are termed prevailing torque type nuts. They are specified in JIS B1056. Various types of prevailing torque type nuts are proposed, for instance, nuts into which some amount of nylon is inserted, nuts whose pitch diameter is gradually reduced, etc. Hexagon bolts with flange and hexagon nuts with flange, specified in JIS B1089 and B1090, respectively, are intended to raise the resistance to return rotation. More specifically, equivalent friction diameter of nut-bearing surface d_{nu} or bolt head-bearing surface d_{hd} is made larger by increasing the bearing surface area. The free spinning type of antiloosening method intends to raise the resistance to return rotation on thread surface and nut-bearing surface [141]. Concretely, this type utilizes the deformations of nuts with special profile or nuts divided into two parts, termed split nuts. Nuts with serrations at bearing surface are categorized as this type.

2) Special bolt—although most antiloosening innovations are made to nuts—is an example of bolts with antiloosening performance. It is proposed to process the bolt thread profile to have left–right asymmetry, thereby making the interference with female threads harder and raising the action of frictional work. A unique method utilizing the difference in thread pitch is proposed, which is sometime called lead difference utilization type [141]. Two kinds of threads with different thread pitch are processed on a single bolt, and the tightening operation is done by using two nuts with different thread pitches, corresponding to the different bolt thread pitches. The point of this mechanism is that, due to the difference in thread pitch between the two nuts, a nut having larger thread pitch, placed in contact with the fastened plates, is not allowed to rotate in the loosening direction.
3) Special washers, such as toothed washer, spring washer, and conical spring washer, have been reported to show no antiloosening effect, or even that special washers are more likely to loosen than ordinary nuts [131]. However, a special washer used with a special nut is proposed, in which the washer and the bearing surface of the nut are machined to be toothed, with a shape whose inclination angles are larger than the lead angle. This type is considered to be a kind of lead difference utilization type, as introduced above.
4) Antiloosening parts are special parts added to the target threaded fastener. The strategy to prevent loosening by adding such parts is called the mechanical antiloosening method. The special parts are attached to threaded fasteners, and then are bent or deformed and put into use. The combination of hexagon castle nut and split pin belongs to this type. Antiloosening mechanisms by use of wire, tongued washer,or claw washer are also proposed. Although the installation work takes time, high antiloosening effect is expected.
5) Other methods include inserting anaerobic adhesives between mating thread surfaces, with which high antiloosening performance can be achieved without adding any special part. In this case, ordinary threaded fasteners are used as they are.

(2) Utilization of tightening condition and joint geometry

When designing jointed components, it is sometimes the case that special-shaped nuts or special parts equipped with antiloosening functionality cannot be used due to various reasons. As a countermeasure against this problem, some strategies are introduced here to effectively enhance the antiloosening performance by small changes of tightening condition or joint geometry.

1) Increase of resistance force against return rotation of nut, as rotation loosening is caused by the relative slip on the contact surface. The resistance force against the return rotation can be raised by two strategies. The first one is to raise bolt stress to the extent that plastic deformations or other problems do not matter. The second strategy is to intentionally increase the coefficient of friction, which has been commonly employed in architecture or its related fields. This strategy is suitable for joints subjected to large shear force.

2) Suppression of relative slip, as bending deformation occurs in bolts when subjected to shear force. By designing the joint shape to have a large grip length, it is possible to use long bolts, by which the occurrence of relative slip can effectively be suppressed due to the low joint stiffness. A slender bolt with reduced shank is also effective because of its low stiffness. In this case, however, care must be paid to the material strength when setting the bolt stress high. Long bolts have another advantage concerning antiloosening performance: for a constant return rotation angle of nut, bolt force reduction is smaller comparing to short bolts, as evident from numerical example 6.1.
3) Double-nut tightening is a traditional loosening prevention method by use of two nuts. The antiloosening mechanism is discussed in detail in [131]. Researches by FE analysis are also reported [142]. The primary factor concerning the antiloosening performance is the locking action due to the compression force generated between the two nuts. According to Yamamoto's theory, there are two types of tightening procedure, in which the upper nut is rotated in the positive direction or the lower nut is rotated in the negative direction, and it is proposed that the latter procedure is more appropriate [131]. Regarding the nut shape, a standard nut should be placed on the upper side and a thin nut on the lower side, from the mechanical point of view. However, the procedure has a work-related problem that it requires two types of spanners. Therefore, in some cases, the two nuts with different thickness are placed upside down or two identical standard nuts are used.

Finally, it should be kept in mind that even if rotation loosening could be prevented by any specific method, fatigue failures may occur in the bolted joints when clamped with insufficient bolt force, owing to the interface separation.

6.3.4 Finite element simulation of rotation loosening caused by repeated shear load

Among various types of rotation loosening mechanisms, the one due to repeated shear load is particularly dangerous, because the external force applied perpendicular to bolt axis often causes a significant loss of bolt force. In this section, the loosening mechanism is examined by three-dimensional finite element (FE) analysis using helical thread models introduced in section 2.5. Fig. 6.3 shows a finite element model constructed to simulate the case of Fig. 6.1A. Detailed descriptions about the modeling scheme, analytical and boundary conditions, etc. are given in [45,46]. The numerical model basically corresponds to Junker's test, except for one point that the effect of friction force at the interface between upper and lower plates is taken into consideration. The action of friction force on the contact surface critically affects the loosening mechanism because the start of slip in a bolted joint is determined by the amount of friction force there. Fig. 6.4 schematically explains the loosening mechanism, in which the relationship between the amount of displacement applied to the lower plate in Fig. 6.3 and the resulting shear load between the two plates is shown. The displacement is forcibly applied to the lower plate repeatedly from side to side. The amount of the displacement is changed between ± 1 mm.

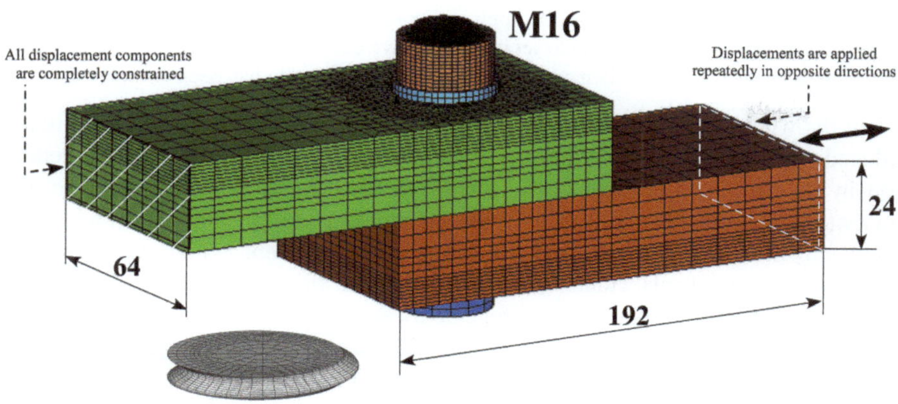

Fig. 6.3 Finite element model of bolted joint for analyzing rotation loosening mechanism.

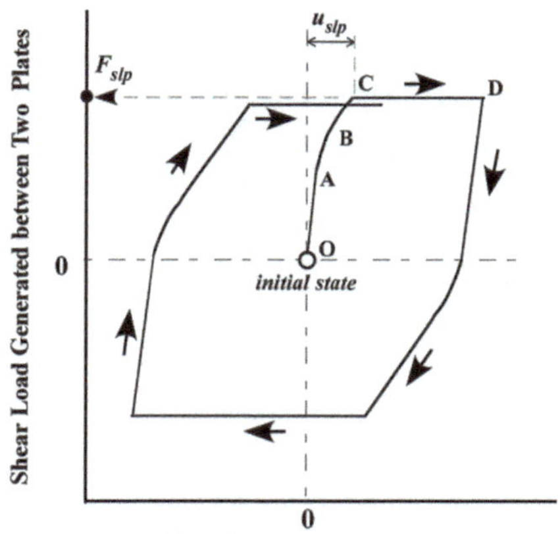

Fig. 6.4 Mechanism of rotation loosening of bolted joint subjected to repeated shear load.

The four points, A, B, C, and D, in the figure represent the critical situations stated below:

O-A: The target bolted joint deforms in an elastic manner like a single body.
Point A: Slip is initiated at the interface between two plates. Shear load starts to behave nonlinearly against the applied displacement.
A-B: The slip on the thread surface starts and it promotes the nonlinear behavior.
B-C: The contact condition on the thread surface makes a transition from partial slip to whole slip. Slip on the nut-bearing surface also starts. Hence, the joint stiffness in the transverse direction is decreased and the nonlinearity appears more noticeably.
Point C: All contact surfaces, including nut-bearing surface, get into the whole slip condition, and the amount of shear load acting on the joint becomes constant. The shear load at this point is defined as critical shear load F_{slip}.
Point D: The direction of applied displacement is reversed.

After point D: In the process of the applied displacement being decreased, the shear load reduces to zero and then it starts to act in the negative direction. After the shear load reaches a constant value, a similar behavior is repeated until one cycle has been completed.

When the whole slip condition occurs by exceeding Point C, bolt force is inevitably reduced. The phenomenon of bolt force reduction can be recognized by a slight reduction in F_{slp} from the original value when one cycle is completed. Therefore, the amount of F_{slp} is an important factor to devise measures for preventing rotation loosening. u_{slp} in the figure is defined as a critical displacement, which represents the threshold value of displacement that starts the whole slip condition. Accordingly, if a constant amount of displacement is repeatedly applied to the bolted joint perpendicular to the bolt axis, the amount of u_{slp} can be used as an index to design bolted joints not loosened by rotation loosening. In any event, the most important aspect in the figure is that the bolt force is slightly reduced after one cycle, once the amount of applied displacement exceeds u_{slp}. That is, although the reduction in bolt force during one cycle is small, it may lead to a significant reduction in, or the complete loss of, bolt force, because the shear load is repeatedly applied.

In Fig. 6.5, the numerical results performed corresponding to Fig. 6.4 are presented, in which coefficients of friction on contact surfaces are changed to 0.1, 0.15, 0.17, and 0.2. As coefficient of friction μ increases, both F_{slp} and u_{slp} are significantly increased, which substantiates that increasing μ can effectively suppress rotation loosening. Although the initial bolt force is constant in this figure, it is predictable that increasing the initial bolt force is also effective for avoiding rotation loosening, since the friction force increases in proportion to initial bolt force. Fig. 6.6 shows how the bolt force is reduced when subjected to repeated shear load, where numerical calculations are performed for three cycles. The ordinate represents the ratio of bolt stress σ_b under

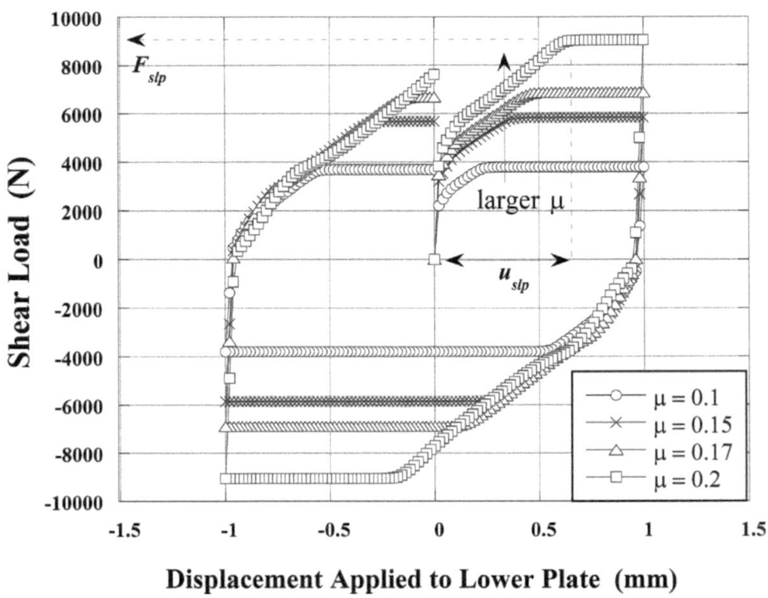

Fig. 6.5 Rotation loosening process of bolted joint evaluated by FE analysis.

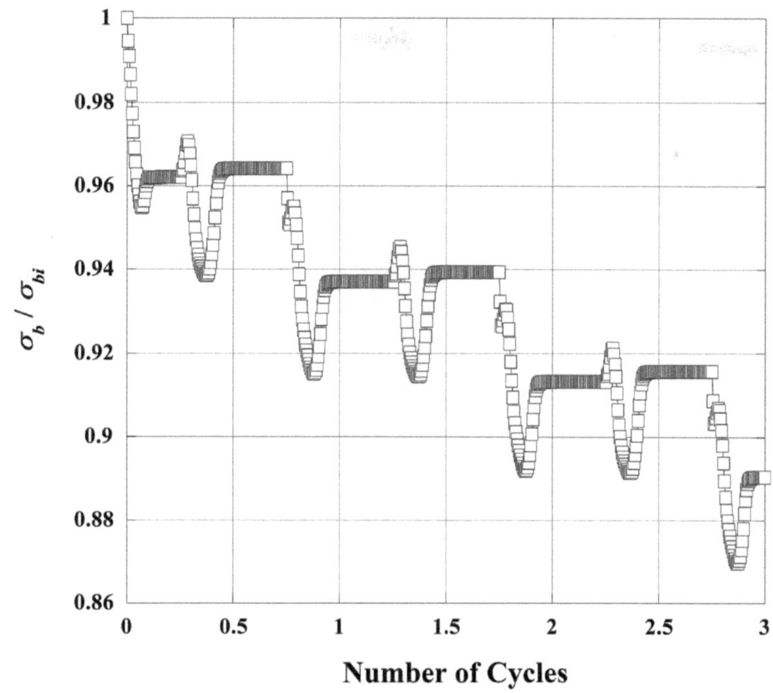

Fig. 6.6 Progress of bolt force reduction due to rotation loosening.

repeated shear load to initial bolt stress σ_{bi}. It is easily predicted that the bolt stress is successively lowered and finally disappears completely after a short time. Fig. 6.7 illustrates a simple strategy to prevent rotation loosening, in which a constant displacement is repeatedly applied perpendicular to the bolt axis. The strategy aims to decrease the amount of displacement generated around the bolt-nut pair by increasing the displacement of the portion away from the bolt axis. Hence, the displacement near the bolt-nut pair predictably becomes less than u_{slp}.

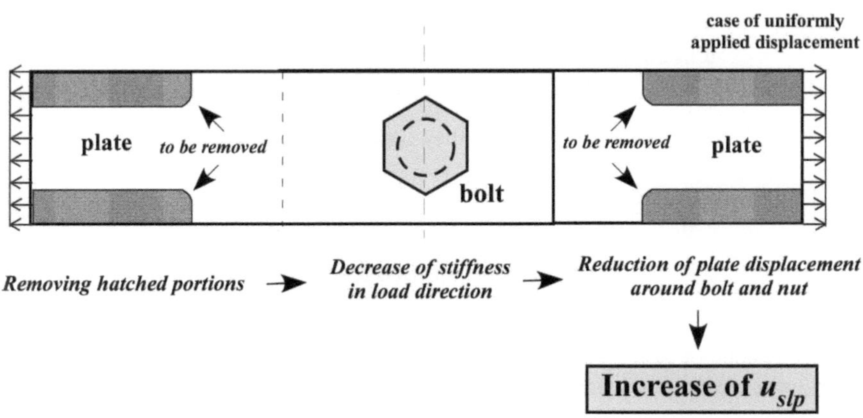

Fig. 6.7 A simple strategy to prevent rotation loosening.

According to Yamamoto's theory, explained in Fig. 6.2, rotation loosening occurs when the bolt body is twisted too much by the applied shear load and then the twisted deformation is released. That is, if this process is repeated, it leads to a significant reduction in bolt force. In Fig. 6.8B, it is shown how the rotation angles of five points

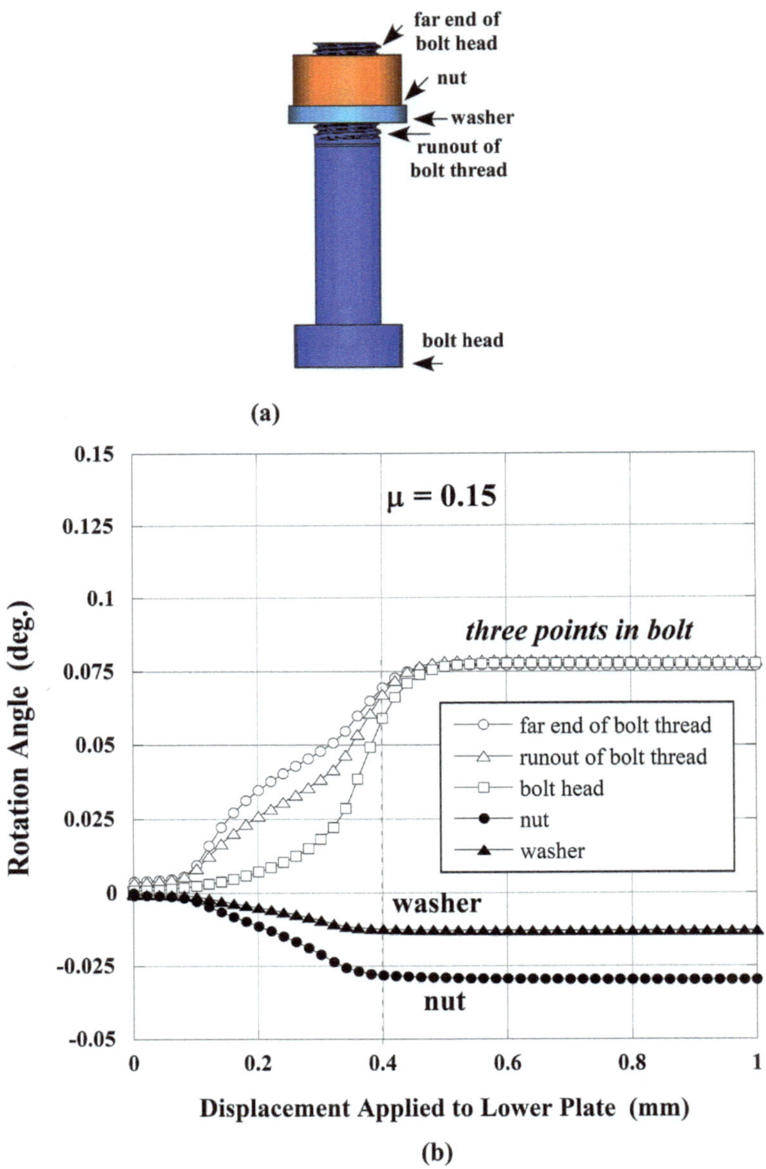

Fig. 6.8 Variation of rotation angles of selected points during the loosening process. (A) Measuring points of rotation angle and (B) progress of rotation angle.

selected on the bolt-nut pair, shown in Fig. 6.8A, progress when shear force is applied. The numerical results correspond to the region O-A-B-C-D in Fig. 6.4. In [45,46], in addition to the detailed considerations of the rotation loosening mechanism, the effects of thread pitch, grip length, etc., are also examined, and then, the effectiveness of the use of larger grip length and smaller thread pitch is quantitatively substantiated. Finally, a point worthy of special mention is given on coefficients of friction: bolt force reduction is primarily influenced by the coefficient of friction on the nut-bearing surface, μ_{nu}, because the whole slip condition causing the bolt force reduction is likely to occur in case of small μ_{nu}. When μ_{th} is small, the slip on the thread surface occurs, but it does not necessarily lead to a successive bolt force reduction. In any event, it is important to avoid the occurrence of the whole slip condition to prevent rotation loosening.

6.4 Bolt force reduction due to nonrotation loosening

6.4.1 Mechanism of nonrotation loosening

Nonrotation loosening occurs due to various causes. In the following, the major causes are listed and the mechanism of each loosening phenomenon is briefly explained [130–132].

(1) Embedment occurring at the interface

There are four kinds of interface, i.e., thread surface, nut-bearing surface, bolt head-bearing surface, and the interface consisting of fastened plates. There exist numerous microprojections on each contact surface, which closely relate to surface roughness. These are deformed plastically when clamping the joint by applying bolt force, so that the mating surfaces composing the interface deform so as to approach each other. This phenomenon of bolt force reduction, due to the plastic deformation, is termed embedment, as introduced in section 2.1.3. Embedment usually progresses with time and can cause a significant reduction in bolt force. In particular, bolt force is largely reduced when the joint is subjected to vibrative external forces.

(2) Plastic deformation around bearing surfaces

High contact pressure exerts on the nut- and bolt head-bearing surfaces owing to the applied bolt force. When the contact pressure exceeds a certain limit, the plate surfaces located under the nut and bolt head are deformed plastically. If such plastic deformations progress with time, loosening occurs without nut rotation due to the depression under the nut- or bolt head-bearing surface. To prevent a depression occurring, the contact pressure should be designed below an appropriate limit value. As introduced in section 2.1.1, limit values of the contact pressure are tabulated in [26]. According to that, the ratio of the limit contact pressure to tensile strength approximately ranges from 70% to 85% for carbon steel, although it may be slightly lower in the case of alloy steel, 35%–50% for stainless steel, and 80%–95% for aluminum and aluminum alloy. In contrast, the ratio ranges from two to four for gray cast iron, reflecting its material property of high compression strength. This means that the limit contact pressure is much higher than the tensile strength.

(3) Progress of plastic deformation due to external force

When subjected to large external force repeatedly, bolt force can be reduced due to the progress of plastic deformation. The mechanism of bolt force reduction is expounded in [130].

(4) Nonrotation loosening due to thermal load

1) Difference in thermal expansion between threaded fastener and fastened plates causes nonrotation loosening. In numerical examples 5.2 and 5.3, concrete calculation examples are given to show how bolt force varies under thermal load. When subjected to thermal load, the joint temperature is raised and the bolt-nut pair and fastened plates are both elongated. In that case, if the amounts of elongation of the two portions are equal, bolt force is unchanged. However, if the elongation of the bolt-nut pair is larger, the bolt force is reduced. In ordinary bolted joints, thermal load is applied to the bolt-nut pair from the surrounding area; hence, there are few cases that the bolt temperature is higher than the temperature of fastened plates. However, if a bolt-nut pair made of stainless steel is used to clamp plates made of carbon steel, bolt force can be reduced even if the bolt-nut temperature is lower than the plate temperature, because the coefficient of linear expansion of stainless steel is about 1.5 times that of carbon steel. When the temperature condition and nominal diameter are identical, loosening is more likely to occur with larger grip length. Nonrotation loosening caused by thermal load is discussed in more detail in Section 6.4.5.

2) Creep and relaxation occurs in bolted joints exposed to high temperature continuously, and loosening due to creep is of critical concern among various causes. In [130], bolt force reduction due to creep is compared between the ones estimated by creep theory and the measured values. From the mechanical point of view, it can be regarded as a kind of residual stress, since recrystallization occurs as the temperature of threaded fasteners is increased. The recrystallization temperature of carbon steel is approximately 500 °C. According to VDI 2230 (2003), relaxation occurs at the temperature above half the recrystallization temperature. Furthermore, some examples are cited on bolt force reduction that the bolt force is significantly reduced above the service temperature of about 240 °C in structural carbon steel and heat-treated steel, and about 160 °C in aluminum alloy.

3) Change in material properties of threaded fasteners can cause nonrotaion loosening. When a bolt-nut pair and fastened plates are made of the same material and are subjected to the same temperature increase, the amounts of thermal expansion generated in the two portions are equal. In this case, unless the temperature is not so high, bolt force is unchanged because equal strains are generated in the two portions. However, if the joint temperature is raised too high, the bolt force is reduced due to the decrease of Young's modulus. Then, the ratio of Young's modulus at raised temperature to the modulus at room temperature is calculated using table 1.4. Calculating the ratios at 200 and 300 °C, they are about 94% and 91% for carbon steel and stainless steel, respectively. At 400 °C, the ratios decrease to 87% for carbon steel and 80% for stainless steel. This indicates that a bolt-nut pair made of stainless steel is more likely to lose bolt force at high temperature.

(5) Stiffness reduction of fastened plates and related nonlinear, hysteresis, and viscoelasticity behaviors

For the purpose of sealing contained fluids, low stiffness gaskets are commonly used when connecting pipe flanges. In sheet gaskets, for example, the relationship between compression force and gasket displacement is non-linear, and hysteresis characteristics are observed between loading and unloading curves. When the flange temperature is raised in the running condition, gasket stiffness is significantly decreased and, therefore, the curve relating gasket contact stress to compression strain significantly shifts toward the lower stiffness side. At the same time, creep strains are generated as in metal materials. These phenomena interact with each other in a complex manner, thereby causing a significant bolt force reduction. Thermal and mechanical behaviors of bolts used for clamping pipe flanges are discussed in Chapter 7.

(6) Others

When clamping plates with painted surfaces by bolt-nut pairs, bolt force reduction can progress with time due to the existence of paint film. The extent of bolt force reduction changes depending on the type of paint, the ratio of paint film thickness to grip length, etc. Specifically, in the case of thin plates thickly painted, the stiffness is greatly different between the paint film and the plates. Therefore, it is predicted that the amount of bolt force reduction largely varies due to a slight plate warpage, the scatter in paint film thickness, etc.

6.4.2 Estimation method of amount of embedment

Embedment is a major cause for nonrotation loosening, in which bolt force is reduced without nut rotation. Since embedment inevitably occurs to some extent, it is highly important, from the viewpoint of joint safety, to predict the amount of embedment δ_z with practical accuracy. Hence, if δ_z is accurately predicted at the design stage, it is possible to estimate the amount of bolt force reduction due to embedment using the spring constants of each part of the bolted joint. In VDI 2230 (1986), an estimating formula for δ_z is proposed:

$$\delta_z = 0.00329\left(L_f/d\right)^{0.34} \tag{6.3}$$

Eq. (6.3) is easy to handle, because δ_z (mm) can be calculated once grip length L_f and nominal diameter d are given. However, it is pointed out that the formula tends to underestimate the amount of δ_z compared to that in actual bolted joints. Therefore, the formula was deleted at the revision of VDI 2230 conducted in 2003. Instead, for three kinds of contact surfaces, i.e., thread surface, nut- and bolt head-bearing surfaces, and plate interface, standard values of embedment δ_z are presented for three levels of surface roughness, although in a slightly passive form [29]. For the actual calculation, external forces are classified into tension, compression, and shear force.

Before the above formula was presented, Junker's proposition had been widely employed, in which the effects of surface roughness, the number of contact surfaces, and the direction of applied external force are taken into account. For instance, in the case of a bolted joint being subjected to axial vibration force, concrete values of embedment δ_z are shown, i.e., 5 μm on thread surface, 4 μm on nut- and bolt head-bearing surfaces and plate interface with relatively large surface roughness, etc. Then, the sum of embedment on each contact surface is regarded as the embedment δ_z of the target bolted joint [143]. In that sense, the standard values given in VDI 2230 (2003) are fairly similar to Junker's proposition.

However, it has been pointed out that Junker's proposition tends to overestimate the amount of embedment. Considering the specific form of contact pressure distribution at the interface, one of the causes for the overestimation could be that, in Junker's proposition, the location of the plate interface existing between nut- and bolt head-bearing surfaces is not taken into account. As an example, supposing that a bolted joint is composed of two plates and grip length is constant, the plate interface is located in the center of the grip length if the thickness of the two plates is identical. Then, the contact pressure necessarily exerts on a wide area. In contrast, for the combination of two plates with different thickness, higher contact pressure is applied to a narrower area as the plate interface approaches the nut- or bolt head-bearing surface. In Fig. 6.9, the relationship between the location of the interface and the form of contact pressure distribution is schematically illustrated.

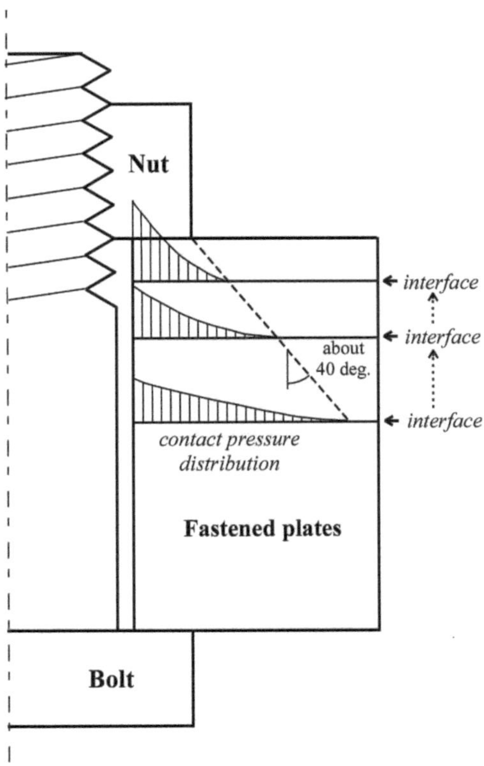

Fig. 6.9 Relationship between location of interface and contact pressure distribution.

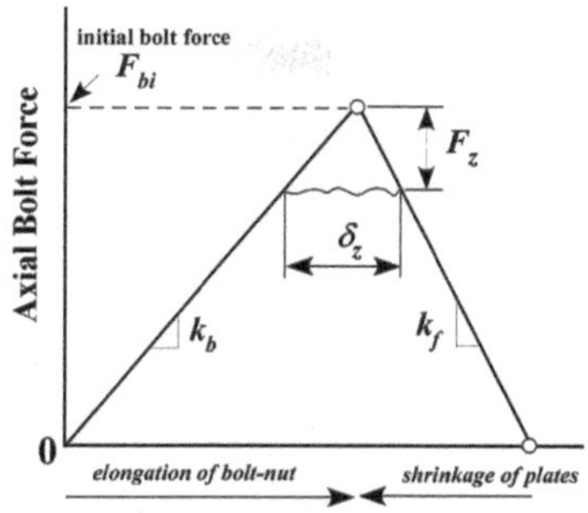

Fig. 6.10 Relationship between bolt force reduction F_z and embedment δ_z.

6.4.3 Bolt force reduction due to embedment

If the amount of embedment δ_z is given, the resulting bolt force reduction F_z can be calculated by using the triangular section of a bolted joint diagram, shown in fig. 4.32. It is designated as a joint triangle. In Fig. 6.10, the relationship between F_z and δ_z is illustrated by use of a joint triangle. In the joint triangle, the base length represents the sum of the elongation of bolt-nut connection δ_b and the shrinkage of fastened plates δ_f. The gradients of left and right sides represent the spring constants of bolt-nut connection k_b and fastened plates k_f in eq. (2.5), respectively. The wavy line in Fig.6.10 shows the amount of δ_z, which is the sum of embedment produced on the four kinds of interface explained in Section 6.4.1. That is, bolt force reduction F_z can be calculated by the total amount of embedment δ_z summed up for the four kinds of interface. Accordingly, since the joint triangle is similar to a triangle whose base represents δ_z, the height of the latter triangle is equal to bolt force reduction F_z. Hence, F_z is calculated by the following equation:

$$F_z = \left(\frac{1}{k_b} + \frac{1}{k_f}\right)^{-1} \cdot \delta_z = Z \cdot \delta_z \tag{6.4}$$

In Eq. (6.4), Z corresponds to the spring constant that connects k_b and k_f in series. It represents the total stiffness of a bolted joint and is termed embedment factor. Therefore, Z is equal to k_{total} in eq. (2.5). Eq. (6.4) provides the important information that if δ_z is constant and bolts and nuts with identical nominal diameter are used, bolt force reduction F_z is decreased as k_b and k_f are decreased. In Fig. 6.11, how F_z varies depending on the amounts of k_b and k_f when δ_z is constant is shown. In a low stiffness joint, where k_b and k_f are both small, bolt force reduction F_z becomes small for the same amount of δ_z. In the figure, the effectiveness of the use of a low stiffness bolt with small k_b is also shown. The above results indicate that the lower the joint stiffness, the smaller the bolt force reduction due to embedment.

On the other hand, since the stiffness of thin plates is high, k_b and k_f are both large. Hence, care must be taken because large bolt force reduction occurs for the same amount of embedment in comparison with thick plates. Additionally, the estimation accuracy of bolt force reduction F_z in thin plates is greatly influenced not only by the

Fig. 6.11 Effect of joint stiffness on bolt force reduction F_z.

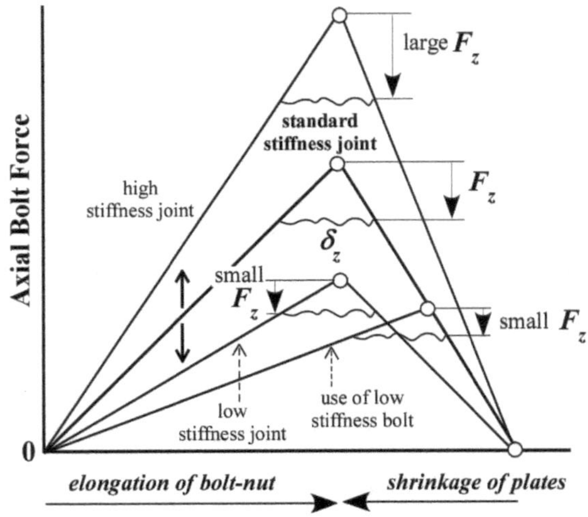

amount of embedment δ_z but also by the evaluation accuracy of joint stiffness, k_b and k_f. In this way, embedment is of critical concern in the bolted joint composed of thin plates, but the accurate prediction of δ_z is quite difficult. The reasons are listed below.

1) In a bolted joint composed of thin plates, k_b and k_f are both large, hence, even a small amount of embedment δ_z produces large bolt force reduction F_z, as evident from Eq. (6.4).
2) Equivalent length of engaged threads L_{th} significantly changes due to the effect of inclination angle existing on the nut-bearing surface. As shown in fig. 2.8B and C, it varies approximately from 0.85*d* to 1.70*d*, where *d* designates nominal diameter. When the spring constant of engaged threads k_b is decreased due to geometric errors existing around the nut-bearing surface, etc., the amount of bolt force reduction F_z can be smaller than the predicted value. This phenomenon particularly appears in the tightening process of thin plates because its spring constant k_b is largely reduced.
3) When tightening warped or distorted thin plates, the actual bolt force reduction F_z becomes smaller than the predicted value, because spring constant of fastened plates k_f becomes smaller than the theoretical value. However, when subjected to vibration forces, the amount of embedment δ_z is increased and the bolt force reduction is progressed, which makes the precise prediction of F_z difficult.
4) According to Eq. (6.4), the amount of bolt force reduction F_z is basically independent of initial bolt force F_{bi}. However, setting F_{bi} higher, the value of k_f can be significantly changed, because thin plates get into close contact so as to cancel the effect of warping or distortion. In contrast, in the case of initial bolt force F_{bi} being low, mating surfaces are in discontinuous contact condition, hence, the embedment is progressed under vibration forces, etc. As a result, bolt force can be significantly reduced.
5) Embedment factor given in Eq. (6.4) represents the total stiffness of a bolted joint and it is almost proportional to nominal diameter *d*. Therefore, for an identical amount of embedment δ_z, bolt force reduction F_z increases nearly in proportion to *d*. However, replacing F_z by bolt stress reduction $\Delta\sigma_b$ and considering that the bolt cylindrical area is proportional to the square of *d*, it is found that $\Delta\sigma_b$ is reduced in almost inverse proportion to nominal diameter

d. In other words, the smaller the nominal diameter of bolts and nuts, the more likely bolt stress is to be reduced due to embedment.

Learning by numerical example 6.2: Effect of embedment and grip length on bolt stress reduction

Using Eq. (6.4), bolt force reduction due to embedment F_z is calculated and the bolt stress reduction $\Delta\sigma_b$ is evaluated. For a bolted joint clamped by a pair of bolt and nut of M16 with coarse thread, assuming δ_z is 12 μm, bolt force reduction is calculated under the same condition as in numerical example 6.1. The ratio of grip length L_f to nominal diameter d, L_f/d, is changed to 1, 2, 4, 6, 8, and 10. The amounts of $\Delta\sigma_b$ are calculated to be 45.8, 30.6, 18.4, 13.1, 10.2, and 8.4 MPa for the ascending order of L_f/d. This indicates that a significant bolt stress reduction occurs due to embedment in the joint consisting of thin plates.

Learning by numerical example 6.3: Relationship between embedment and nominal diameter

For a bolted joint of $L_f/d=1$, nominal diameter of a bolt-nut pair is changed to M12, M16, M24, and M36. The joint configurations are all supposed to be similar. Assuming δ_z is 12 μm, bolt stress reduction $\Delta\sigma_b$ is calculated to be 59.4, 45.8, 30.6, and 20.7 MPa in the order of increasing nominal diameter. The result demonstrates the statement given in 5) that larger bolt stress reduction occurs with smaller nominal diameter for an identical amount of embedment.

Learning by numerical example 6.4: Relationship between embedment and equivalent length of engaged threads

For a bolted joint of $L_f/d=1$ clamped by a pair of bolt and nut of M16, bolt stress reduction $\Delta\sigma_b$ is compared for the cases of equivalent length of engaged threads L_{th} being $0.85d$ and $1.5d$. For $L_{th}=0.85d$, $\Delta\sigma_b$ is 45.8 MPa, as calculated above. When L_{th} is increased to $1.5d$ due to geometric error existing around nut-bearing surface, etc., $\Delta\sigma_b$ is reduced to 36.5 MPa. This calculation example shows the difficulty in the estimation of bolt stress reduction in thin plates.

6.4.4 Prevention methods of nonrotation loosening

It is essentially impossible to completely prevent nonrotation loosening, unlike rotation loosening, because its loosening phenomenon is influenced by microscopic plastic deformations on the mating surfaces. In other words, nonrotation loosening inevitably occurs to a greater or lesser extent. In the following, some countermeasures effective in suppressing nonrotation loosening are listed [130,131].

(1) Use of low stiffness bolt

When a certain amount of embedment δ_z is inevitably generated, it is effective to reduce embedment factor Z for suppressing bolt force reduction F_z. Hence, thin and long bolts are used in order to decrease the spring constant of bolt-nut connection k_b. Another method is to use bolts and nuts made of materials with low Young's modulus. Except for the issues concerning cost, strength, and thermal properties, titanium-based materials and aluminum alloys are good candidates from the viewpoint of low stiffness materials. The reason is that Young's modulus is about half in the former case and is about one third in the latter case, comparing to the modulus of carbon steel.

(2) Reducing the stiffness of fastened plates

It is also effective to lower the spring constant of fastened plates k_f for suppressing bolt force reduction F_z. When grip length L_f is fixed, k_f can be lowered by reducing the diameter of fastened plates around the jointed portion or inserting a spacer, as shown in the right illustration of fig. 4.58B. In this case, k_b is also effectively reduced because of the use of a long bolt. However, attention should be paid that the usage of spacer increases the number of contact surfaces. One way to avoid the increase of the number of contact surfaces is to employ the joint shape in the left illustration of fig. 4.58B.

(3) Suppression of embedment

In order to suppress the amount of embedment, it is effective to reduce the surface roughness of fastened plates. Especially in thin plates, it is important to decrease the warping and distortion and improve the flatness. Since the amount of embedment δ_z increases as the number of contact surfaces is increased, the target jointed portion should preferably be designed to have fewer contact surfaces.

(4) Suppression of reduction rate of bolt force

Bolt force reduction due to embedment F_z is basically constant regardless of the amount of initial bolt force F_{bi}, as found from Eq. (6.4). Hence, raising F_{bi} to the extent possible, the reduction rate of bolt force can be lowered.

(5) Prevention of depression of bearing surfaces

When requiring a high bolt stress in the joint consisting of a bolt-nut pair made of high strength material and fastened plates made of low strength material, the contact pressure can be lowered by expanding the bearing surface area by use of washers or hexagon bolts and nuts with flange.

(6) Suppression of wear at the interface

In a bolted joint subjected to vibration force, materials with high wear resistance are applied to the portions composing the interface for preventing fretting wear.

(7) Suppression of the difference in thermal expansion

Considering the material combination of threaded fasteners and fastened plates and the heat flow pattern around the joint, the difference in thermal expansion between a bolt-nut pair and fastened plates, which is the major cause of bolt force variation, should be suppressed to as little as possible. In regard to this, the application of long bolts, introduced in (1), to the joint subjected to thermal load can increase the difference in thermal expansion, since the amount of elongation is proportional to its length. Therefore, when using long bolts aiming at the suppression of embedment, attention must be paid to the differences in temperature and coefficient of linear expansion between threaded fasteners and fastened plates.

6.4.5 Loosening due to difference in thermal expansion

It is explained in (4) and (5) of Section 6.4.1 how bolt force reduction occurs when bolted joints are subjected to thermal load. In this section, the mechanism of nonrotation loosening, caused by the difference in thermal expansion occurring in the joint, is studied referring to basic theory of thermal stress and numerical results by FEM, which were introduced in Chapter 5.

1) Effect of joint materials

With smaller thermal conductivity of joint materials, the differences in temperature and thermal expansion between a bolt-nut pair and fastened plates are increased; hence, bolt force variation is increased. When bolted joints are heated, bolt stress of the joint made of stainless steel becomes higher than those made of carbon steel and aluminum alloy, as shown in fig. 5.9. This also suggests that the joint made of stainless steel is more likely to loosen under cold energy.

2) Effect of the difference in coefficient of linear expansion

In a bolted joint consisting of a bolt-nut pair and fastened plates made of different materials, nonrotation loosening tends to occur when the coefficient of linear expansion of the former is larger than that of the latter, as shown in fig. 5.11. The amounts of coefficient of linear expansion vary depending on the service temperature. For the commonly used materials of carbon steel, stainless steel, and aluminum alloy, the values at room temperature are 11.8×10^{-6}, 17.3×10^{-6}, and 23.2×10^{-6}, in the above order, according to table 1.5. When dealing with the joints consisting of different materials shown above, particular attention must be given to the bolt force variation due to the difference in thermal expansion. Also important is that even for the combination of the same kind of materials, bolt force can be reduced to some extent or more under large temperature increase, if a certain amount of difference exists between the coefficients of the two materials.

Meanwhile, the coefficient of linear expansion for austenitic stainless steel is cited from [23]. On the other hand, the coefficient given in [24] is 13.6×10^{-6}. In this text, the former value is employed, because this value was newly measured just before its publication. If using the latter value, the calculated bolt force reduction can be considerably smaller than the actual one. Incidentally, the coefficient of linear expansion of titanium is very small, as shown in table 1.5. Utilizing its characteristics, a research paper has been published [25] that proposes a design procedure for a joint that is hard to loosen against thermal load.

3) Effect of grip length

The amount of bolt force reduction is increased with increasing grip length, as found from eqs. (5.7)–(5.9). Numerical results by three-dimensional FEM indicate the same tendency. When using long bolts to improve the fatigue strength, attention should also be given to the bolt force variation due to thermal expansion, if the target joint is subjected to thermal load at the same time. In many cases, the advantage brought by the use of long bolts seems to surpass the disadvantage caused by larger expansion due to thermal load.

Learning by numerical example 6.5: Nonrotation loosening due to thermal load and coefficient of linear expansion

Assuming that the coefficient of linear expansion of stainless steel is, tentatively, 13.6×10^{-6}, as given in [24], under the same condition as in numerical example 5.2, bolt stress reduction $\Delta\sigma_b$ is calculated to be 15.0, 18.0, 19.3, 20.0, and 20.4 MPa for bolted joints of $L_f/d = 2, 4, 6, 8$, and 10. Comparing these results to those obtained by assuming the coefficient to be 17.3×10^{-6}, $\Delta\sigma_b$ is approximately reduced to about one-third. Although these calculation results can be predicted to some extent from eq. (5.9), it is of significant importance to use precise values of coefficient of linear expansion when designing bolted joints subjected to thermal load.

6.5 Inspection of bolt tightening state by torque measurement [144]

6.5.1 Inspection methods of bolted joint integrity

The integrity of a bolted joint is sometimes checked after some running time, i.e., whether the target bolt is properly tightened with prescribed bolt force within the acceptable error. Fig. 6.12 shows the summary of three types of inspection methods

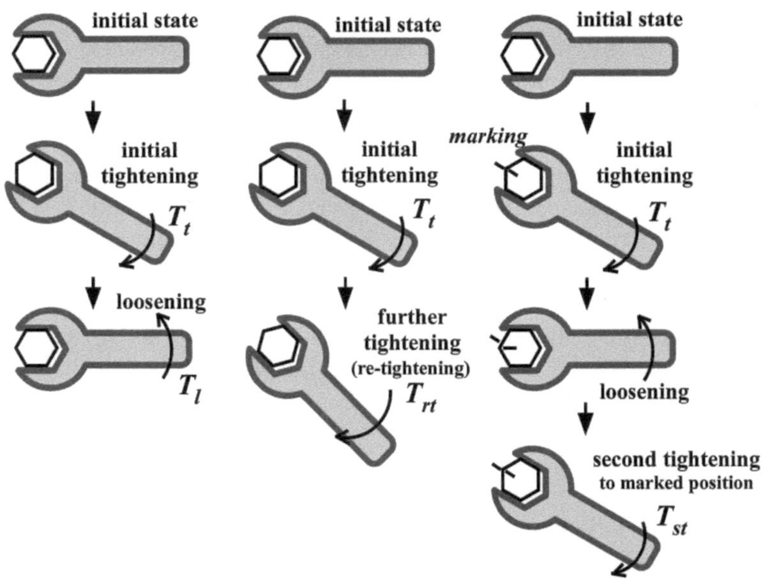

Fig. 6.12 Summary of commonly used inspection methods. (A) Release torque method, (B) re-tightening torque method, and (C) marking method.

commonly used, i.e., release torque method, retightening torque method, and marking method. In all cases, the inspection is carried out based on the torque measurement.

(1) Release torque method

The judgment is made by the magnitude of loosening torque T_l required for loosening the target bolt in the clamped state. For instance, the allowable range of the ratio of T_l to the initial tightening torque T_t is set to be between 0.6 and 0.9, based on experience. The torque ratio T_l/T_t varies according to nominal diameter d, thread pitch P, and coefficient of friction μ. Although the target bolt needs to be retightened, it is rather widely applied to small bolts because of its relatively easy work.

(2) Retightening torque method

The target bolt in the clamped state is further tightened for inspection. The judgment is made by the magnitude of applied torque when the nut rotation starts, which is termed retightening torque T_{rt} herein. T_{rt} tends to be larger than the initial tightening torque T_t, because the coefficient of friction in the former is likely to be larger than the latter, due to the difference between static friction and dynamic friction. The allowable range of T_{rt}/T_t is set to be, for instance, between 0.9 and 1.2.

(3) Marking method

When the tightening operation is completed, a matching mark is drawn between plate surface and nut or bolt head. The target bolt is once loosened and tightened again to the marked point. The judgment is made by the magnitude of the torque required at that time, which is termed second tightening torque T_{st} herein. There is a case that the allowable range of T_{st}/T_t is set to be between 0.9 and 1.1.

In the next section, the characteristics of the three methods are examined using the torque–bolt force relationship derived in section 3.2, under the following calculation conditions:

Nominal diameter d*: M4, M6, M8, M10, M12, M16, M24, M36, M48, and M64.*
Thread pitch P*: coarse screw thread, fine screw thread, and ultra-fine screw thread.*
Coefficient of friction μ, μ^**: 0.05, 0.07, 0.1, 0.15, 0.2, and 0.3.*

It is assumed that the coefficients of friction on thread surface μ_{th} and nut-bearing surface μ_{nu} are equal. Coefficient of friction without asterisk, μ, is the one in the initial tightening process. The coefficient with asterisk, μ^*, represents the one for the cases other than the initial tightening, because its amount varies to some extent from the initial value μ.

6.5.2 Release torque method

The relationship between tightening torque T_t and bolt force F_b is given in eq. (3.15).

$$T_t = \frac{1}{2} F_b \left\{ d_2 \tan\left(\rho'_{th} + \beta\right) + \mu_{nu} d_{nu} \right\} \qquad \text{(3.15 reshown)}$$

The equation for loosening torque T_l is given in eq. (3.19), and it is rewritten to express the loosening process in the release torque method.

$$T_l = \frac{1}{2}F_b^*\left\{d_2 \tan\left(\rho_{th}'^{\,*} - \beta\right) + \mu_{nu}{}^* d_{nu}\right\} \tag{6.5}$$

Asterisks are added to bolt force F_b and friction angle of thread surface ρ'_{th} besides μ_{nu}, which may be changed due to the change in coefficients of friction. In the following calculations, the bolt forces are supposed to be unchanged for simplicity, between the initial tightening process and the inspection stage, i.e., $F_b = F_b^*$.

Fig. 6.13 shows the relationship between the torque ratio T_l/T_t and nominal diameter d. The parameter is coefficient of dynamic friction μ relating to the initial tightening torque T_t. The calculations are performed by equating μ^* to μ. It is found that the amount of T_l/T_t is significantly influenced by μ, especially in the range of small μ. This result can be predicted by the calculation results in numerical example 3.5 in section 3.2.4. Furthermore, T_l/T_t is changeable in small bolts, say, M12 and smaller. The calculation results seem to deny the conventional method, i.e., release torque method has commonly been applied to small bolts. Although not shown here, the effect of pitch size of bolts with identical nominal diameter is not so significant on the inspection accuracy, compared to that of coefficient of friction. The amount of T_l/T_t changes for varying thread pitch by up to about 10%.

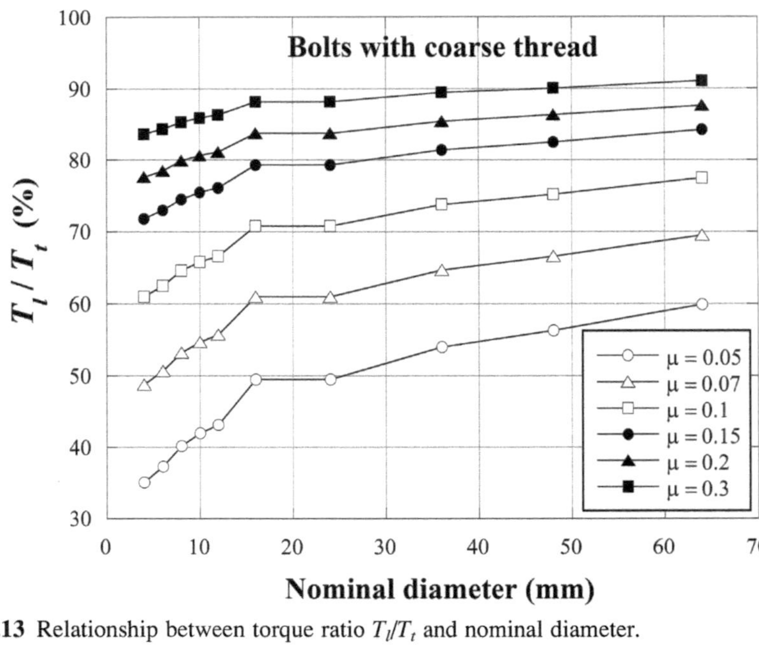

Fig. 6.13 Relationship between torque ratio T_l/T_t and nominal diameter.

Table 6.1 Ratio of retightening torque to initial tightening torque in case of $\mu = 0.15$ (%).

μ^*/μ	0.9	1.05	1.1	1.2	1.3	1.4	1.5
M6	91.3	104.4	108.7	117.5	126.2	135	143.7
M16	91	104.5	109	118.1	127.1	136.1	145.2
M36	90.9	104.6	109.1	118.3	127.4	136.5	145.7
M64	90.7	104.6	109.3	118.5	127.8	137.1	146.3
M16 $P=1.5$	90.7	104.6	109.3	118.5	127.8	137.1	146.3
M16 $P=1.0$	90.5	104.7	109.5	119	128.5	138	147.5

6.5.3 Retightening torque method

The relationship between retightening torque T_{rt} and bolt force F_b^* thus obtained is given in the same form as eq. (3.15):

$$T_{rt} = \frac{1}{2} F_b{}^* \left\{ d_2 \tan\left(\rho'_{th}{}^* + \beta \right) + \mu_{nu}{}^* d_{nu} \right\} \tag{6.6}$$

The ratio of T_{rt} to initial tightening torque T_t, given in eq. (3.15), tends to be larger than unity due to the difference between coefficients of static friction μ^* and dynamic friction μ, as stated in Section 6.5.1. In this section, the ratio μ^*/μ is tentatively changed to 0.9, 1.05, 1.1, 1.2, 1.3, 1.4, and 1.5. In Table 6.1, the calculation results of torque ratio T_{rt}/T_t are given for varying μ^*/μ, in which μ is set to be 0.15. It shows that the effects of nominal diameter d and thread pitch P are negligible on the inspection accuracy, comparing to that of coefficient of friction. Fig. 6.14 represents the relationship between T_{rt}/T_t and coefficient of dynamic friction μ in the initial tightening process. It is found that the amount of T_{rt}/T_t is fairly close to the value of μ^*/μ. Accordingly, it is concluded that the inspection accuracy depends on how accurately the amount of μ^*/μ is estimated.

6.5.4 Marking method

The relationship between second tightening torque T_{st} and bolt force is given in the same form as Eq. (6.6) just by changing T_{rt} into T_{st}. The inspection accuracy is examined by the following equation, which is derived by the equations for T_{st} and T_t:

$$\frac{F_b^*}{F_b} \cdot \frac{T_t}{T_{st}} = \frac{d_2 \tan\left(\rho'_{th} + \beta\right) + \mu_{nu} d_{nu}}{d_2 \tan\left(\rho'^*{}_{th} + \beta\right) + \mu_{nu}^* d_{nu}} \tag{6.7}$$

Using the above equation, the sensitivity of μ^*/μ to F_b^*/F_b is evaluated. The ratio μ^*/μ is changed to 0.7, 0.8, 0.9, 1.1, 1.2, and 1.3, considering the scatter in coefficient of friction. Fig. 6.15 shows the numerical results for varying coefficient of friction μ. The ratio F_b^*/F_b is to be obtained by dividing the calculated value shown in the figure by the measured value of T_t/T_{st}. It is found from the figure that the product of F_b^*/F_b and T_t/T_{st}

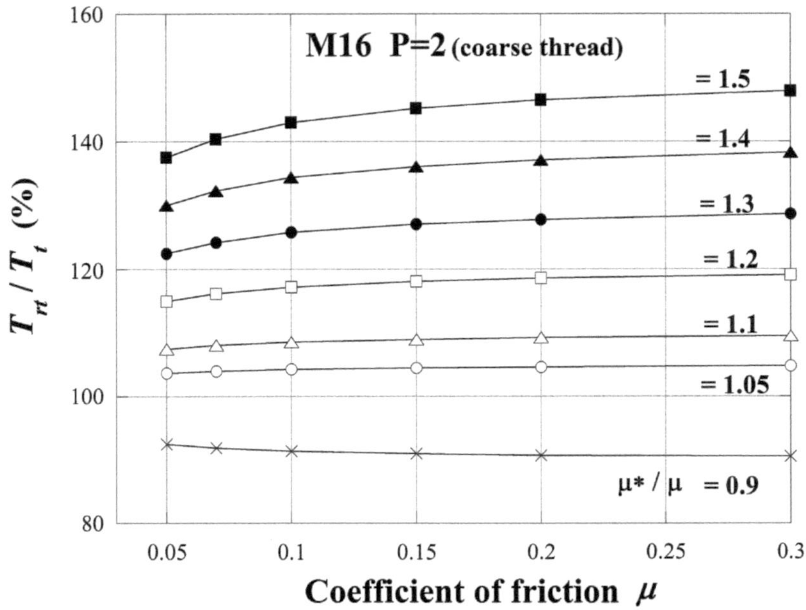

Fig. 6.14 Relationship between torque ratio T_{rt}/T_t and coefficient of friction.

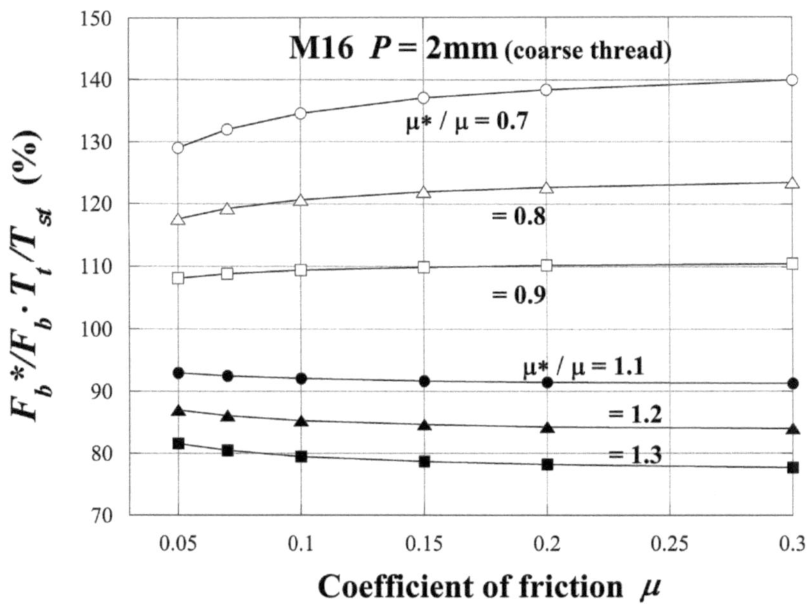

Fig. 6.15 Characteristics of marking method for varying coefficient of friction.

is nearly equal to the reciprocal of μ^*/μ. The allowable rage of T_t/T_{st} being from 0.9 to 1.1, which is determined empirically, roughly corresponds to the range of μ^*/μ, being from 0.9 to 1.1, if the bolt forces are supposed to be almost unchanged, $F_b = F_b^*$.

6.5.5 Verification of inspection methods by experiment and inspection guidelines

In this section, the three inspection methods discussed in the foregoing sections are examined by experiments. Test equipment shown in fig. 3.12 is used for experimentation [71], and 20 pairs of bolt and nut of M16 are prepared for each inspection method. Initial bolt stress is set to be 100 MPa. Fig. 6.16 shows the experimental results of measured torque ratios, T_l/T_t for release torque method, T_{rt}/T_t for retightening torque method, and T_{st}/T_t for marking method. The average values and standard deviation σ of T_l/T_t, T_{rt}/T_t, and T_{st}/T_t are presented in the figure. The former values are found to be about 0.79, 1.02, and 0.98 for each inspection method, all of which stay within the allowable range empirically determined, i.e., from 0.6 to 0.9

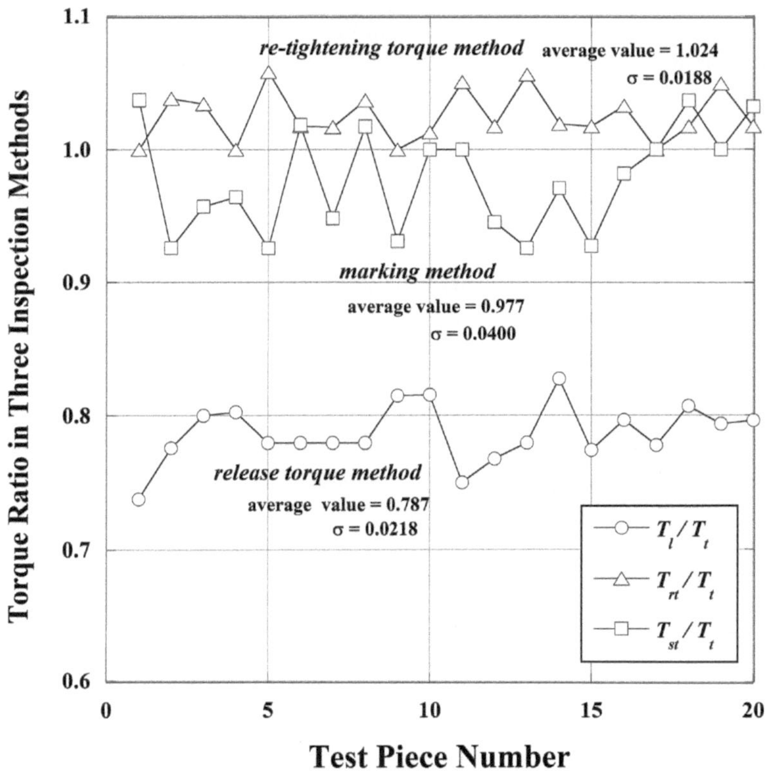

Fig. 6.16 Experimental results of torque ratios in three types of inspection methods.

Table 6.2 Statistical values of measured coefficients of friction.

	μ_{th}	μ_{nu}	μ_{th^*}	μ_{nu^*}
Release torque method				
Average value	0.146	0.125	0.144	0.137
σ	0.0082	0.0122	0.0088	0.0143
2σ/average (%)	11.2	19.6	12.2	21.0
Retightening torque method				
Average value	0.142	0.124	0.140	0.129
σ	0.0111	0.0136	0.0108	0.0152
2σ/average (%)	15.6	21.8	15.4	23.6
Marking method				
Average value	0.136	0.128	0.140	0.138
σ	0.0087	0.0133	0.0107	0.0192
2σ/average (%)	12.8	20.7	15.4	27.7

for T_l/T_t, from 0.9 to 1.2 for T_{rt}/T_t, and from 0.9 to 1.1 for T_{st}/T_t. As for the scatters generating in the three methods, the ratios of 2σ to average value are calculated to be 5.6%, 3.7%, and 8.2%, in the order of introduction.

The statistical results suggest that retightening torque method gives a relatively high inspection accuracy. One more concluding remark is that release torque method is preferably applied to the bolts, say, M16 and larger while avoiding the contact conditions with very small coefficient of friction.

In Table 6.2, measured values of coefficients of friction are compared among the three methods. The ratio of 2σ to average value is calculated for evaluating the scatter. The values of coefficient of friction on thread surface, μ_{th} and μ^*_{th}, are about 0.14, and no significant change is observed between the initial tightening process and the inspection stage. Those on nut-bearing surface, μ_{nu} and μ^*_{nu}, are about 0.13 and the scatter is larger than the case of μ_{th} and μ^*_{th}. Next, the variation of coefficients of friction is compared in detail between the initial tightening process and the inspection stage. The variation rates of μ_{th} and μ_{nu} are about 1.7% and 9.5% for release torque method, 1.5% and 3.9% for retightening torque method, and 2.9% and 7.6% for marking method. This shows that the variation in μ_{nu} becomes relatively large in release torque method and marking method.

Thermal and mechanical behaviors of pipe flange connections

7

7.1 Thermal and mechanical behaviors inherent to pipe flange connections

(1) Gaskets and bolt force govern the performance of pipe flange connections

Pipe flange connections are important machine elements used for connecting pipelines. Ordinarily, pipeline sealing performance is ensured by inserting a low stiffness gasket between a pair of flanges. This situation can be expressed by simple spring models shown in fig. 2.4. That is, a low stiffness spring is inserted between two high stiffness springs, in which the former and latter springs correspond to a gasket and a pair of flanges, respectively. Therefore, the mechanical behavior of the entire pipe flange connection is dominated by the compression characteristics of the low stiffness gasket, because the springs are connected in series. For instance, widely used sheet gaskets exhibit a nonlinear relationship between gasket contact stress and strain, in addition to hysteresis characteristics between loading and unloading curves. Furthermore, the gasket compression characteristics vary with temperature. Thus, it is very difficult to predict the bolt force variation in pipe flange connections. Concretely, it is extremely difficult to evaluate the transition process from the initial state to the running state and then to the shutdown operation.

(2) Tightening process of the bolts used for clamping a pipe flange connection and its sealing performance

For clamping pipe flange connections, a number of bolts equal to a multiple of four are commonly used. Each bolt is usually tightened sequentially, by which gasket contact pressure is generated. If the bolt force is reduced due to various reasons, such as poor management of initial bolt force, the action of external force, or thermal load in the running condition, etc., the leakage of contained fluids can occur. On the other hand, too high a bolt force can cause a failure or rupture of the gasket or excessive plastic deformations in the pipe flange body. Therefore, the bolts used for clamping pipe flange connections do not need to have high fatigue strength, but must have the characteristics that the bolt forces are unlikely to change when subjected to mechanical external force or thermal load.

Sealing performance is controlled by gasket contact pressure. The magnitude of gasket contact pressure at a selected point on the flange interface is determined not only by a single bolt force but also by the bolt forces generated in the multiple bolts located around the objective point. This means that the contact pressure at a certain point is determined by the action of multiple bolts. In that sense, a certain amount of bolt force scatter is permissible if the scatter in gasket contact pressure stays within an

The Mechanics of Threaded Fasteners and Bolted Joints for Engineering and Design.
https://doi.org/10.1016/B978-0-323-95357-3.00007-1

allowable range. Accordingly, the accuracy required for the bolt force management is relatively lower than that for the bolts used in machines and structures, in which fatigue strength is of primary importance. However, since multiple bolts are tightened sequentially in pipe flange connections, there is a difficult problem of elastic interaction, introduced in section 3.7, by which large bolt force scatter can be generated.

(3) Effect of type of gasket

Gasket compression characteristics are largely influenced by the gasket materials. In the case of metal gaskets, the relationship between compression stress and strain is linear, but the relationship is nonlinear in widely used spiral wound gaskets and joint sheet gaskets.

7.2 Gasket compression characteristics and flange rotation

Fig. 7.1A schematically shows the relationship between gasket contact stress and strain of sheet gaskets and spiral wound gaskets, in which loading curves are comparatively smooth and convex downward. Spiral wound gaskets are usually used at higher temperature and pressure, comparing to sheet gaskets. On the other hand, in the case of gaskets that contain PTFE (polytetrafluoroethylene), the relationship between gasket contact stress and strain shows a very distinctive shape, as shown in Fig. 7.1B.

Incidentally, when high pressure fluids flow in the pipeline, flange rotation occurs in the bolt tightening process, as explained in section 3.7.2, because large or small raised face flanges, illustrated in fig. 3.42, are used. Hence, the gasket contact pressure distributes in a specific pattern, shown in fig. 3.43A, where it steeply increases from inside to outside. Accordingly, pipe flange connections behave in a very complicated manner when subjected to inner pressure and thermal load, because of the specific pressure distribution pattern and the compression characteristics of the objective gasket.

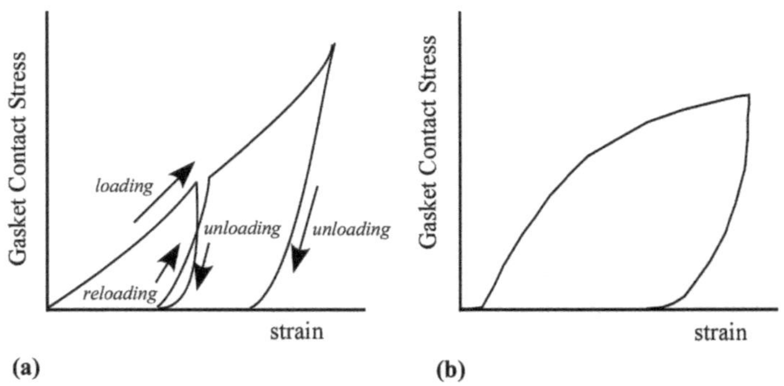

Fig. 7.1 Nonlinear relationship between gasket contact stress and strain. (A) Sheet gasket, spiral wound gasket and (B) PTFE gasket.

Nonlinear and hysteresis properties, shown in Fig. 7.1, are formulated in various manners. In the case of spiral wound gaskets, for instance, gasket contact stress in the loading and unloading curves can be expressed by sixth degree function and exponential function of compression strain [89], respectively. As for sheet gaskets, it is proposed that loading and unloading curves are both formulated as exponential function of compression strain [90]. Meanwhile, some high-performance commercial structural analysis programs provide a special element, termed gasket element. To use the special elements, loading and unloading curves need to be measured in advance by experiment, and then the relationship between gasket contact stress and strain can be reproduced as continuous functions by using the discrete data of the measured values.

7.3 Temperature dependency of gasket compression characteristics

In many cases, high or low temperature fluids flow in pipelines. Accordingly, flange and gasket temperatures are significantly changed during the three stages, i.e., immediately after the start of plant operation, in the steady state operation, and after plant shutdown. During these processes, compression characteristics of sheet gaskets and spiral wound gaskets are greatly varied depending on the gasket temperature. In order to evaluate the changes in bolt forces, which govern the sealing performance of pipe flange connections in the process from the start of plant operation to the plant shutdown, the temperature dependency of compression characteristics of the target gasket must be taken into consideration. Test equipment, shown in Fig. 7.2, has been developed for measuring the relationship between gasket contact stress and strain for varying gasket temperature, aiming at covering the service temperature of sheet gaskets [145].

The measured results of compression characteristics of aramid sheet gaskets obtained by the test equipment are shown in Fig. 7.3A and B. Gasket temperatures are set to be at room temperature and 100 °C, respectively. It is found that the compression curves greatly shift towards the low stiffness side at elevated temperature of 100 °C. The stiffness of the target gasket is significantly decreased in the range from room temperature to 75 °C. Then, the reduction rate of gasket stiffness is decreased above 75 °C. The experimental results are explained in detail in [145]. Since the gasket stiffness is significantly influenced by the surrounding temperature in this manner, it is quite difficult to evaluate the mechanical behavior of pipe flange connections under thermal load with high accuracy. In the next section, the relationship between gasket contact stress and strain, measured by the test equipment shown in Fig. 7.2, is incorporated into FE analysis. Then, the numerical results are presented on the variations of bolt temperature and bolt force of the pipe flange connection with time, corresponding to the three stages, i.e., immediately after the start of plant operation, in the steady state operation, and after the plant shutdown.

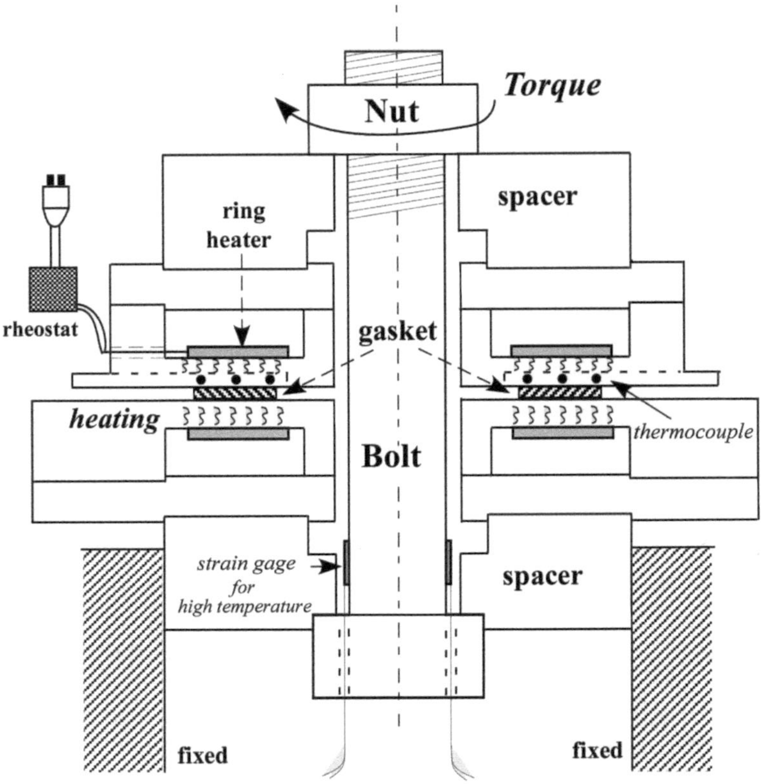

Fig. 7.2 Test equipment for measuring the relationship between gasket contact stress and strain at elevated temperature.

7.4 Analysis of thermal and mechanical behaviors in running condition and during shutdown operation

From the viewpoint of the safety management of pipelines, the sealing of high temperature fluids is the most important issue. Plant assembly is usually carried out at room temperature. Then, when the plant operation is started, a difference in thermal expansion occurs among the parts composing the pipe flange connection due to the action of high temperature fluids flowing in the pipeline. As a result, bolt forces that clamp pipe flange connections necessarily vary to a greater or lesser extent. Meanwhile, the dimensions, shapes, and materials of bolts, nuts, and pipe flanges used for the target pipeline are determined according to fluid type, flow rate, fluid pressure and temperature, etc. Even for pipe flanges made of the same material, their shapes are not similar to each other, with varying nominal size, as explained in section 3.7. In general, the larger the flange size, the less the flange stiffness.

Fig. 7.4 shows a finite element model of the relatively small pipe flange that was used for obtaining fig. 3.45B, with large raised face of inner diameter of 50 mm and nominal pressure 40 K specified in JIS. Its stiffness is considerably higher than that of the large pipe flange of 20 in. in inner diameter, shown in fig. 3.44. Accordingly, in order to evaluate thermal and mechanical behaviors of pipe flange connections under thermal load with practical accuracy, it is required to introduce a proper numerical method that can consider the effect of nominal size. Since there is a limit to evaluation by experiment, a growing number of research papers by FE analysis have been published in the conference held by Pressure Vessel and Piping Division of ASME. In the following, the numerical results by FEM are presented, in which gasket compression characteristics at high temperature are taken into account.

Fig. 7.5A and B illustrates the shape of the target flange specified in JIS, nominal pressure 20 and inner diameter of 65 mm, and the corresponding axisymmetric FE model [146,147]. It has been confirmed that axisymmetric models can be used

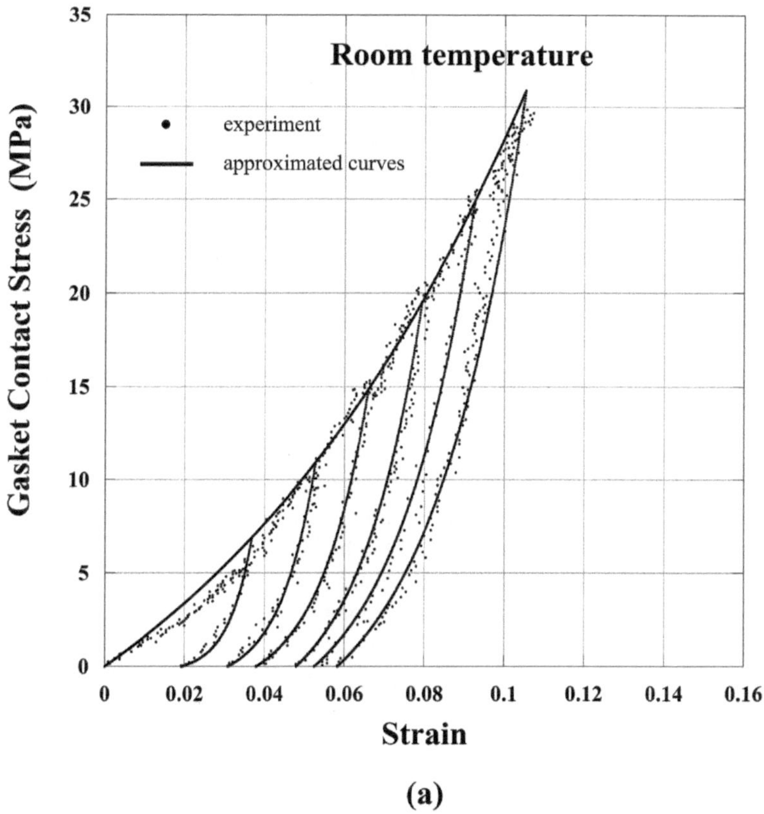

Fig. 7.3 Measured results of compression characteristics of aramid sheet gaskets. (A) At room temperature and (B) at 100°C.

(Continued)

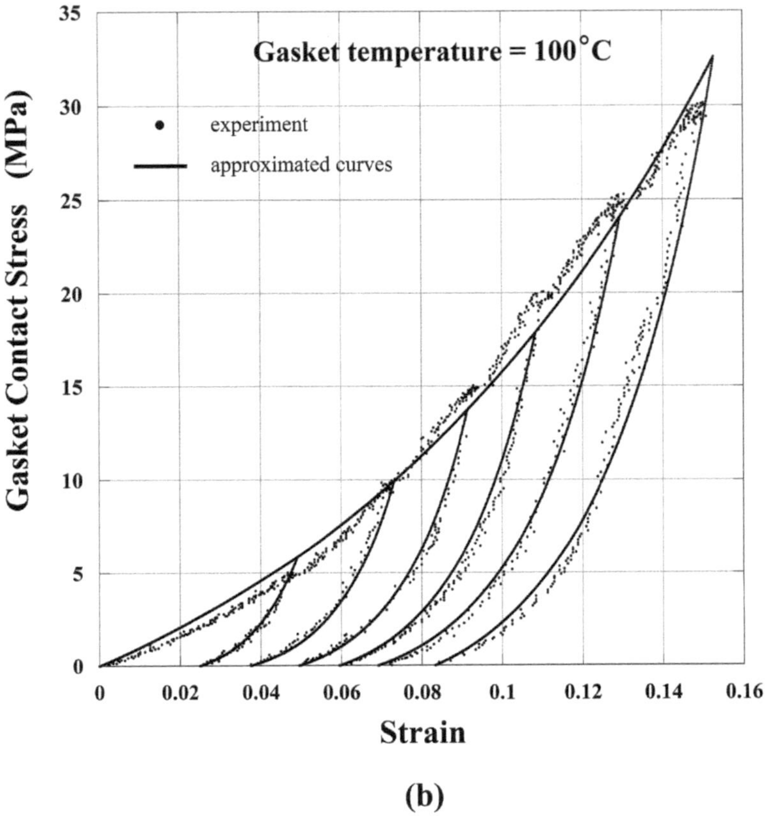

Fig. 7.3, cont'd

effectively if the primary purpose is the evaluation of bolt force variation due to thermal load. In this case, however, there are two geometric problems, i.e., bolt shape and the bolt hole existing in the flange. These problems are solved by adjusting the relevant stiffness by changing Young's modulus of the bolt body and bolt hole portion. Details of FE models and boundary conditions are given in [146,147], together with the procedure for how to treat the temperature dependency of gasket compression properties. In Fig. 7.6, variations of bolt force and bolt temperature with time are presented in the process from the start of plant operation to the plant shutdown 30,000 s later and in the process of returning to room temperature for another 30,000 s [147]. Bolt force is presented in the dimensionless form by dividing it by the initial bolt force. It is found that bolt force is decreased to nearly 70% of the initial value in the steady state operation. After the plant shutdown, the bolt force is further decreased to about 40%.

There are some differences between the experimental results and numerical ones calculated by FEM, which are particularly observed in the bolt forces just after the start of plant operation. It is presumed that the difference is not caused by the use of axisymmetric FE models, but it is primarily caused by not taking account of the

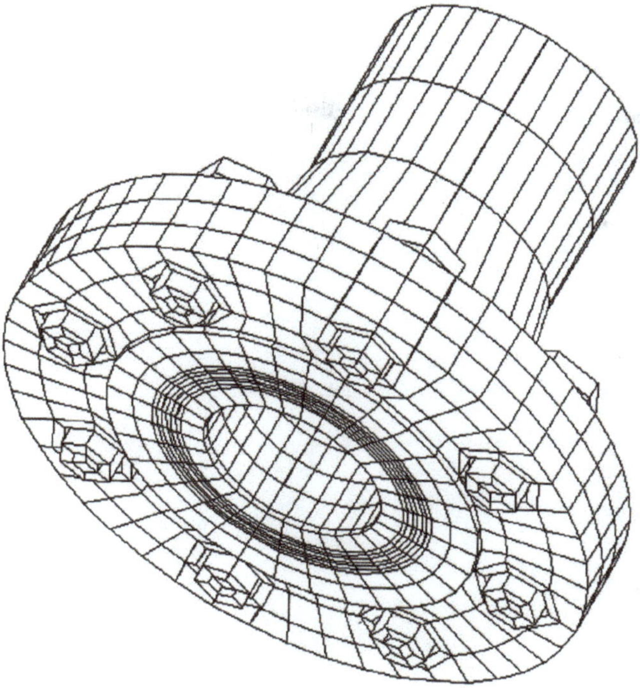

Fig. 7.4 Finite element model for a small pipe flange specified in JIS.

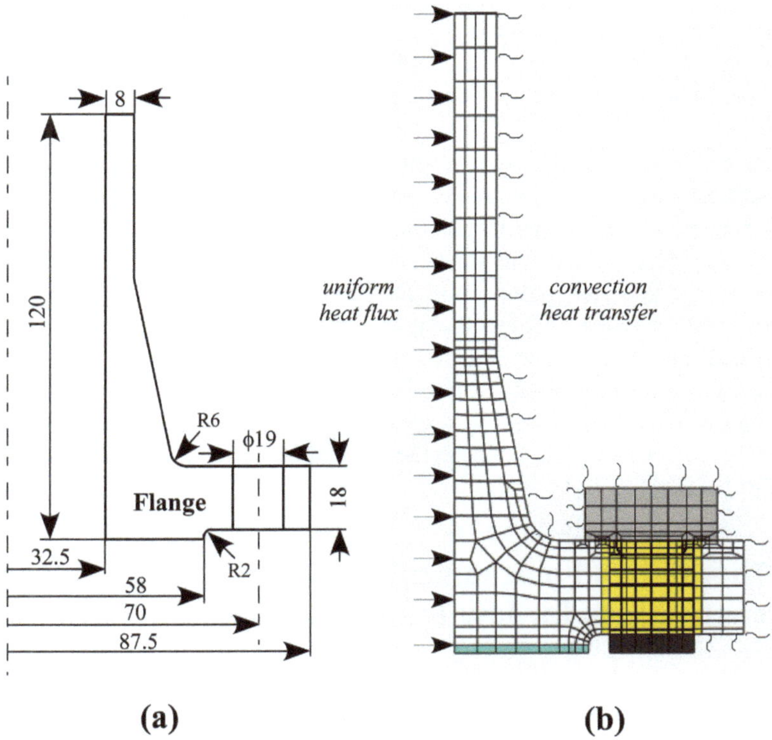

Fig. 7.5 Detailed dimensions and axisymmetric FE model of target pipe flange. (A) Detailed dimensions and (B) FE model.

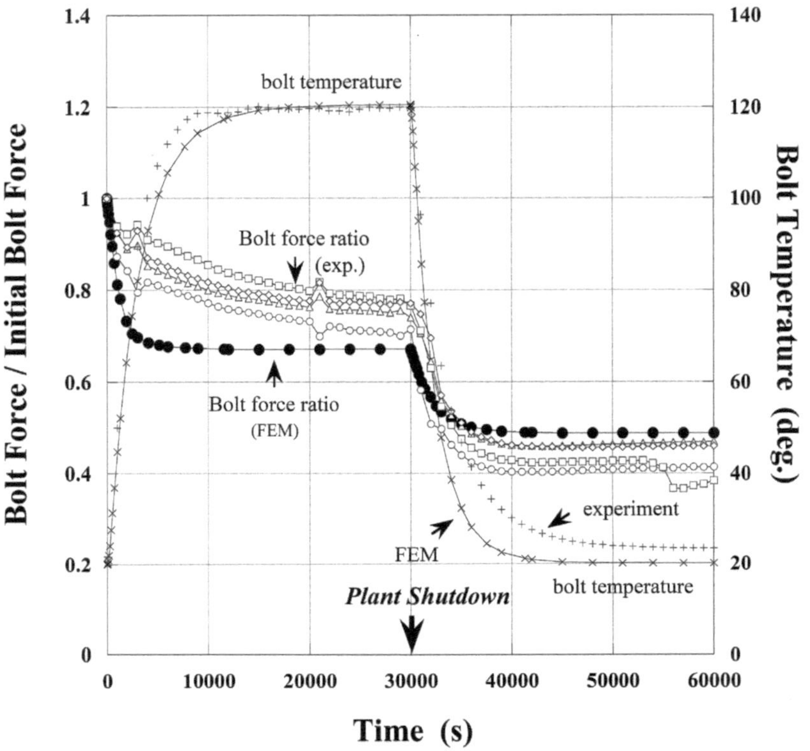

Fig. 7.6 Variations of bolt force and bolt temperature with time.

effect of creep in FE analysis. In any event, regardless of making various assumptions, the amounts of bolt forces in the steady state and after the plant shutdown are in fairly good agreement with the experimental ones, which indicates the effectiveness of the proposed numerical method. The fundamental cause for the specific behavior of bolt forces, by which pipe flange connections are clamped, is a decrease in gasket stiffness. In other words, the gasket stiffness that is originally low is reduced when subjected to thermal load, and it is further reduced in the shutdown process. The relationship between gasket contact stress and strain of sheet gaskets is nonlinear, hence, its stiffness cannot be expressed simply by use of Young's modulus. However, the approximate values can be considered to be in the order of a few hundredths of that of carbon steel [146,147].

7.5 Thermal and mechanical behaviors of pipe flange connections for low temperature fluids [148,149]

Leakage accidents of low temperature fluids are sometimes reported in accordance with the rapid increase of the demand for LNG (liquefied natural gas) and other low temperature fluids. When dealing with high temperature fluids, a decrease in

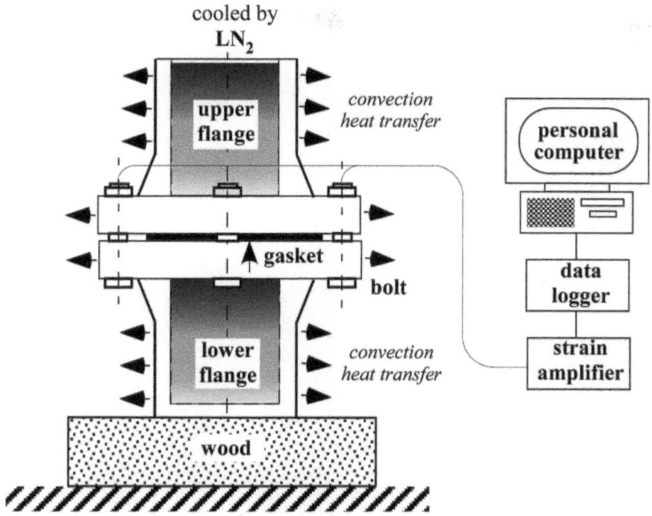

Fig. 7.7 Experimental setup for pipe flange connection used for low temperature fluids.

gasket stiffness is one of the major causes for the leakage accident of contained fluids, as explained in the previous section. In contrast, in the case of low temperature fluids, a significant decrease or change in gasket stiffness does not occur. In this section, the leakage mechanism specific to low temperature fluids is explained based on experimental and numerical results.

Fig. 7.7 shows an experimental setup for examining thermal and mechanical behaviors of pipe flange connections in which low temperature fluids flow. One end of a pair of pipe flanges is closed with a thick plate made of the same stainless steel as the flange material. The target pipe flange is the one specified in JIS, with nominal pressure 20 and inner diameter of 65 mm. A pair of pipe flanges is clamped with eight bolts of M16. Liquid nitrogen LN_2 is used as a low temperature fluid. Its temperature is -196 °C. In the experiments, LN_2 is continuously poured into the test flange connection, because LN_2 continues to evaporate wildly until reaching a steady state. Variations of bolt temperature and bolt force are measured by thermocouples and strain gages attached to four bolts. Fig. 7.8 shows the status of the test equipment in the steady state. A lot of frost is attached around it.

Fig. 7.9 depicts the FE model used for numerical analysis. One-eighth of the target flange connection is modeled considering the geometric symmetry. Coefficient of heat transfer on the inner surface of the pipe flange is set to be 8000 W/m^{2}K based on the experimental results, where the flange surface is assumed to be under boiling heat transfer state. In Fig. 7.10, numerical results of the bolt temperature and bolt force variations with time are compared to the experimental ones. Bolt force variation is represented as the ratio of bolt stress at that time σ_b to initial bolt stress σ_{bi}, which is termed bolt stress residual rate herein. Both results are in fairly good agreement, which substantiates the validity of the proposed numerical procedure. The experimental results of bolt temperatures and stresses in the figure are the averaged ones

Fig. 7.8 Status of test flange during experimentation.

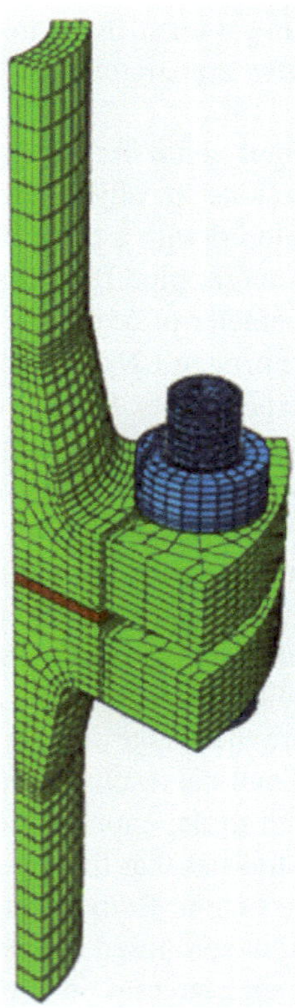

Fig. 7.9 Finite element model used for numerical analysis.

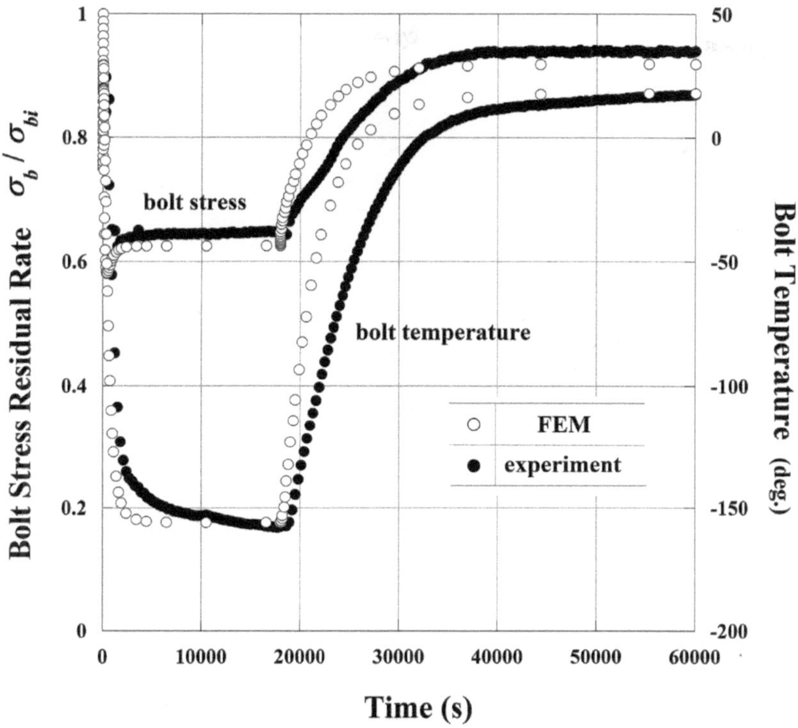

Fig. 7.10 Comparison of numerical and experimental results of bolt temperature and bolt stress variations with time.

measured in the four bolts. It is found that the bolt stress is reduced to about 65% of the initial value just after the start of cooling. This value is significant from the viewpoint of sealing performance.

Next, to examine the bolt force reduction phenomenon in more realistic conditions, a finite element model, shown in Fig. 7.11, is introduced to simulate the process of the fluid surface being moving up and down. The rising and falling of the fluid surface occur when starting the transfer of low temperature fluids and just before completing the transfer work. Considering the geometric symmetry, half of the pipe flange connection is modeled, and four bolts are identified by adding the numbers from the bottom to the top. Engaged threads are modeled by simple equivalent models with cylindrical shape. Parametric calculations are conducted by changing the time required to fill the cross section of the pipe flange with low temperature fluids, i.e., 20, 80, 320, and 1280 s. In the surface falling process, the same conditions are adopted. Presented in Fig. 7.12A and B are the variations of bolt temperature and bolt stress with time in the surface rising process. Bolt temperature is decreased to about −160 °C in the steady state. Bolt stress residual rate ranges between a bit below 60% and a slightly above 70% in the steady state. It is worth noting that the residual rate of bolt1 is reduced to 37% of the initial state in 250 s after the start of experimentation, for the case of required rising time of 320 s. Although not shown here, the minimum bolt stress residual rate is about 60% in the surface falling process.

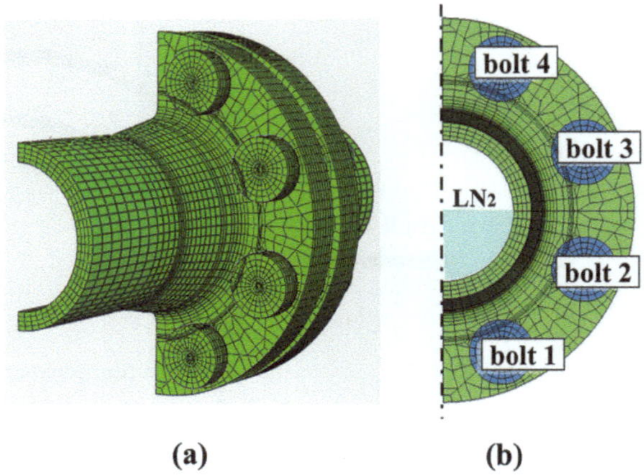

Fig. 7.11 Finite element model for simulating the surface rising and falling processes.
(A) Overview and (B) cross section of FE model.

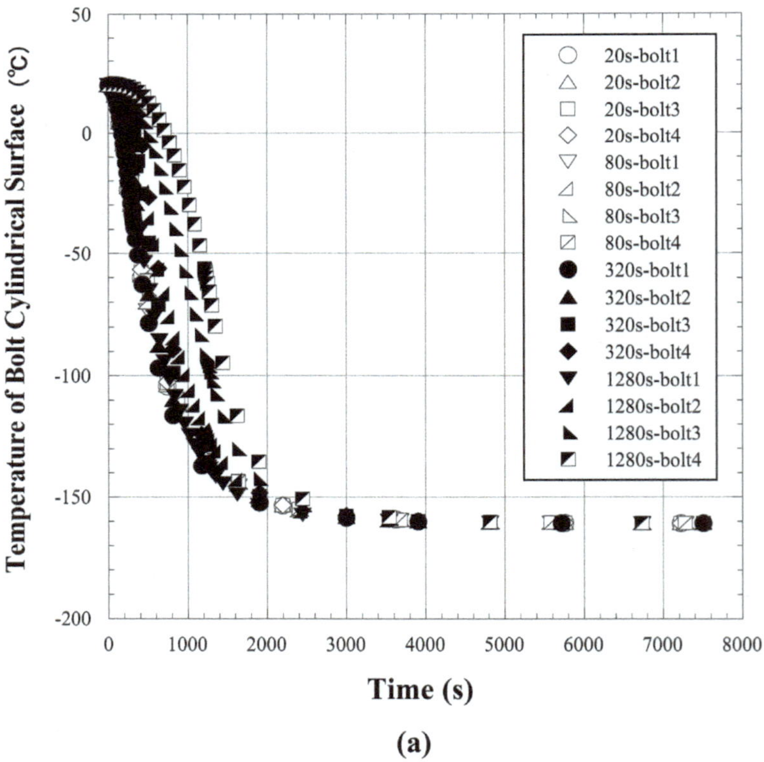

Fig. 7.12 Numerical results of bolt temperature and bolt stress variations with time. (A) Bolt temperature and (B) bolt stress residual rate.

(Continued)

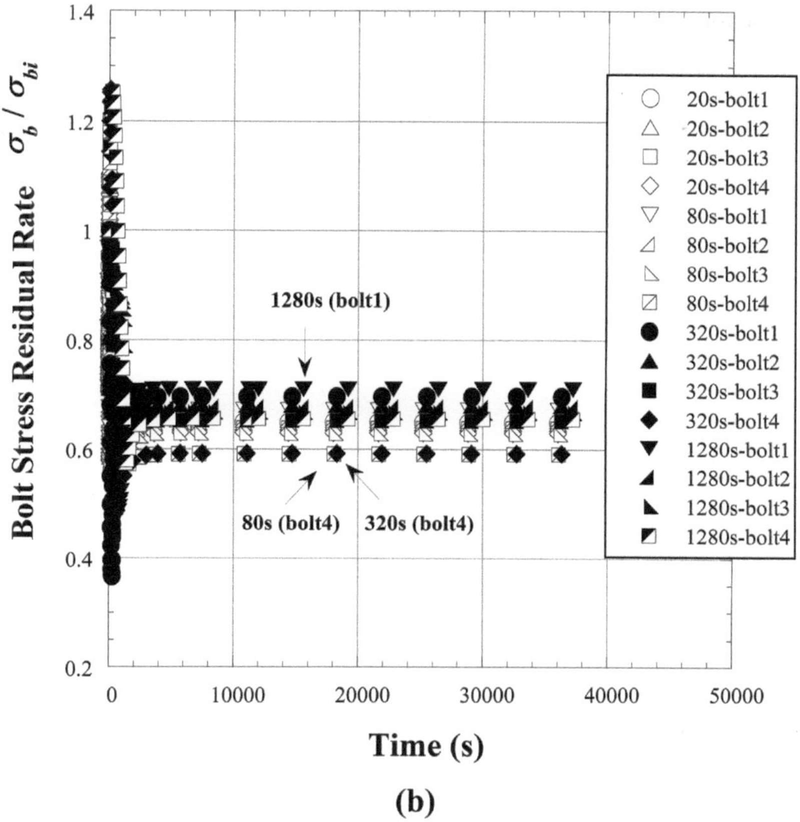

Fig. 7.12, cont'd

LNG is the most commonly used low temperature liquid. Its temperature is about −163 °C. Bolt stress residual rate is calculated in the same manner as the case of LN_2. Assuming that no boiling occurs, the same FE models are used for numerical calculations. Coefficient of heat transfer on the flange inner surface is estimated by a well-known equation for turbulent flow in a smooth tube:

$$Nu = 0.023\,Re^{0.8}Pr^{0.4} \tag{7.1}$$

where *Nu*, *Re*, and *Pr* are the Nusselt number, Reynolds number, and Prandtl number, respectively.

Then, the coefficient of heat transfer is set to be 1000 W/m²/K, based on the calculated results for the common range of flow velocity. In this case, bolt stress residual rate is reduced to about from 72% to 76% of the initial state and, just after the start of cooling, it temporarily drops to 66% of the initial state. In any event, special attention must be given to the bolt force variation in the rising and falling processes of the fluid surface.

Learning from problems and accidents with threaded fasteners

8

8.1 Introduction

Problems and failures of threaded fasteners occur in various forms. The complex geometry of threaded fasteners and the diversity of applied external forces are considered to be the essential causes. Aiming at providing some hints to cope with these problems, in this chapter, a variety of problems occurring in bolted joints and characteristic mechanical behaviors inherent to threaded fasteners are introduced.

In Section 8.2, wheel falling-off accidents of large vehicles are treated, which are caused by bolt ruptures due to metal fatigue. Although the recommended torque value for tightening wheel bolts has been raised to prevent wheel falling-off accidents, which were occurring rather frequently, it is concluded that it is possible that occurrence of the problem would become very low just by controlling the tightening torque with practical accuracy. In Section 8.3, the mechanism of fatigue failure occurring in the threaded portion of the axle of a rollercoaster is explained from the viewpoint of strain energy. In Section 8.4, bearing type joints clamped with multiple bolts are treated, which are used for supporting large shear force, and it is shown that the magnitude of shear force supported by each bolt, arranged along the load direction, can be calculated by the elementary theory of solid mechanics. In Section 8.5, mechanical behavior of reamer bolts used for clamping rigid flanged shaft couplings is examined, in which the effects of the fitness of reamer bolts, bolt force scatter, and assembly error on the joint strength are comprehensively clarified by FE analysis. In Section 8.6, bolt force reduction occurring in reamer bolts clamped by cooled fitting is treated, and a simple formula is derived by use of spring models to calculate the amount of reduction.

In Section 8.7, sealing performance of the plugs used for hydraulic equipment is discussed, to which oil pressure is applied dynamically. It is shown that one-dimensional spring models work well to express this dynamic behavior. In Section 8.8, vibration problems inherent to bolted joints are treated, in which a numerical method by FEM is proposed that can quantitatively evaluate the effect of surface roughness on the natural frequency and natural mode of vibration of bolted joints. In Section 8.9, to clarify the mechanical behavior of bolted joints by FEM, some tips are provided to effectively perform the numerical calculations, for instance, a modeling scheme considering the geometric symmetry or periodic symmetry of the target bolted joint. In addition, mesh patterns for engaged threads are concretely illustrated that can offer numerical results with sufficient practical accuracy.

The Mechanics of Threaded Fasteners and Bolted Joints for Engineering and Design.
https://doi.org/10.1016/B978-0-323-95357-3.00006-X

Although each subject treats a specific problem, such as stress analysis, low temperature thermal load, modal analysis, etc., it is expected that the contents introduced here would give some tips to solve various problems concerning threaded fasteners and bolted joints.

8.2 Fatigue failure of wheel bolts for large vehicles specified in JIS

8.2.1 Summary of wheel falling-off accidents

Wheel falling-off accidents from large vehicles has become a serious social problem and threatens the safety of civilian life. Unfortunately, such accidents occur rather frequently. For instance, in April 2008, a huge wheel fallen off a large trailer bounced into the windshield of a sightseeing bus. According to statistics, most of the wheel falling-off accidents occur in rear wheels, and two thirds of them originate from left-side rear wheels [150]. The major cause of the accidents is fatigue failures of threaded fasteners, which are used for installing the wheel to the hub attached to the end of axle. One fundamental cause for wheel falling-off accidents is that the rear wheels of large vehicles specified in JIS have a dual-tire structure, and the tightening operation is done in two stages using dedicated threaded fasteners called wheel bolts. In the actual operation, an inner wheel is first tightened followed by an outer wheel.

When tightening wheel bolts, impact wrenches have commonly been used because a large number of target bolts need to be tightened. At the present time, it is mandatary in Japan to clamp the wheel bolts with tightening torque of about 600 Nm to cope with the frequently occurring accidents. However, the problem is that the rear wheels of large vehicles have a very complex structure. Therefore, it is difficult to say that a reliable formula for wheel bolts, which relates the tightening torque to bolt force to be generated, has been prepared and a proper procedure for tightening inner and outer wheels in two stages has been established.

Meanwhile, since some fundamental problems regarding JIS-style wheels had been clarified by a series of related researches, ISO-style wheels came to be used in large vehicles manufactured after 2008, in which inner and outer wheels are tightened together at the same time using 10 wheel bolts. In contrast, eight bolts are used in JIS-style tire wheels. Accordingly, it was strongly expected in Japan to eradicate the accidents or achieve a significant reduction of them by changing the joint style into ISO-style. However, the number of wheel falling-off accidents does not show a monotonic decrease. In addition, it has been reported that a considerable number of wheel falling-off accidents, due to fatigue failure of wheel bolts, occur in the USA and Europe, except in Germany [150]. The situation suggests that a more essential cause for the accidents is too low a bolt force due to insufficient torque management. In the remaining part of Section 8.2, mechanical behavior of the wheel bolts is explained in detail, based on the research results obtained in the course of performing the contract research with JRTT (Japan Railway Construction, Transport and Technology Agency).

8.2.2 Joint configuration, tightening method, and fatigue failure of threaded fasteners

Fig. 8.1A shows the standard structure of a rear wheel of a large vehicle and the tightening sequence. Inner and outer wheels are attached to a hub in two stages by use of dedicated threaded fasteners consisting of wheel bolt, inner nut, and outer nut. Fig. 8.1B shows the actual threaded fasteners. Among them, an inner nut is a specific part that characterizes the tightening operation of the rear wheel of large vehicles. Fine screw threads of M20 and M30 are processed inside and outside a bell-shaped portion, which works as a female thread and a male thread when tightening inner and outer wheels, respectively. Bearing surfaces of the inner and outer nuts are shaped like spherical washers with small areas. Fig. 8.2 depicts the structure around the jointed portion when the tightening operation of inner and outer wheels has been completed. A standard number of wheel bolts in JIS-style is eight, while the number is six for wheels with small diameter. Wheels clamped by eight bolts are treated here, because they are employed in most large vehicles. The tightening procedure is explained in the following, corresponding to Fig. 8.1A:

Step 1: Bolt hole portions of an inner wheel are mounted to wheel bolts sticking out of a hub.

Step 2: Inner nuts are attached to wheel bolts and then an inner wheel is tightened to a hub by applying torque to the square portion of inner nut.

Step 3: An outer wheel is mounted to wheel bolts and then outer nuts are screwed into the male threads processed on the outside surface of the inner nut.

As shown above, inner and outer wheels used in the rear side of large vehicles are tightened in two stages. A specified tightening torque is 600 Nm for both inner and outer wheels. As for the tightening number of wheel bolts, it is equal to the number of bolts in front wheels having single-tire structure, while it is twice the number of

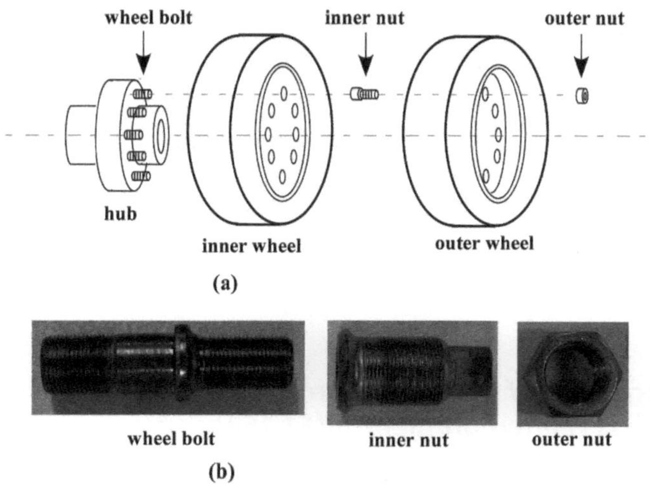

Fig. 8.1 (A) Tightening sequence of rear wheel and (B) dedicated threaded fasteners for large vehicles.

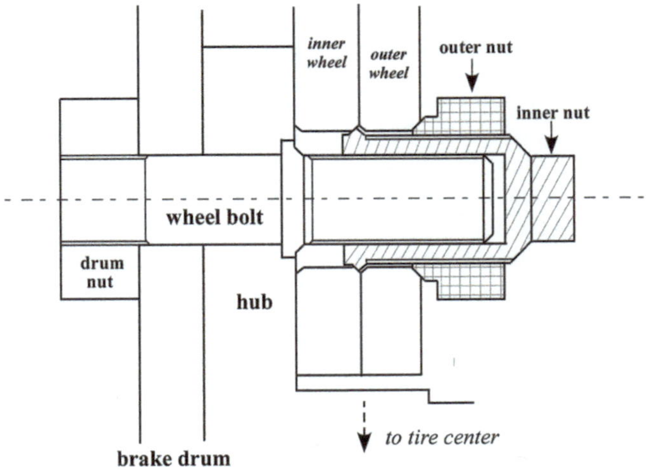

Fig. 8.2 Joint configuration of rear wheel at the completion of tightening operation.

(a) (b)

Fig. 8.3 Examples of ruptured (A) wheel bolt and (B) inner nut due to metal fatigue.

bolts in rear wheels having dual-tire structure. Consequently, even in the minimum case of large vehicles with one axle shaft in the front and two axle shafts in the rear, the total tightening number is 80, and the number is increased to 112 for three axle shafts in the rear.

It should be noted that, in Japan, traffic drives on the left-hand carriageway. This could be one major reason why most of the wheel falling-off accidents occur in rear wheels and two-thirds of them originate from the left-side rear wheels. Shown in Fig. 8.3 are examples of a wheel bolt and an inner nut ruptured due to metal fatigue [150]. Referring to the joint configuration in Fig. 8.2, it is found that both threaded fasteners are ruptured around the first thread root nearest to the nut-bearing surface, as in the case of ordinary bolt-nut connections.

8.2.3 Mechanics of tightening process of wheel bolts and its related problems [151,152]

In Fig. 8.4, the relationship between torque and bolt force is schematically shown in the two-stage tightening operation. When tightening the inner and outer wheels with the specified torque of 600 Nm, bolt force produced in a wheel bolt becomes about 123 kN at the completion of inner wheel tightening, and the value is increased by

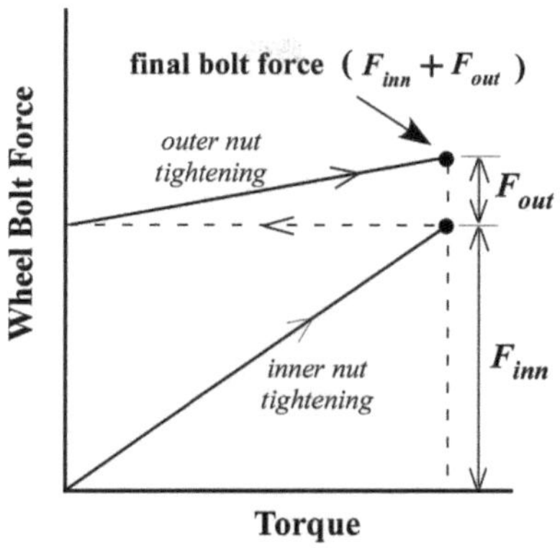

Fig. 8.4 Torque-bolt force relationship in two-stage tightening operation of rear wheel.

as much as about 30 kN when tightening the outer wheel, assuming the coefficient of friction to be 0.15. Thus, the final bolt force is about 153 kN. As discussed in section 3.2.2, the magnitude of bolt force produced by a constant torque is inversely proportional to nominal diameter. Considering that the nominal diameters of screw threads used for tightening inner and outer wheels are M20 and M30, the bolt force generated by the outer-wheel tightening will be approximately two-thirds of the one generated in the inner-wheel tightening process. Specifically, the increase in bolt force is expected to be about 80 kN, i.e., two-thirds of 123 kN. The difference between 80 kN and the actual increase of 30 kN in bolt force is traced to inherent characteristics of the two-stage tightening operation.

Since wheel bolts are tightened in two stages, two nut factors of K_{inn} and K_{out} are necessary as a tightening guideline, corresponding to the inner-wheel and outer-wheel tightening processes, respectively. Using K_{inn} and K_{out}, the tightening torques required for inner nut, T_{inn}, and outer nut, T_{out}, are calculated, and then the two components of bolt force, F_{inn} and F_{out}, are obtained. As a result, the final bolt force produced in a wheel bolt becomes the sum of F_{inn} and F_{out}:

$$\begin{aligned} T_{inn} &= K_{inn} F_{inn} d \\ T_{out} &= K_{out} F_{out} d \\ F_b &= F_{inn} + F_{out} \end{aligned} \tag{8.1}$$

Fig. 8.5 shows an FE model of the jointed components of wheels, which is constructed to derive the equation relating tightening torque to bolt force. Assuming axisymmetric geometry, the shapes of inner and outer wheels, hub, and brake drum are simplified to the extent that the mechanical behavior of actual joints is not impaired. Details of the analytical procedure and conditions are given in [152]. Two nut factors, K_{inn} and K_{out}, defined in Eq. (8.1) are expressed as a function of coefficient of friction, as in the case of ordinary bolt-nut connections:

$$\begin{aligned} K_{inn} &= 0.6004\mu_{ith} + 0.8423\mu_{inu} + 0.02120 \\ K_{out} &= 3.179\mu_{oth} + 0.6198\mu_{onu} + 0.08163 \end{aligned} \tag{8.2}$$

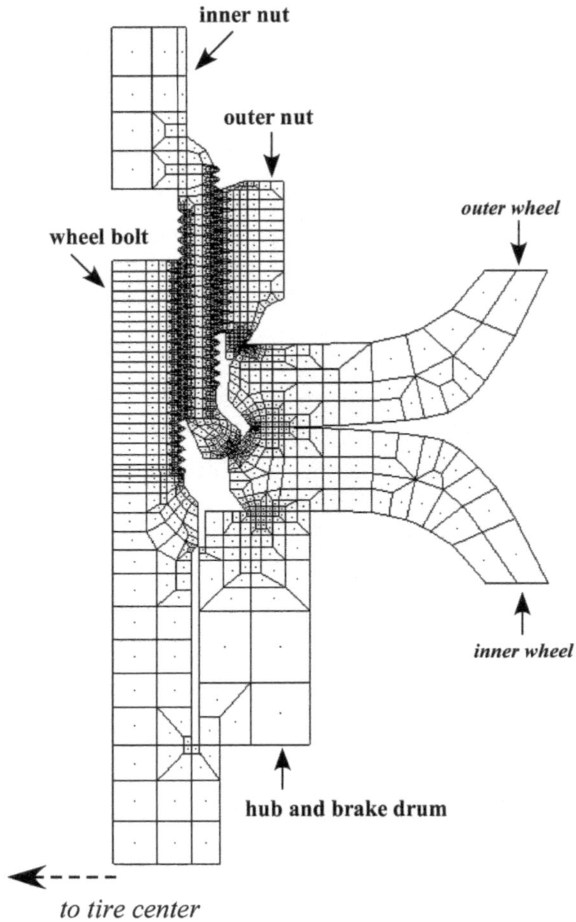

Fig. 8.5 FE model of the jointed components of a wheel.

where μ_{ith}, μ_{inu}, μ_{oth}, and μ_{onu} represent coefficients of friction on thread surface and nut-bearing surface in the inner-wheel and outer-wheel tightening processes, respectively; and inner nut and outer nut are identified by the first subscript, *i* and *o*.

The magnitude of K_{inn} is considerably smaller than K_{out}, corresponding to the fact that large bolt force is produced when tightening the inner wheel, as shown in Fig. 8.4. This means that the tightening characteristics of wheel bolts are largely influenced by K_{inn}. Comparing K_{inn} with nut factor K for ordinary bolt-nut connections, given in eq. (3.18), coefficient for μ_{inu} in Eq. (8.2) is 0.8423, which is significantly larger than 0.65 for μ_{nu}. It can be said, therefore, that μ_{inu} has a relatively larger effect on the nut factor, comparing to the case of ordinary bolt-nut connections. In other words, when coefficients of friction are varied due to the repetition of assembly and disassembly of jointed components, wheel bolt forces are more easily changed than those in ordinary bolt-nut connections.

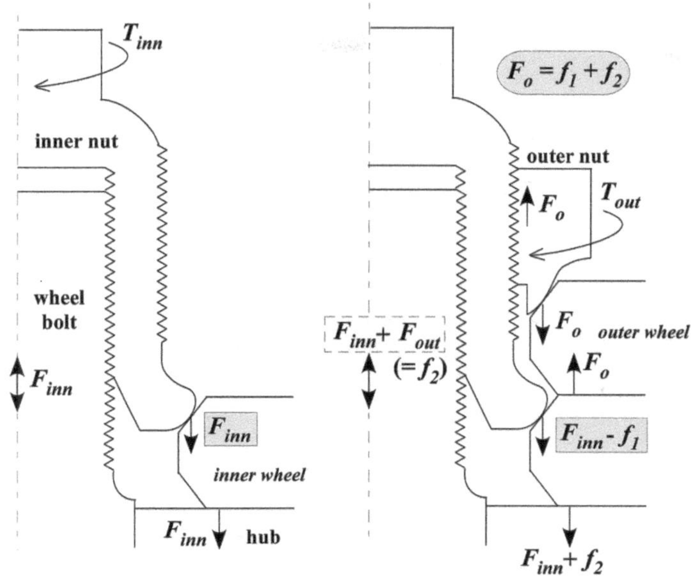

Fig. 8.6 Equilibrium of forces acting on wheel bolt in two-stage tightening operation of rear wheel. (A) Inner nut tightening and (B) outer nut tightening.

Fig. 8.6 illustrates the equilibrium of forces acting on each part of the wheel bolt when tightening rear wheels in two stages [70]. The mechanism of bolt force generation of F_{inn} in the inner-wheel tightening process is the same as that in ordinary bolt-nut connections. However, the bolt force increase in wheel bolts, F_{out} ($=f_2$), in the outer-wheel tightening process is approximately 30% of the total axial force F_0. The remaining force f_1—nearly 70% of F_0—acts so as to push down the inner wheel, thereby lowering the contact stress on the inner nut-bearing surface. As a result, the bolt force produced in the inner-wheel tightening process, F_{inn}, accounts for a large percentage of the final bolt force, F_b, at the completion of tightening work. According to the numerical results, the ratios of f_1 and f_2 to F_0 show a small variation, less than 2%, when the coefficient of friction is varied from 0.1 to 0.2. Then, referring to the ratios calculated for coefficient of friction being 0.15, f_2 becomes only 28% of F_0 and the remaining 72% is occupied by f_1. Next, using the ratios for f_1 and f_2 and Eqs. (8.1) and (8.2), it is examined in Fig. 8.7 how each coefficient of friction on the contact surfaces affects the compression force ($=F_{inn}-f_1$) acting on the inner nut-bearing surface at the completion of tightening work.

Tightening torque is set to be 600 Nm in all cases. The abscissa represents coefficients of friction, μ_{ith} and μ_{inu}, in the inner-wheel tightening process and the parameters are μ_{oth} and μ_{onu} in the outer-wheel tightening process. The figure indicates that if F_{inn} is small and F_0 is large, there are cases in which the contact pressure on the inner nut-bearing surface almost vanishes. In other words, such a state occurs when μ_{ith} and

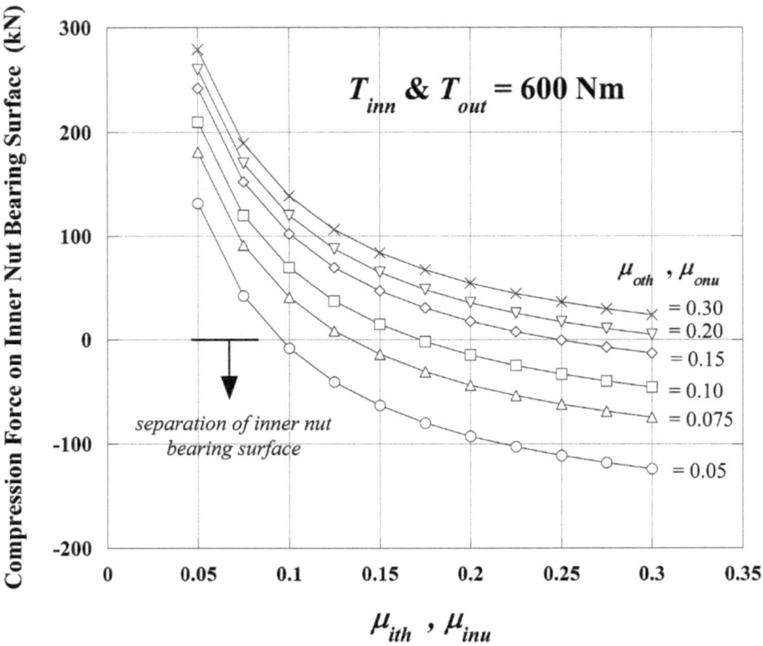

Fig. 8.7 Effect of coefficients of friction on the compression force acting on inner nut-bearing surface.

μ_{inu} are large and μ_{oth} and μ_{onu} are small. In that case, the movements of inner and outer wheels are constrained only by the compression force acting on the interface between the two wheels. Therefore, it is presumed that the contact conditions around the jointed portion consisting of two wheels fall into a highly unstable state.

8.2.4 Scatters in bolt force and friction coefficient and its reduction strategy

Fig. 8.8 shows an experimental setup constructed by use of the actual parts of hub, brake drum, and other peripheral components [70]. Fig. 8.8A represents the state before installing wheels, in which the parts sticking out of the hub surface are wheel bolts. Shown in Fig. 8.8B is the state of the installation of inner and outer wheels being completed. Threaded portions of wheel bolts are partly removed and machined into a cylindrical shape, and two strain gages are attached, each one 180 degrees apart, to measure axial bolt force, as shown in Fig. 8.8C. When measuring coefficients of friction on thread surface and nut-bearing surface separately, two crossed gages are additionally attached for measuring the magnitude of torque exerted on the bolt cylindrical portion.

Shown in Fig. 8.9 are the measured results of coefficients of friction on thread surface μ_{ith} and inner nut-bearing surface μ_{inu} when tightening an inner wheel. The experiments are repeated five times using machine oil as a lubricant. The tightening torque

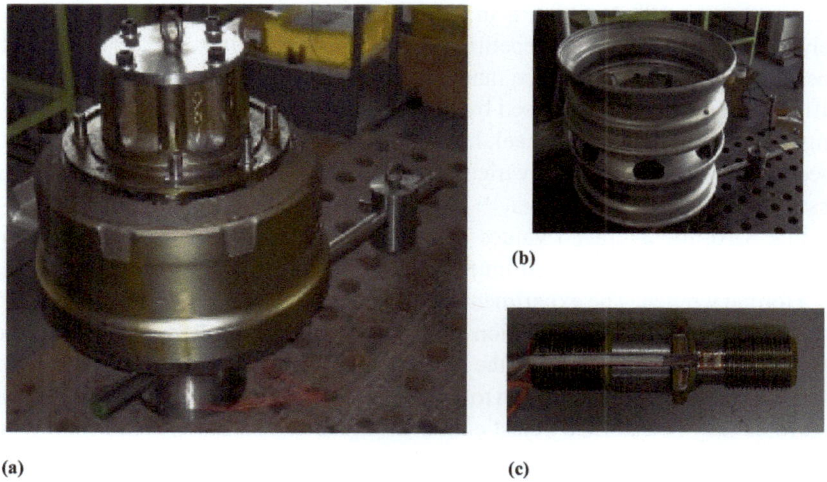

Fig. 8.8 Experimental setup by use of actual parts of wheel joint. (A) Experimental setup for examining tightening process, (B) installing inner and outer wheels, and (C) test specimen of wheel bolt.

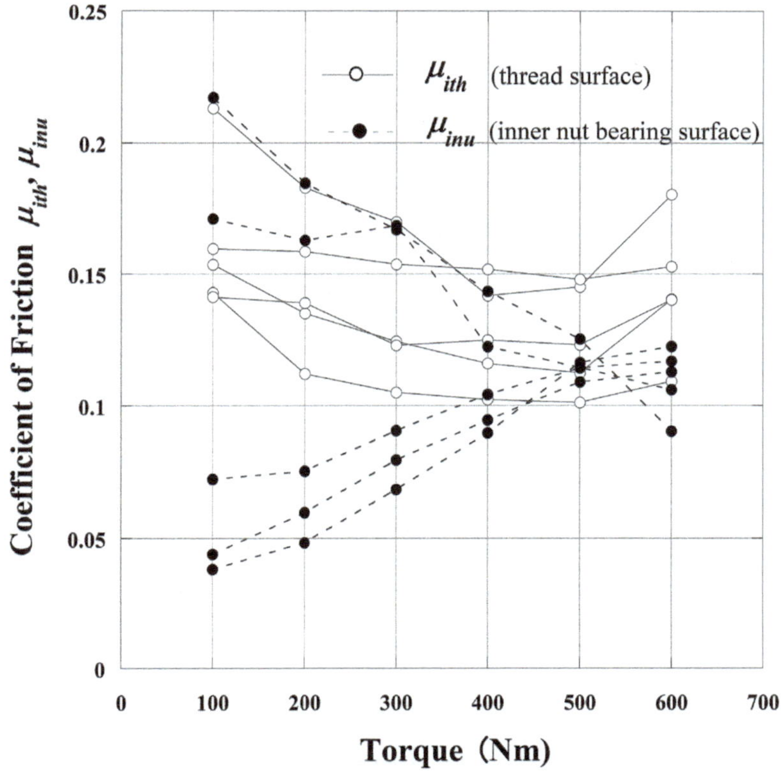

Fig. 8.9 Measured results of μ_{ith} and μ_{inu} in the tightening process of inner wheel.

is increased from 100 to 600 Nm in increments of 100 Nm. Both μ_{ith} and μ_{inu} are largely varied according to the repetition of tightening and disassembly. The variation of coefficients of friction is more markedly observed in μ_{inu}. It is presumed that the significant change in μ_{inu} is caused by the fact that the mating bearing surfaces, consisting of inner nut and inner wheel, have a spherical shape and the contact condition between the two surfaces is varied owing to the repetition of tightening and disassembly.

In Fig. 8.10, the averaged values of bolt force and its scatter generated in eight wheel bolts are shown, in which inner and outer wheels are tightened in two stages using a torque wrench. The experiments are repeated five times and the bolt force scatter is expressed by standard deviation $\pm 1\sigma$. A considerable scatter is observed among the eight bolts. Also significant is the scatter generated in each bolt during the experiment's five repetitions. In addition to the scatter in coefficients of friction, the scatter in bolt force may be caused by elastic interaction among bolts, as discussed in section 3.7, when tightening them sequentially. However, it is confirmed by the numerical results using FEM and experimental results that the effect of elastic interaction can practically be negligible [70]. The major reason for this is that the thicknesses of inner and outer wheels are thin and the plate surfaces composing the interface are almost

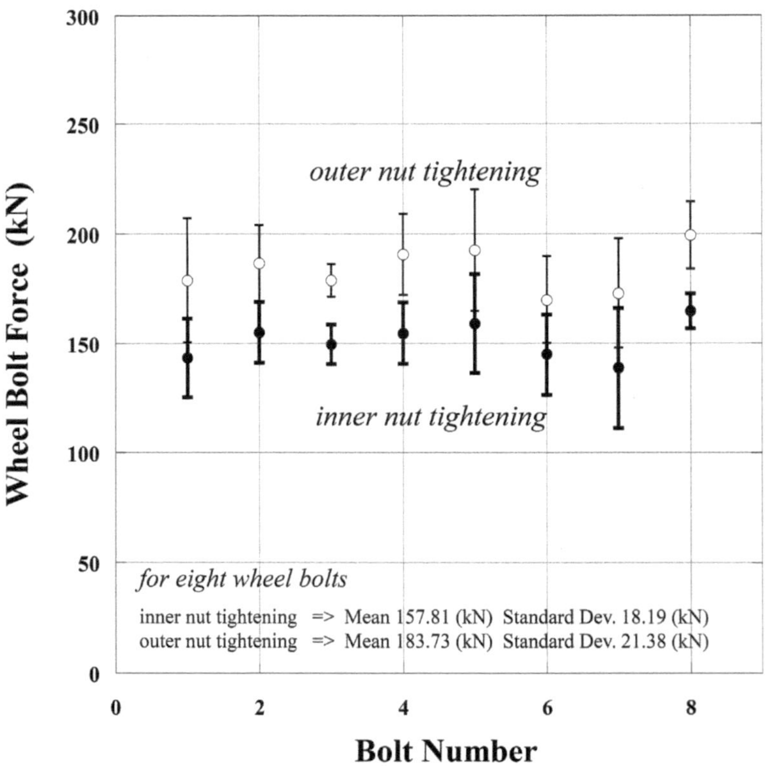

Fig. 8.10 Averaged values of bolt force and its scatter generated in eight wheel bolts.

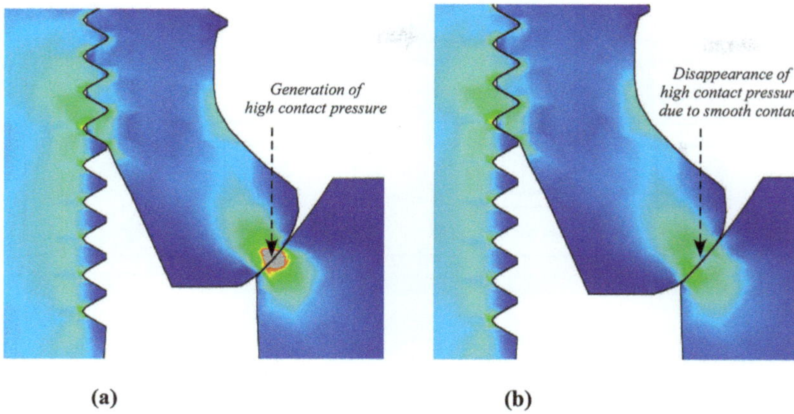

Fig. 8.11 von Mises stress distributions around inner nut-bearing surfaces with (A) original shape and (B) modified shape.

flat, thereby making the joint stiffness very high because of the close contact. This result provides important information on elastic interaction. That is, in the tightening process of a bolted joint by use of multiple bolts arranged around the circumference, the effect of elastic interaction is almost negligible when the stiffness of fastened plates is high and low stiffness parts like sheet gaskets are not included.

According to the experimental results shown in Fig. 8.9, in order to reduce the bolt force scatter in wheel bolts, it is considered effective to lower the scatter in the coefficient of friction on the inner nut-bearing surface μ_{inu}. Then, by constructing FE models whose dimensions correspond to the ones of the actual inner nut and inner wheel, contact pressure distributions around the inner nut-bearing surface are evaluated in detail. In Fig. 8.11, von Mises stress distributions around the inner nut-bearing surface, obtained by the FE model having the original shape of the actual jointed portion, are compared to those obtained by the model having a slightly modified shape. In the latter, aiming at the smooth contact according to theory of Hertzian contact stress, the bearing surfaces of inner nut and inner wheel are modified to be in contact as convex and concave spherical surfaces. It is found that the maximum stress occurring around the inner nut-bearing surface is significantly decreased by a slight modification of bearing surface shape. A similar effect can also be obtained regarding the outer nut-bearing surface. Accordingly, for the purpose of reducing the bolt force scatter in wheel bolts, it is effective that the curvature radius of the inner nut-bearing surface is processed to be slightly smaller than that of the mating inner wheel-bearing surface. As a result, smooth contact can be achieved between the two surfaces like the Hertzian contact problem.

8.2.5 Measurement of stress amplitude of wheel bolts [153]

Fig. 8.12 illustrates the construction of a stress amplitude measurement equipment. The equipment is assembled by diverting the rear portion of an actual trailer with

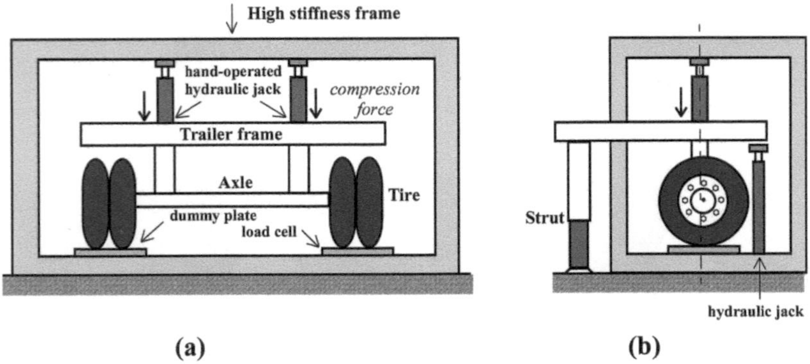

Fig. 8.12 Construction of stress amplitude measurement equipment. (A) Front view and (B) side view.

maximum loading capacity of 200 kN. The magnitude of repeated stress, produced in a wheel bolt in accordance with tire rotation, is measured in a state close to the actual running condition. In the experiments, the external forces corresponding to the live load of a trailer are applied by hand-operated hydraulic jacks. The magnitude of the compression force is measured for one side of the rear wheels with dual-tire structure by load cells exclusively prepared for measuring the live load of large vehicles.

Wheel bolts are wholly threaded. To cope with this problem, the part of threaded portion that is not engaged with an inner nut is removed and processed into a cylindrical shape. Then, four strain gages, each one 90 degrees apart, are attached to the cylindrical surface in the circumferential direction, as shown in Fig. 8.13A. Four positions of the strain gages are identified by IN, OUT, R, and L, where IN and OUT designate the inside and outside of the wheel in the radial direction, and R and L the right side and left side as viewed from the front of the wheel. Fig. 8.13B shows a test specimen of wheel bolt with four strain gages being attached, in which four narrow grooves are machined to pull out a lead wire of each strain gage. JIS-style wheels for large vehicles are tightened with eight wheel bolts. Hence, the bolt located at the farthest point from the ground is named bolt-1, which is the reference point. Other bolts are named clockwise as bolt-2, bolt-3, etc. The measurements of stress amplitude are carried out by the strain gage attached to bolt-1.

Fig. 8.14 shows an example of experimental results, in which the relationship between axial bolt stress at IN and tire rotation angle is presented. The parameter is a live load. The stress amplitude is defined by dividing the difference between the maximum and minimum stresses, occurring during one rotation, by two. The relationship between the stress amplitude and live load when bolt force is reduced to 30% of the standard value is shown in Fig. 8.15. Live load of 25 kN corresponds to the maximum loading capacity. Although not shown here, when the bolt force is reduced to 50% of the standard value, the stress amplitudes vary negligibly with live load, i.e., between about 5 and 8 MPa. However, there is a significant increase in stress amplitude in Fig. 8.15. In addition, when the experiments are repeated for the bolt force being 30% of the standard value, the measured bolt forces show a fairly large

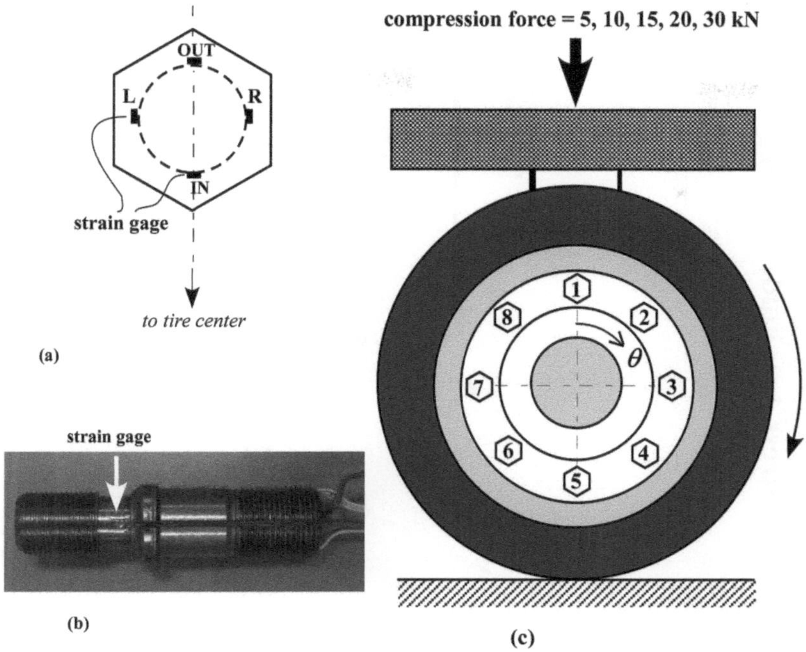

Fig. 8.13 Measuring method of stress amplitude and test specimen of wheel bolt. (A) Measurement position of bolt strain, (B) test specimen of wheel bolt, and (C) bolt number.

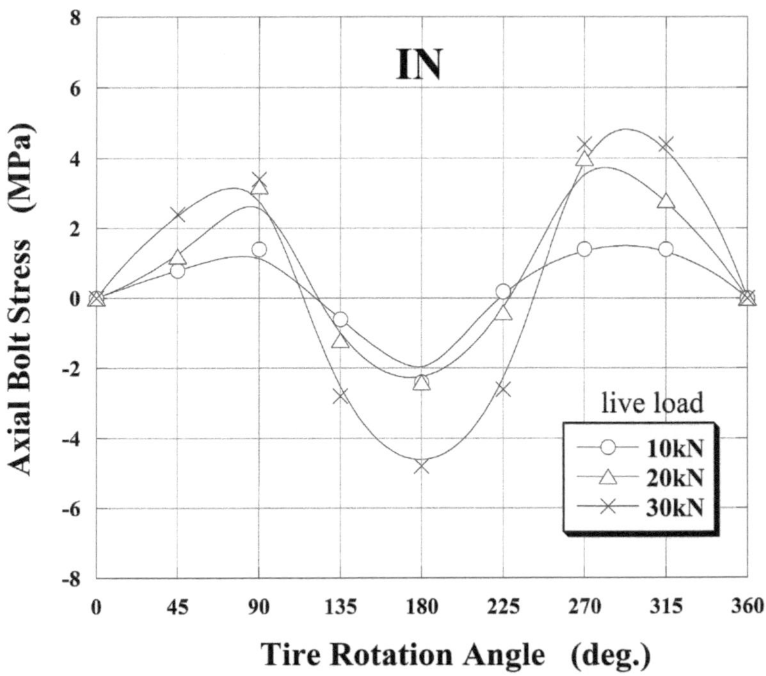

Fig. 8.14 Relationship between axial bolt stress at IN and tire rotation angle.

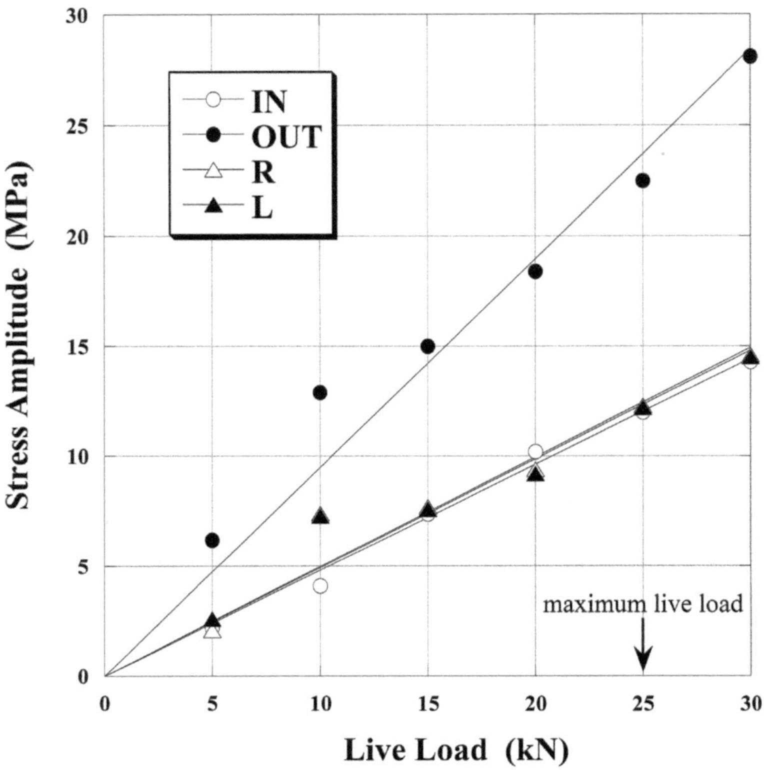

Fig. 8.15 Relationship between stress amplitude and live load under reduced bolt force.

dispersion, which means that the wheels are being clamped in a rather unstable state due to insufficient bolt force. As a result, a practically important guideline is derived that the stress amplitude of wheel bolts can be sharply increased when the bolt force is reduced to about 30% of the standard value.

The prescribed tightening torque for wheel bolts is 600 Nm. This value was raised from the former value of 400 Nm to cope with the frequent occurrence of wheel falling-off accidents. However, according to the experimental results explained above, the stress amplitude shows a negligible increase when the bolt force is reduced to 50% of the standard value. Therefore, if the tightening operation were performed with the former prescribed torque of 400 Nm, it is expected that fatigue failures of wheel bolts rarely occur, even if a certain extent of bolt force scatter occurs due to a variety of factors.

8.2.6 Analysis of stress amplitude of wheel bolts by FEM [154]

Fig. 8.16A shows an FE model of the whole joint and Fig. 8.16B and C represent the detailed mesh patterns of wheel bolt and inner nut. Considering the complex geometry around the joint and computational efficiency, the numerical models constructed here

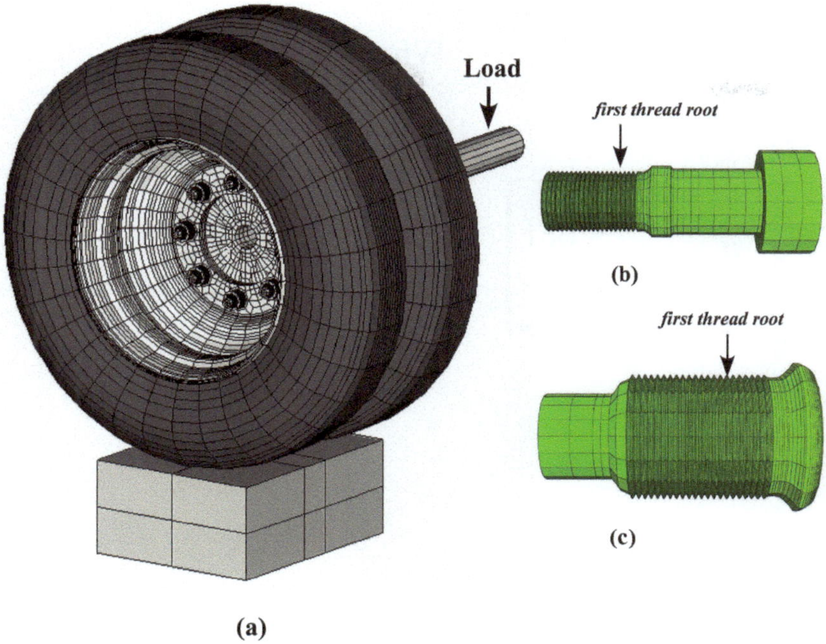

Fig. 8.16 FE models of (A) whole joint, (B) wheel bolt, and (C) inner nut for stress amplitude evaluation.

are all assumed to be linearly elastic, including the tire and the ground. Shown in Fig. 8.17A and B are the relationship between the stress amplitude, occurred at the first thread root of wheel bolt, and live load. In each case, the bolt force is maintained in the normal state and is reduced to 30% of the standard value, respectively. Coefficient of friction is set to be 0.15. In the latter case, the stress amplitude shows a steep increase with increasing live load at the position of OUT on the wheel bolt surface. When coefficient of friction is increased, although not shown here, high stress amplitude is produced in a similar manner to the case of bolt force being significantly decreased. This result suggests that particular attention should be paid to the surface roughness of contact surfaces related to the tightening and disassembly processes, together with the measured results of coefficient of friction on the inner nut-bearing surface, shown in Fig. 8.9.

8.2.7 Development of tightening apparatus equipped with multiple shafts and torque control mechanism [155]

The tightening torque recommended for wheel bolts is 600 Nm. In order to tighten the bolts by human power using a dedicated wrench of 7.2 kg mass and 1.2 m arm length, the magnitude of applied force is approximately 500 N. The work can possibly be achieved only by a person with significant physical strength. Therefore, a special

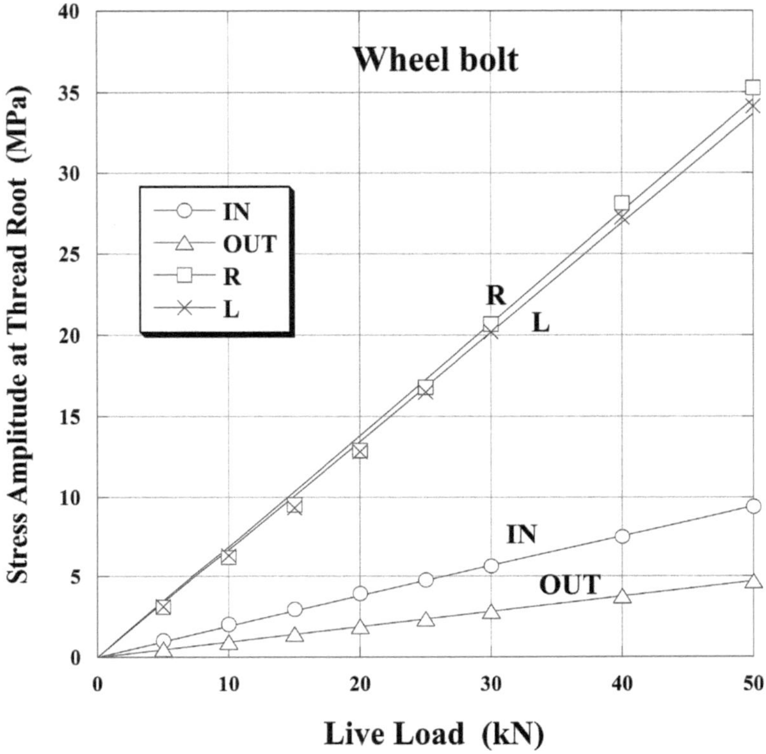

(a)

Fig. 8.17 Relationship between stress amplitude and live load in (A) normal and

(Continued)

torque wrench has been developed that needs smaller tightening torque of 300 Nm due to the design of the mechanical mechanism. In any event, when repeating the tightening operation of wheel bolts and inner nuts, for example, 80 times in total, a very large tightening energy is required. It is practically impossible for one worker to complete the work in a short time. Alternatively, if an impact wrench is employed to improve the working efficiency, which is not equipped with torque control function, it leads to a serious problem of insufficient bolt force or a failure of threaded fastener due to overtightening. As a countermeasure for the forgoing problem, it is recommended to use tightening equipment with torque control function.

Fig. 8.18 represents the measured results of bolt force when tightening operation is carried out by air-powered wrench with torque control function. Comparing to the experimental results in Fig. 8.10 performed by manual torque wrench, the scatter in bolt force becomes very small. When using manual wrenches, the work requires

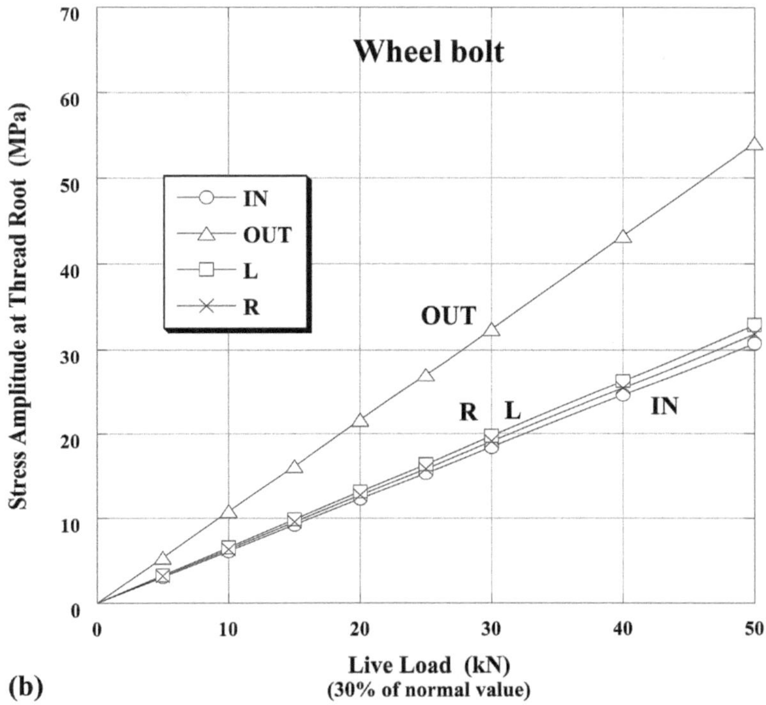

Fig. 8.17, cont'd (B) reduced bolt force states.

all the strength of the worker to tighten with the specified torque of 600 Nm. Hence, it is difficult to progress the tightening work at a constant speed, thus leading to a large scatter in bolt force. In contrast, smooth tightening operation can be achieved by use of a powered wrench equipped with torque control function. That may be the major cause of the small scatter in bolt force, shown in Fig. 8.18. Fig. 8.19 depicts tightening equipment especially developed to tighten the wheels of large vehicles, by which four wheel bolts can be tightened simultaneously, because four air-powered torque wrenches are incorporated into the equipment. The tightening operation is stopped when the applied torque reaches a prescribed value of 600 Nm. The whole equipment is supported by an arc-shaped arm, which enables it to be rotated clockwise or anti-clockwise by 45 degrees from a horizontal position. Therefore, by repeating the simultaneous tightening of four bolts twice, the tightening operation of a set of eight wheel bolts or eight inner nuts can be completed.

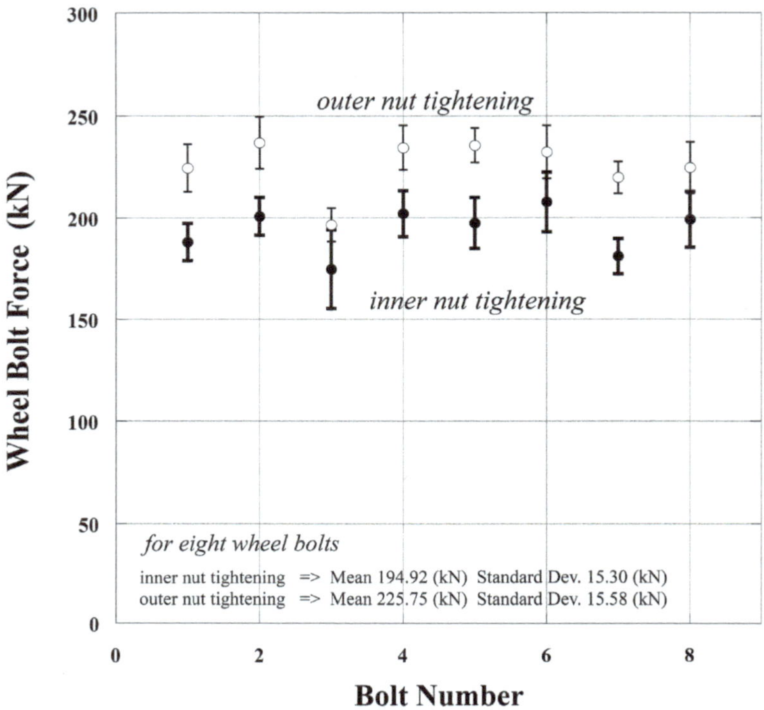

Fig. 8.18 Examination of tightening accuracy of air-powered torque wrench.

Fig. 8.19 Newly developed equipment to simultaneously tighten four wheel bolts or inner nuts.

8.3 Fatigue failure of threaded portion of rollercoaster axle

A terrible rollercoaster accident occurred in May 2007 at Expo Land, located in Osaka, Japan. It has been reported that the cause of the accident was the rupture of the threaded portion processed at the end of an axle, due to metal fatigue. In the following, it is discussed how the shape of the threaded portion affected the strength.

1) Strain energy stored in a stepped bar and the shape of the axle of the rollercoaster

Slender bolts with reduced shank are customary used as a crankpin bolt in large-scaled internal combustion engines. The use of slender bolts is a well-known strategy, by reducing the bolt stiffness, for raising the fatigue strength of jointed components subjected to large external forces repeatedly. On the other hand, from the viewpoint of strain energy, a stepped bar can absorb a lower amount of strain energy than a straight bar.

In Fig. 8.20A, strain energies U stored in straight and stepped bars are compared [156]. As the length of the thick portion increases for the total length, the absorbing capacity of strain energy is decreased because the deformation of the total bar is concentrated around a slender portion. Fig. 8.20B illustrates the original shape of the ruptured rollercoaster axle. It is like a stepped bar consisting of three bars with different diameters. The threaded portion has a small diameter and its length is short for the total length of the axle. Accordingly, when the axle is subjected to axial force, the total absorbing capacity of strain energy necessarily becomes small, since the deformation is concentrated on this slender portion.

2) On the fatigue failure of the axle

When axles or bars are ruptured due to metal fatigue, the configuration of fracture surface varies depending on the way external forces are applied. In the case of propeller shafts, in which the shear force due to twisting moment is a dominant factor, the rupture commonly occurs in a 45 degrees orientation relative to the axis. It coincides with the direction of the principal stress exerted on the shaft. In contrast, in case

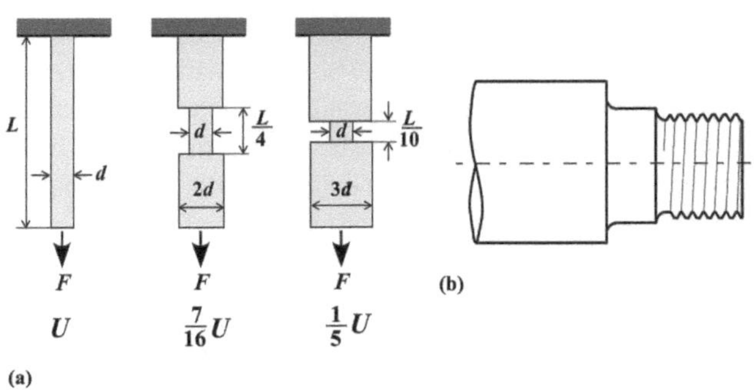

Fig. 8.20 Strain energy stored in straight and stepped bars and the end configuration of the ruptured axle. (A) Comparison of strain energy and (B) stepped bar with threaded portion.

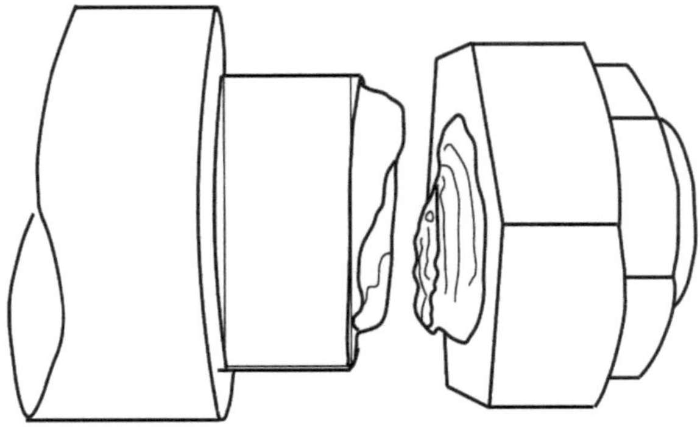

Fig. 8.21 Configuration of the fracture surface of the rollercoaster axle.

of bolt-nut connections under repeated loading, the rupture usually progresses in the cross section almost perpendicular to the bolt axis. The ruptured axle of the rollercoaster had been subjected to various types of loadings, i.e., bending due to the weight of the rollercoaster and passengers acting upward and downward, axial loading due to the turning motion of the rollercoaster, etc.

The fracture surface of the axle is almost perpendicular to the axis, as shown in Fig. 8.21. The configuration of the fracture surface is similar to the one occurring in the pulsating fatigue test of bolt-nut connections. In addition, a comment has been issued that the axial force applied repeatedly is a dominant factor in this accident [157], since fine striped patterns inherent to fatigue failure are observed. One and a half months after the accident, in the process of accident investigation, serious cracks were found at the same position of another rollercoaster, which had been operated as a pair. It indicates that there were possibly fundamental problems on the axle configuration and the management of tightening operation of the axles.

8.4 Mechanical properties of multibolted joints under shear loads [158]

8.4.1 Multibolted joints clamped by friction type and bearing type

When a joint is clamped by multiple bolts, the bolts are usually arranged perpendicular to the load direction so that the applied external force is equally shared among them. On the other hand, there are cases in which, due to various reasons, bolts are arranged along the load direction. When the strength calculation is carried out for this type of joints, it is customary to assume that the applied external force is equally supported among all bolts. In fact, however, each bolt shares a different percentage of the applied load.

In multibolted joints, friction type joints tightened with high strength bolts are widely used, as depicted in Fig. 8.22A. Mechanical behavior of the friction type joints is almost equal to that of monolithic structures, unless slip occurs on contact surfaces. In bearing type joints, shown in Fig. 8.22B, the applied shear load is supported by

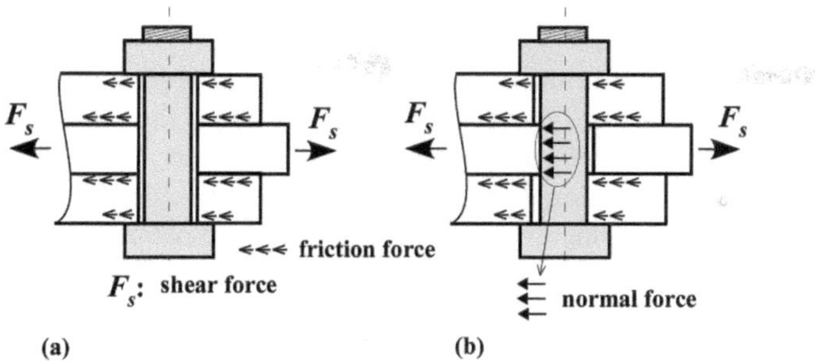

Fig. 8.22 (A) Friction type joint and (B) bearing type joint.

bringing the bolt cylindrical surface into contact with the bolt hole of fastened plates. At that time, a certain percentage of applied load is also sustained by the friction force on the contact surface. In bearing type joints, therefore, it is difficult to quantitatively evaluate the percentages of the shear load sustained by the friction force on the contact surface and the bearing force acting on the bolt hole surface. Fig. 8.23A and B illustrates chain and zigzag types of multibolted joints; *m* and *n* represent parallel and perpendicular row numbers to the direction of applied load. For the case of a joint being subjected to shear load, it has been pointed out that regarding the perpendicular row number *n*, the percentages of the sustained load become large at both ends, but no quantitative evaluation had not been performed [159]. In order to apply bearing type joints, which are less expensive than friction type joints, safely and effectively, it is necessary to establish a calculation method that can quantitatively evaluate the joints' mechanical behavior.

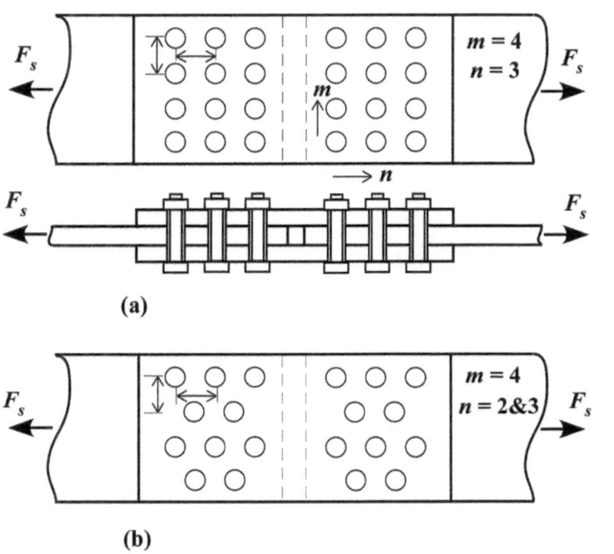

Fig. 8.23 (A) Chain type bolted joint and (B) zigzag type bolted joint.

8.4.2 Mechanism of shear force transfer in bearing type joints

Fig. 8.24 shows the equilibrium of forces in a bearing type joint of perpendicular row number $n=4$. Bolt number i is labeled from the center of the joint toward its edge at which shear force is applied. Shear force F_s, which is the applied external force, is transferred as the combination of bearing force on the bolt cylindrical surface and friction force at the plate interface. For simplicity, the case of parallel row number $m=1$ is treated here. Let the bearing and friction forces sustained by bolt i be F_i^N and F_i^μ. Then, the sum of F_i^N and F_i^μ for all bolts is naturally equal to F_s:

$$\sum_{i=1}^{n}\left(F_i^N + F_i^\mu\right) = F_s \tag{8.3}$$

Next, R_i^N and R_i^μ are defined by dividing F_i^N and F_i^μ by F_s:

$$R_i^N = \frac{F_i^N}{F_s}, \quad R_i^\mu = \frac{F_i^\mu}{F_s} \tag{8.4}$$

In Eq. (8.4), R_i^N and R_i^μ are termed shear load-bearing ratios by bearing force and friction force. The sum of them, R_i, is simply called shear load-bearing ratio of bolt i:

$$R_i = R_i^N + R_i^\mu \tag{8.5}$$

The sum of R_i for all bolts becomes unity as a natural consequence:

$$\sum_{i=1}^{n} R_i = \sum_{i=1}^{n}\left(R_i^N + R_i^\mu\right) = 1 \tag{8.6}$$

Contact surfaces existing in a bolted joint are classified into four types, i.e., thread surface, nut-bearing surface, bolt head-bearing surface, and plate interface. Among them, it is predicted that the plate interface has a dominant effect on shear load-bearing

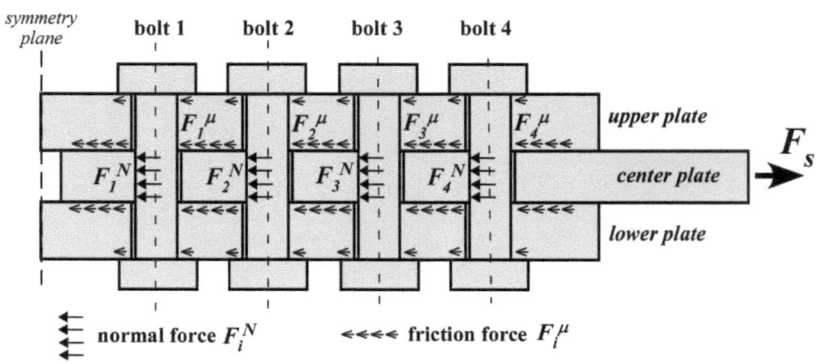

Fig. 8.24 Equilibrium of normal and friction forces in a bearing type joint.

ratio R_i, because its contact area is the largest. In the conventional design procedure, F_i^{μ} is assumed to be zero for bearing type joints, while F_i^N is set to be zero for friction type joints.

8.4.3 Finite element analysis of multibolted joints clamped by bearing type

The analytical object is a butt-type chain bolted joint shown in Fig. 8.23A. Three plates with identical thickness of 12 mm are clamped by M16 bolts with coarse thread. The bolt hole diameter is 17.5 mm. With regard to the bolt arrangement, the distances between neighboring bolts in the load direction and the distance perpendicular to the load direction are both 65 mm. The distance from the center of the outmost bolts to the plate edge is set to be 28 mm.

Other than the joint geometry, bolt force F_b, applied shear force F_s, and coefficient of friction μ are the factors affecting the mechanical behavior of the joint. In the numerical calculations, F_s and μ are varied for a constant bolt stress σ_b. The magnitude of shear force F_s is determined on the basis of mean shear stress τ exerted on a single shear plane of each bolt. Since there are two shear planes in a joint consisting of three plates, shear force F_s is calculated by the following equation for the case of perpendicular row number n and parallel row number $m=1$:

$$F_s = \frac{\pi}{4} d^2 \tau \times 2 \times n \tag{8.7}$$

where d designates bolt nominal diameter.

The standard analytical condition is selected so that the joint contact condition is in the bearing type, i.e., $\sigma_b = 100\,\text{MPa}$, $\tau = 30\,\text{MPa}$, $\mu = 0.07$. Fig. 8.25A and B shows the finite element models, the case of $m=1$ and $n=5$, and the case of $m=3$ and $n=5$, respectively. Fig. 8.26 represents the numerical results for varying perpendicular row number n from 2 to 5 under the standard condition of $\mu = 0.07$. The ordinate represents shear load-bearing ratios, R_i^N and R_i^{μ}, and the abscissa shows the bolt number

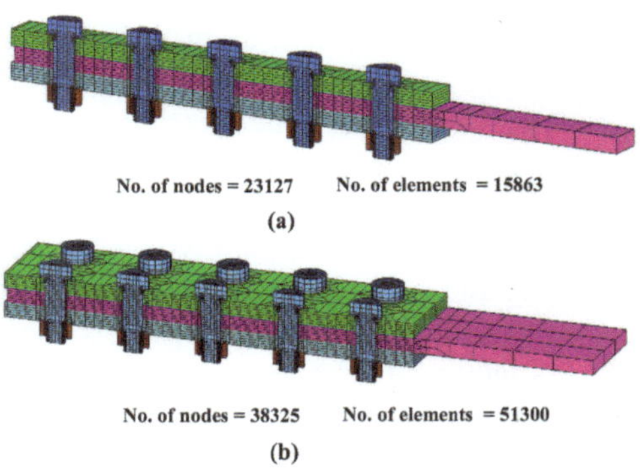

Fig. 8.25 Finite element models of multibolted joints. (A) $m=1$, $n=5$ and (B) $m=3$, $n=5$.

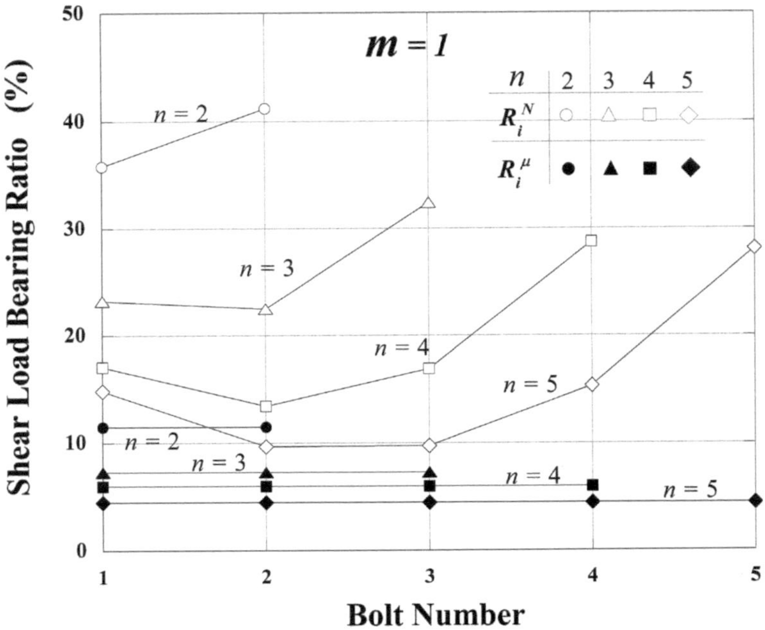

Fig. 8.26 Effects of perpendicular row number and coefficient of friction on shear load-bearing ratio.

i defined in Section 8.4.2. Shear load-bearing ratio R_i, which is the sum of R_i^N and R_i^μ, becomes relatively large at both ends, i.e., in bolt-1 and bolt-n. This result coincides with the load distribution pattern stated in [159], which supports the indication that the fractures are likely to occur in the bolts located at both ends.

Next, how R_i^N and R_i^μ contribute to R_i is compared. Fig. 8.26 shows that R_i^μ is almost equal for all bolts. The result is consistent with the conventional design procedure for friction type joints, which states that each bolt equally sustains the applied shear load. It is therefore concluded that the characteristic distribution patterns of R_i are caused by the bearing forces acting on bolt cylindrical surface. Although not shown here, the amounts of R_i^μ increase as μ is increased. In that case, R_i^μ is almost equal for the bolts arranged in the load direction, as in the case of Fig. 8.26.

For a given joint geometry, the determination of the joint type, bearing type or friction one, depends on the factors such as applied shear force, bolt force, and coefficient of friction. The conditions that determine the joint type are studied using the joint model of $m=1$ and $n=5$. In Fig. 8.27, it is shown how the joint type falls into bearing type or friction type for varying loading and surface conditions, where the ordinate is τ and the abscissa is the product of μ and bolt stress σ_b. When τ is constant, the joint gets into friction type as $\mu\sigma_b$ is increased. The gradient of the right straight line, which divides bearing type and friction one, is about 1.1. On the contrary, as τ is relatively increased or $\mu\sigma_b$ is decreased, slip occurs at the interface.

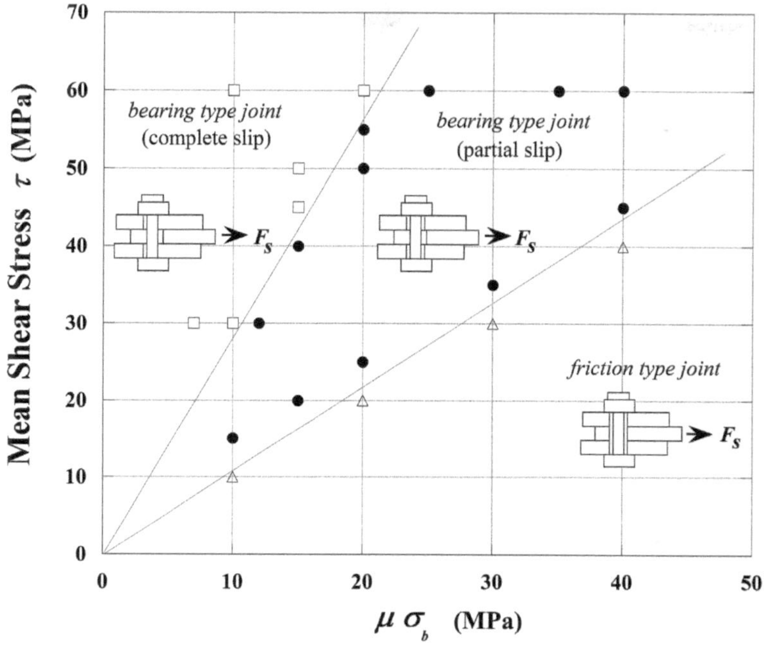

Fig. 8.27 Judgment conditions for determining joint type.

In the case of the joint consisting of three plates treated here, the center plate first starts to slip, then the left surface of the bolt body is brought into contact with the plate hole surface. As τ is further increased or $\mu\sigma_b$ is decreased, the bolt starts to slip against the upper and lower plates. Then, the right cylindrical surface of the bolt comes into contact with the hole surfaces of upper and lower plates. These situations are both regarded as bearing type joints. The former is designated as partial slip and the latter is complete slip in the figure. The gradient of the straight line that divides the two kinds of bearing type joints is about 2.8. From the foregoing explanation, it is found that the transition from friction type to bearing type progresses in two steps.

In actual joints, it is possible to prescribe the applied shear force F_s and bolt force F_b. However, the joint surface conditions change due to various factors. That is, the joint type in service condition is to be determined depending on coefficient of friction μ. It is worth noting here that the use of Fig. 8.27 makes it possible to obtain the critical value of μ for a constant τ or F_s that changes the joint status from friction type to bearing type of partial slip.

8.4.4 Analysis of shear load-bearing ratio using spring models

It has been substantiated that the numerical results of R_i obtained by three-dimensional finite element analysis, for the case of parallel row number $m = 1$, can favorably be

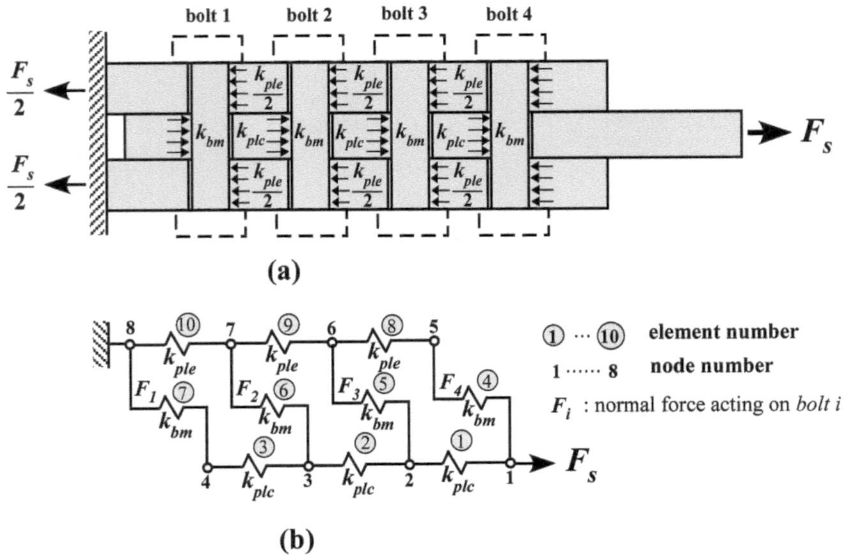

Fig. 8.28 Spring models for calculating shear load-bearing ratio of bearing type joints. (A) Analytical model and (B) spring model.

applied to the cases of $m=2$ and larger. For shear load-bearing ratio by bearing force, R_i^N, it has been shown that R_i^N distributes in a similar manner to the case of $m=1$, regardless of coefficient of friction μ, and the amount of R_i^μ is almost equal for each bolt, as shown in Fig. 8.26. These results indicate the importance of shear load-bearing ratio for the case of $m=1$, for which perpendicular row number n is systematically varied. Hence, in this section, a simple method based on elementary solid mechanics is proposed to predict the shear load-bearing ratio under the condition of $\mu=0$.

Fig. 8.28A illustrates a simplified joint model corresponding to Fig. 8.24, in which a bolt is modeled into a simple cylinder and the effect of friction is not considered. k_{bm}, k_{plc}, and k_{ple} are spring constants that represent the stiffness of bolt body, center plate, and a set of upper and lower plates for one bolt pitch length. Using the three spring models, the whole structure can be modeled as shown in Fig. 8.28B. The equation of k_{bm} is derived by combining the two solutions for a cantilever beam and a beam fixed at both ends, both of which are subjected to uniformly distributed load. The equations of k_{plc} and k_{ple} are derived by assuming a plate under uniform tension. The evaluation method of k_{bm}, k_{plc}, and k_{ple} is explained in detail in [158]. Fig. 8.29 shows the calculated results of shear load-bearing ratio R_i, where perpendicular row number n is changed from 3 to 5. In the figure, numerical results by FEM are also presented for the case of a very small coefficient of friction, $\mu=0.001$. Both results agree quite well. It is concluded, therefore, that for the case of coefficient of friction being small, shear load-bearing ratio can be evaluated by the elementary method explained above.

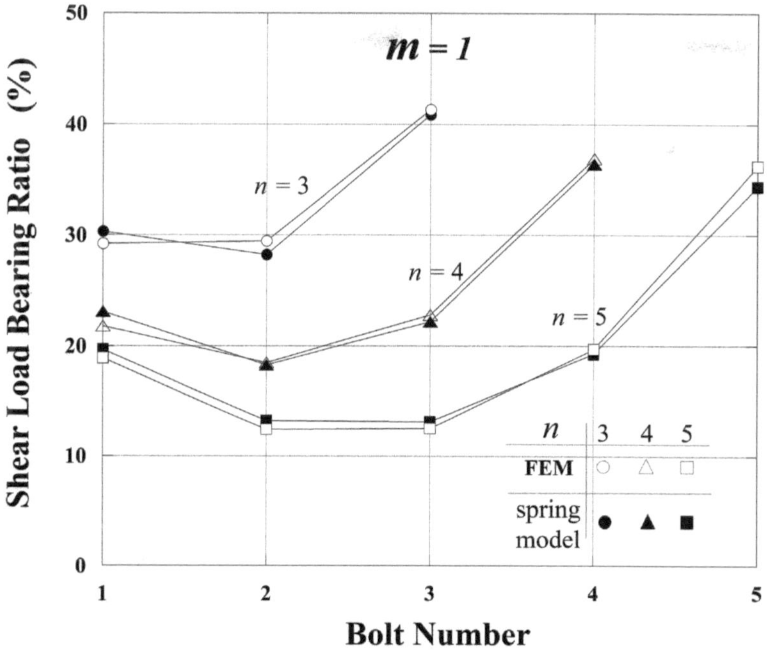

Fig. 8.29 Comparison of shear load-bearing ratios obtained by elementary method and FEM.

8.5 Strength and load capacity of reamer bolts used for rigid flanged shaft couplings [160,161]

8.5.1 Configuration and rupture phenomena of reamer bolts

A reamer bolt has a characteristic shape whose cylindrical diameter is equal to the bolt hole diameter, as shown in Fig. 8.30A. It is used to support large shear forces exerted on the joint by attaching its cylindrical surface to the bolt hole surface. In actual joints, however, since the contact condition of the two diameters being perfectly equal does not exist, the jointed portion is either in interference fit or in clearance fit, as shown in Fig. 8.30B. Regarding the amount of fit δ_c, it is reported that there exists a fair amount of dimensional error, up to $\pm 20\,\mu m$, between the diameters of bolt cylindrical portion and bolt hole, referring to the measured data of the rigid flanged shaft coupling of ship propeller shafts [162]. In the assembly process of actual joints, most joints are prepared in a slightly interference fit condition. According to the accident reports on reamer bolts provided by ClassNK, which is a representative ship classification society, the position of rupture occurrence is not at the interface between the two flanges but on the reamer surface shown in Fig. 8.31, which is somewhat away from the interface toward the ship propeller.

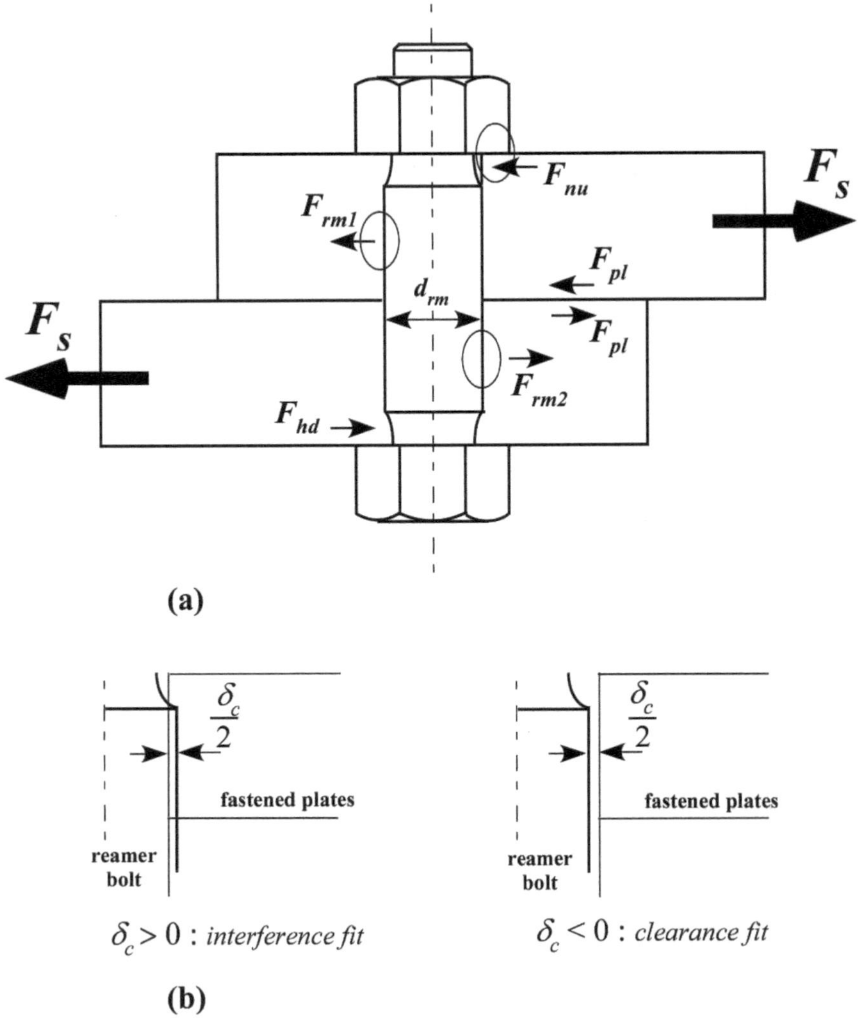

Fig. 8.30 Force equilibrium and fit in reamer bolted joint. (A) Shear force and friction force acting on joint and (B) interference fit and clearance fit.

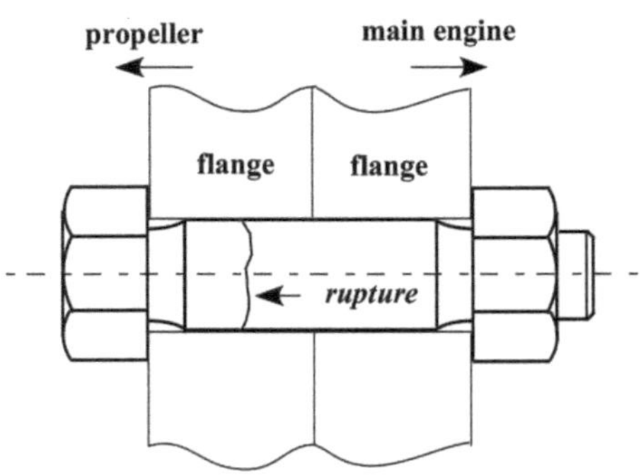

Fig. 8.31 Rupture of reamer bolt used for shaft coupling of ship propeller.

8.5.2 Mechanical properties of reamer bolts and its related problems

Fig. 8.30A shows the equilibrium of forces around the joint clamped by a reamer bolt when subjected to shear force due to external force like shaft torque. As is evident from the figure, the forces acting on the joint must be evaluated by the sum of the force perpendicular to the reamer surface and the friction force at the plate interface. In contrast, a certain amount of clearance exists in ordinary bolts; hence, they are called loose bolts. In loose bolts, the shear force applied to the joint is transmitted only by friction forces acting on the plate interface, and nut- and bolt head-bearing surfaces. When using the high strength bolts introduced in Section 8.4.1, which are commonly used in the field of architecture, the roughness of the plate surface is intentionally raised to increase coefficient of friction. The ingenuity makes it possible to sustain large shear forces by clamping the bolt with high bolt stress. Specifications of reamer bolts are given in JIS B1451 "Rigid Flanged Shaft Couplings." Fig. 8.32 schematically shows the configuration of the flanged shaft coupling. According to JIS, the shear strength of reamer bolts is evaluated by the following equation:

$$\tau = \frac{T}{\frac{B_p}{2} \times \frac{n}{2} \times \frac{\pi {d_{rm}}^2}{4}} \tag{8.8}$$

where τ and T represent mean shear stress at the cross section of reamer bolt and applied torque, and B_p, n, and d_{rm} designate diameter of bolt pitch circle, number of bolts, and reamer bolt diameter, respectively.

The torque transmitted from the rotating shaft is delivered from the driving side flange to the driven side one through reamer bolts that clamp the two flanges. At that time, the delivered torque acts as a normal load on the cylindrical surface of reamer bolt. In Eq. (8.8), it is assumed that the applied torque is sustained by half the number of actually used bolts, $n/2$, considering the dimensional accuracy of bolt hole and bolt location and the assembly error of flanges, etc. In addition, Eq. (8.8) does not take

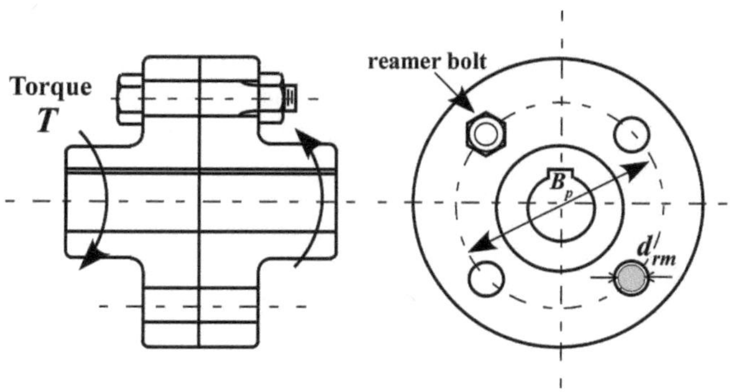

Fig. 8.32 Rigid flanged shaft coupling specified in JIS B1451.

account of the positive effect of friction force at the interface between mating flanges. The problems when using reamer bolts are listed below:

1) In JIS calculation formula, Eq. (8.8), it is assumed that the delivered torque is only sustained by the forces perpendicular to the cylindrical surface of reamer bolt. That is, the effect of friction force is not considered. Alternatively, the effects of various factors like geometric and assembly errors, etc., are treated by adjusting the number of clamping bolts from n to $n/2$. However, the rationale of this assumption is not clear.
2) Since it is virtually impossible to perfectly match the diameters of reamer bolt and bolt hole, a practical calculation procedure should be established, which can evaluate the strength of shaft coupling by considering the dimensional errors to the extent that occurs in the actual joints.
3) When the clearance between reamer bolt and bolt hole varies from plus-minus a few μm to 10 μm, it is substantially impossible to comprehensively clarify the mechanical behavior of the shaft coupling by experimental approaches.
4) A certain amount of dimensional error inevitably occurs in the location of bolt holes of the mating flanges, unless the two bolt holes are processed by stacking the two flanges that make up the shaft coupling. However, in the marine engineering field, components, e.g., main engines and propeller shafts, are usually manufactured by different companies. Therefore, it becomes an unavoidable problem.

In order to solve the aforementioned four problems and elucidate the related mechanical phenomena, it is effective to introduce numerical analysis methods represented by FEM.

8.5.3 Shear force transfer ratio and bending stress

When two plates with identical thickness are tightened by a single reamer bolt, as shown in Fig. 8.30A, shear force applied to the joint, F_s, is carried by the combination of the force perpendicular to reamer surface and friction force. Considering the equilibrium of forces in the upper plate, the following equations are obtained [160]:

$$
\begin{aligned}
R_{rm} &= F_{rm1}/F_s \\
R_\mu &= \left(F_{nu} + F_{pl}\right)/F_s \qquad (8.9)\\
&\text{where } F_s = F_{rm1} + F_{nu} + F_{pl}
\end{aligned}
$$

Similar equilibrium equations hold for the lower plate. R_{rm} and R_μ represent the ratios of shear load transmitted through reamer surface and friction force, respectively. They are collectively termed shear force transfer ratio, which is abbreviated as SFTR. The sum of R_{rm} and R_μ naturally becomes unity. F_{rm1} and F_{rm2} are normal forces acting on reamer surface. F_{nu}, F_{hd}, and F_{pl} are friction forces acting on nut- and bolt head-bearing surfaces and plate interface, respectively. The ratio of R_{rm} to R_μ varies depending on the fit around reamer bolt, coefficient of friction, bolt stress, the magnitude of applied shear force, and joint geometry. If the effects of these factors on the ratio of R_{rm} to R_μ can be evaluated quantitatively, it becomes possible to revise the conventional guideline for the strength evaluation of reamer bolts and construct a new guideline considering both safety and efficiency.

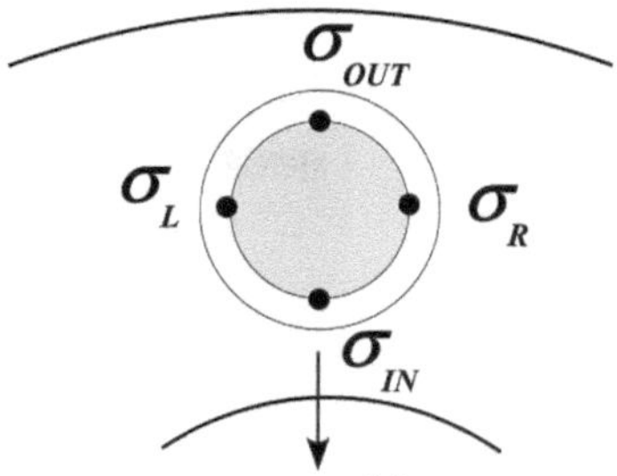

Fig. 8.33 Bending loads exerting on reamer bolt in circumferential and radial directions.

The cylindrical portion of the reamer bolt is subjected to bending stress induced by the applied shear load. Referring to Fig. 8.30, in the case of interference fit, the bending load acts so as to deform the cylindrical portion of the reamer bolt into an S-shape against the interface consisting of two plates. As a natural consequence, how the applied load acts on the joint is significantly changed, depending on the fit between the bolt cylindrical portion and the bolt hole. The cylindrical portion of reamer bolts, used for rigid flanged shaft couplings, is subjected to bending loads in the circumferential and radial directions, as shown in Fig. 8.33. Circumferential bending stress is defined by dividing the difference between the axial stresses exerted on the left and right sides of the cylindrical portion, $\sigma_L - \sigma_R$, by two [161]. Radial bending stress is defined in the same manner using σ_{IN} and σ_{OUT} shown in the figure. When the objective joint is a rectangular-shaped one, which is treated in the next section, its bending stress component corresponds to the bending stress in the circumferential direction by assuming the shear load direction being circumferential.

8.5.4 Effects of fit, friction coefficient, and bolt stress

Fig. 8.34A and B show finite element models of a rectangular-shaped joint and a rigid flanged shaft coupling. In the latter FE model, the engaged thread portions are replaced by simple cylinders aiming at enhancing calculation efficiency. The magnitude of shear force is expressed by the mean shear stress τ exerted on a single shear plane of the target reamer bolt. In the case of the joints shown in Fig. 8.34, the number of shear plane is one, because the fastened plates are composed of two plates. When the joints are composed of three plates, the number of shear plane becomes two. The numerical results for the rectangular-shaped joint are shown in this section [160].

In Fig. 8.35A and B, shear force transfer ratio is shown for the bolt stress σ_b of 100 and 300 MPa, respectively. The abscissa represents the fit δ_C between reamer bolt and bolt hole, and the parameter is coefficient of friction μ on contact surfaces. Mean shear stress τ acting on the reamer bolt is set to be 50 MPa. It is found that as the fit δ_C is increased, R_{rm} increases and R_μ decreases. Regarding coefficient of friction μ, as a natural consequence, R_μ increases and R_{rm} decreases with increasing μ. The effect of μ becomes small in the range of $\delta_c > 0$, i.e., interference fit. In contrast, the effect of δ_C is rather small in the range of $\delta_c < 0$, i.e., clearance fit, and then the values of R_{rm}

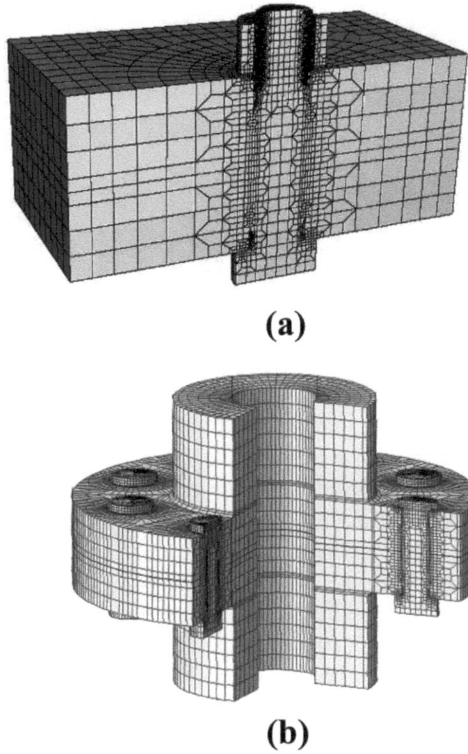

Fig. 8.34 Finite element models of rectangular-shaped joint and rigid flanged shaft coupling.

and R_μ become almost constant. That is because the occurrence of slip brings the reamer surface newly into contact with the bolt hole. Hence, the ratio of R_{rm} to R_μ becomes almost constant. In Fig. 8.35B, when μ is 0.2 and δ_C is less than $-5\,\mu\text{m}$, R_{rm} is zero and R_μ is unity, because contact surfaces are completely in stick condition. In summary, the ratio of R_{rm} to R_μ significantly varies depending on the fit δ_c, coefficient of friction μ and bolt stress σ_b.

The numerical results of bending stress generated in reamer bolts are shown in Fig. 8.36A and B. In the figures, two components are also shown of axial bolt stress on the reamer surface located 180 degrees apart from one another against the shear load direction. The amount of fit δ_C is set to be $-10\,\mu\text{m}$. Coefficient of friction μ is changed to 0.2 and 0.1. In the case of $\mu = 0.2$, the magnitudes of bending stress, depicted by black circles, are all small. However, in the case of $\mu = 0.1$, the bending stress significantly increases and shows the maximum value in the bolt head side away from the interface. This location almost coincides with the one where the rupture of reamer bolts frequently occurs, shown in Fig. 8.31. Although not shown in the figures, in the case of $\delta_c = 0\,\mu\text{m}$ and $\mu = 0.1$, the maximum bending stress is produced around the interface between two flanges.

As a result, it is concluded that the rupture of reamer bolts is likely to occur under large shear load, when δ_C is less than zero, i.e., clearance fit, and μ is small.

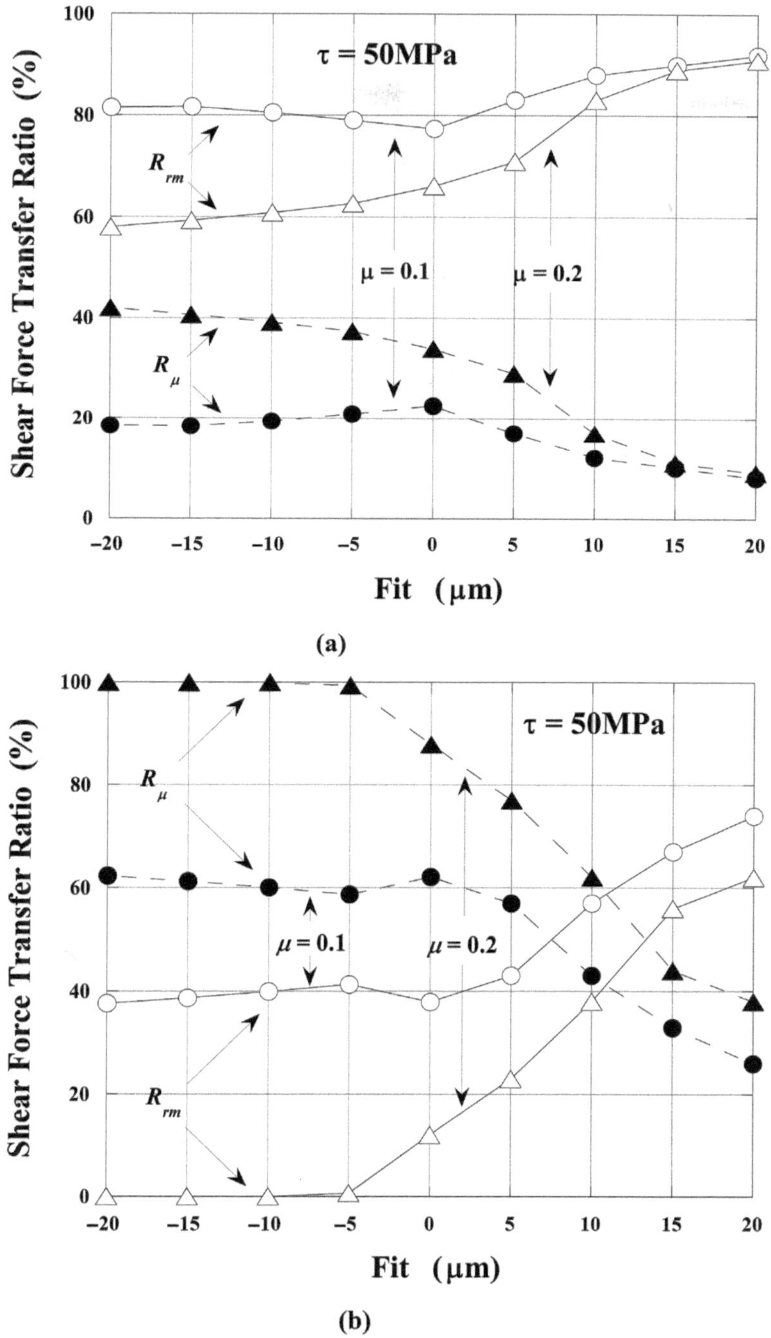

Fig. 8.35 Effects of various factors on shear force transfer ratio. (A) $\sigma_b = 100\,\text{MPa}$ and (B) $\sigma_b = 300\,\text{MPa}$.

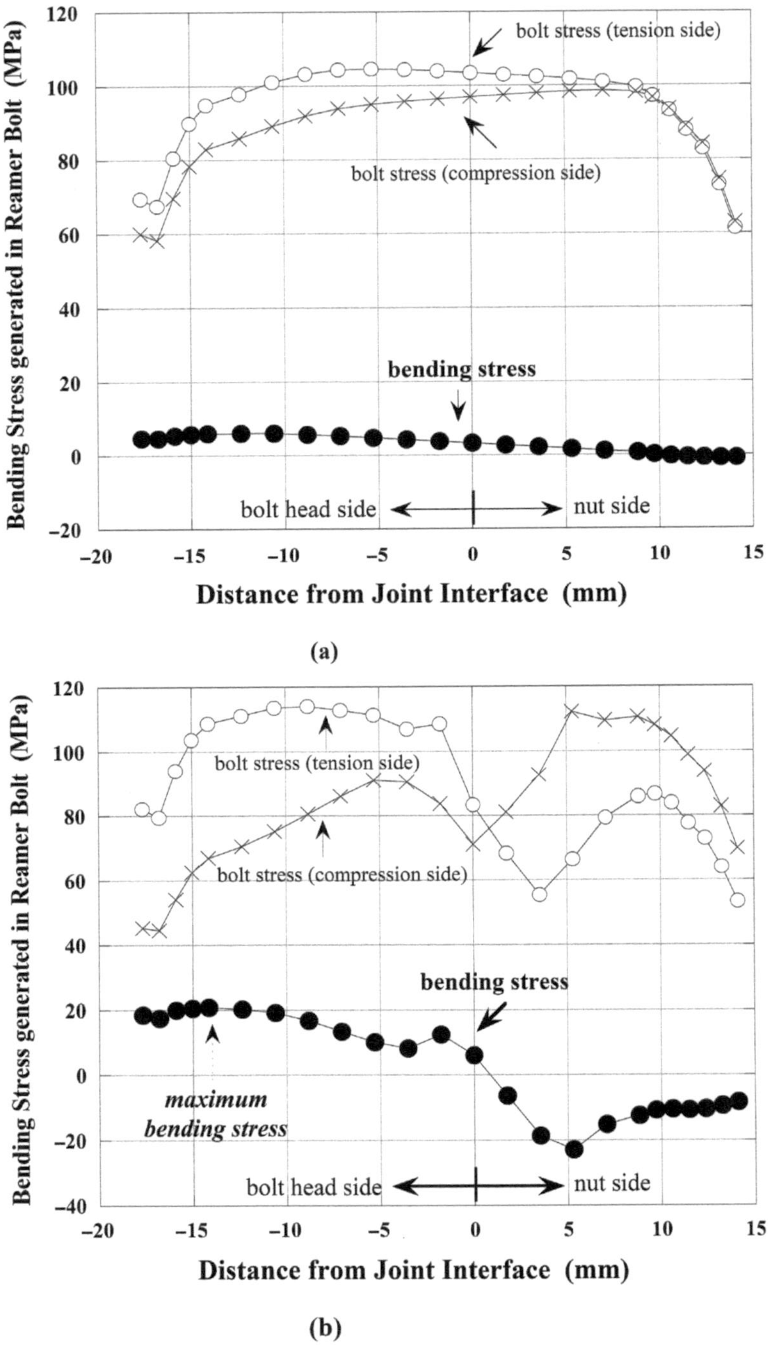

Fig. 8.36 Numerical results of bending stress generated in reamer bolts. (A) $\delta_c = -10\,\mu\text{m}$, $\mu = 0.2$ and (B) $\delta_c = -10\,\mu\text{m}$, $\mu = 0.1$.

This suggests that special attention must be paid to cases like ship propeller shafts, in which the direction of rotation frequently changes, particularly in the harbor area. In that case, even if the contact condition is initially interference fit, it may change into clearance fit due to the repetition of twisting moments applied in the clockwise and anticlockwise directions. In addition, there is a high probability that coefficient of friction is also decreased, owing to the rubbing action at the interface. Then, large bending moment occurring on the reamer surface can cause the fatigue failure of reamer bolts.

8.5.5 Effects of bolt force scatter and alignment error [161]

Since shaft couplings are clamped by multiple bolts, it is considered that the scatter in bolt force among them and the misalignment existing in the shaft system significantly affect the mechanical behavior of the couplings. In order to clarify the phenomena, systematic numerical analyses were conducted using the entire model of the coupling shown in Fig. 8.34B. Regarding the bolt forces, the numerical results have shown that the scatter to the extent usually occurring has little effect on shear force transfer ratio and the bending stress exerting on reamer bolts. As an extreme example, when the bolt force is completely lost in one bolt, it has been reported that in clearance fit, most of the applied shear load is supported on its own reamer surface. On the other hand, the effect of the bolt force loss in one bolt is limited in interference fit.

When connecting two shafts using a pair of rigid flanged shaft couplings, it is practically impossible to make the center axes of driving and driven shafts completely coincide with each other while bringing the mating flange surfaces into close contact. In actual shaft couplings, there exist some types of misalignments, as shown in Fig. 8.37. The three types of misalignments are commonly termed offset or parallel misalignment, endplay or axial misalignment, and angular misalignment [163].

Since the mechanical behavior of shaft couplings is influenced by the stiffness of the entire joint, it varies depending on the bolt stress and the distance between the joint and bearing, as well as the amount of misalignment. Regarding parallel and angular misalignments, in order to avoid the generation of high bending stresses in the running

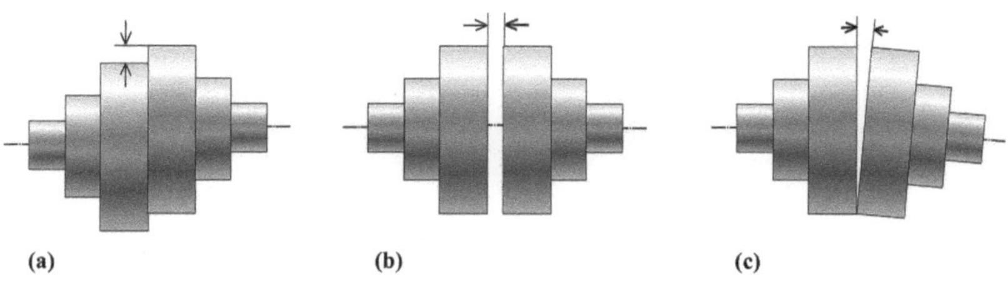

Fig. 8.37 Three basic types of misalignments in shaft couplings. (A) Parallel misalignment (offset), (B) axial misalignment (endplay), and (C) angular misalignment.

condition, the shaft system is aligned very carefully so as to make the center axes of both shafts precisely coincide with each other. However, in shaft systems like ship propeller shafts, a certain amount of parallel and angular misalignment is inevitably produced, owing to the shaft weight and the deflection of bearing foundation. In contrast, endplay or axial misalignment becomes an issue less frequently in actual shaft couplings. In any event, it is important to clarify the threshold of the misalignment beyond which it would affect the mechanical behavior of shaft couplings.

According to the numerical results, the effect of parallel misalignment on shear force transfer ratio and bending stress is small as long as the misalignment stays within one 1/100th of the shaft length. As for endplay or axial misalignment, if the amount is so large that the mating flange surfaces are not in contact each other in the initial tightening state, large bending stresses are produced in the bolt body in both radial and circumferential directions. However, when the amount of misalignment is not so large, its effect can be mitigated by raising the bolt stress.

Angular misalignment is the most troublesome for shaft couplings. Fig. 8.38 shows the effect of angular misalignment on shear force transfer ratio of reamer surface, R_{rm}. The line for 0 degrees indicates the status without angular misalignment. The effect of misalignment is remarkably observed beyond about 0.03 degrees. Although not shown in the figure, bending stresses are also increased beyond 0.03 degrees, corresponding to R_{rm}. If such a large amount of angular misalignment exists, large bending stresses are produced in the bolt body in the radial and circumferential directions. It should be kept in mind that this undesirable phenomenon can scarcely be mitigated even if raising the bolt stress.

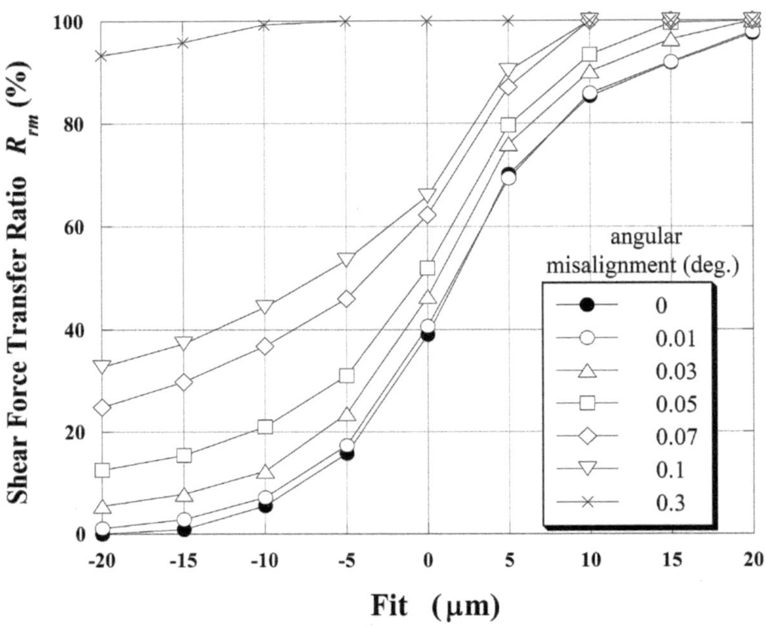

Fig. 8.38 Effect of angular misalignment on shear force transfer ratio.

8.6 Tightening process of reamer bolt by cooled fitting [164]

8.6.1 Tightening procedure and its related problems

Diameter of reamer bolt is basically processed to be equal to bolt hole diameter. However, a certain amount of difference inevitably occurs between the two diameters due to processing errors. To cope with any dimensional error, reamer bolts are cooled in the actual tightening work to reduce the diameter using dry ice or liquid nitrogen, and then are inserted into the bolt hole. After that, a usual tightening operation is carried out. As a consequence, the following problems arise:

1) If the tightening operation is completed while the reamer bolt temperature is still quite low, the bolt force produced at that time is going to be reduced with time, due to bolt elongation occurring while the bolt temperature increases to the ambient one.
2) To avoid the foregoing problem, if the tightening operation is started after the whole joint temperature almost reaches the ambient one, it requires much longer working time.
3) When the fit between the reamer bolt body and bolt hole is in the state of interference at room temperature and the shrinkage by cooled fitting is insufficient, there is a possibility of the seizure occurrence on the reamer surface in the tightening process by torque control method.
4) If the reamer bolt body is in contact with the bolt hole during the tightening process, a part of the tightening torque is consumed as friction work, hence, the bolt force is decreased to below the target value.

When carrying out cooled fitting, in order to cope with the first problem, the tightening work should be started after the joint temperature almost reaches the ambient one [165]; however, this idea causes the second problem of long working time and the third problem of seizure with respect to the fit or the problem of insufficient bolt force. Therefore, if the amount of bolt force reduction due to the temperature increase after the completion of tightening work can be predicted, three problems, from 2) to 4), occurring in the reamer bolts become avoidable while conducting the tightening operation at low temperature.

8.6.2 Estimation method of amount of bolt force reduction due to temperature increase

Fig. 8.39 schematically shows the tightening process of reamer bolt, which consists of three steps, as shown below:

Step 1: A reamer bolt cooled by dry ice or liquid nitrogen is inserted into the bolt hole. It is assumed that a micro-gap exists between the reamer bolt body and bolt hole at this moment. When a cooled reamer bolt is attached to the fastened plates, the bolt temperature starts to increase and the plate temperature decreases below room temperature, because the heat is taken away by the reamer bolt.

Step 2: Mean temperatures of the reamer bolt and fastened plates are supposed to be T_b and T_f at the moment when bolt force F_b is produced by applying tightening torque to the reamer bolt.

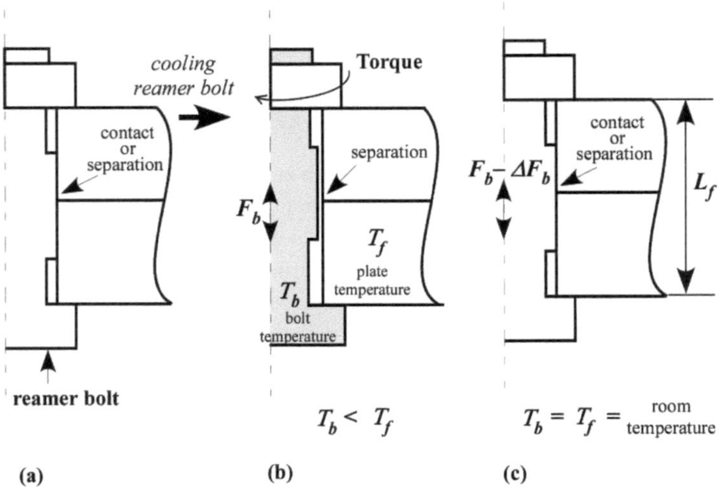

Fig. 8.39 Tightening process of reamer bolt by cooled fitting. (A) Initial state, and (B) tightening by applying torque, and (C) completion of tightening operation.

Step 3: When the tightening operation is completed and then the temperatures of reamer bolt and fastened plates reach room temperature, bolt force is reduced by ΔF_b. This indicates that the cause of bolt force reduction is the temperature difference between T_f and T_b, denoted by ΔT, at the moment of completing the application of torque to the reamer bolt.

When the tightening work is completed and the joint temperature wholly returns to room temperature, the temperature difference ΔT between reamer bolt and fastened plates becomes zero. This situation corresponds to the case that the reamer bolt temperature is relatively raised by ΔT against the plate temperature. As a result, the reamer bolt is elongated by an amount of $\alpha_b \Delta T L_f$, which is the product of coefficient of linear expansion α_b, grip length L_f and ΔT. $\alpha_b \Delta T L_f$ is equal to the sum of elongations of each part of the jointed portion, which is expressed as the product of bolt force reduction ΔF_b times the reciprocal of the spring constant of entire bolted joint k_{total}:

$$\alpha_b \Delta T L_f = \Delta F_b \left(\frac{1}{k_{th}} + \frac{1}{k_s} + \frac{1}{k_{cyl}} + \frac{1}{k_{hd}} + \frac{1}{k_f} \right) = \frac{\Delta F_b}{k_{total}} \tag{8.10}$$

Eq. (8.10) is made identical to eq. (3.40) by removing the terms relevant to approaches of interface in the latter equation, which represents the equilibrium of displacements in the tightening process of thermal expansion method in section 3.5. That is, once the geometry and material of the target joint are determined, bolt force reduction ΔF_b can be estimated for a given temperature difference ΔT, because the spring constants, grip length L_f, and α_b in Eq. (8.10) are known quantities.

In the case of a constant bolt stress, the bolt force increases in proportion to the square of nominal diameter. Therefore, the variation of clamping force is evaluated by bolt stress variation $\Delta \sigma_{rm}$, not by bolt force variation. For screw threads with coarse

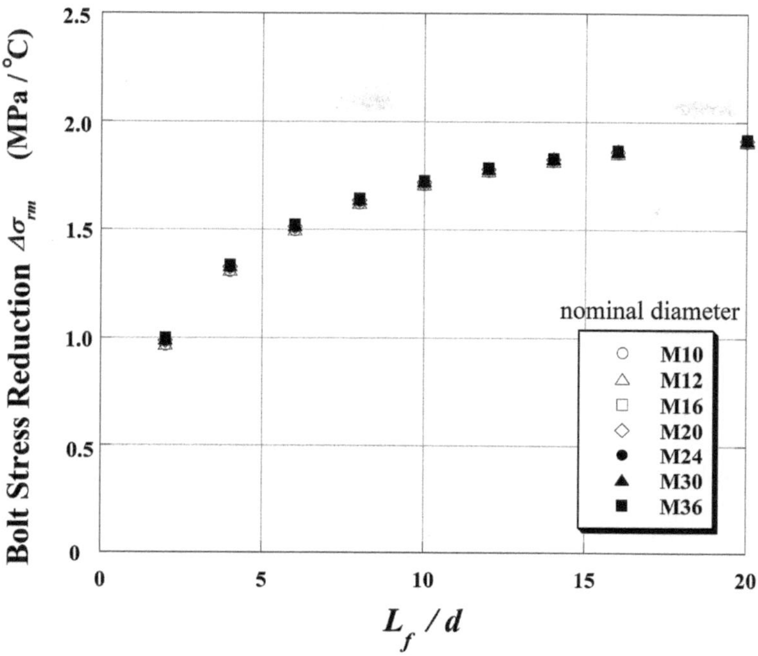

Fig. 8.40 Bolt stress reduction per unit temperature difference.

thread of M10, M12, M16, M20, M24, M30, and M36, bolt force reduction ΔF_b is calculated by Eq. (8.10), then $\Delta\sigma_{rm}$ is obtained by dividing ΔF_b by cross-sectional area for each nominal diameter.

Fig. 8.40 shows the numerical results for the case of ΔT being 1 °C. The abscissa represents a dimensionless ratio of grip length L_f to nominal diameter d. The effect of nominal diameter, which is the parameter in the figure, is negligible. Bolt stress reduction $\Delta\sigma_{rm}$ per unit temperature difference is increased from about 1 to 2 MPa with increasing grip length L_f. Then, $\Delta\sigma_{rm}$ gradually approaches a saturation value. The effect of nominal diameter d and grip length L_f on $\Delta\sigma_{rm}$ can be explained by considering the relationship among d, L_f, and the five spring constants composing a bolted joint, as in the case of thermal expansion method. It is found from Fig. 8.40 that bolt force reduction occurring in reamer bolts tightened by cooled fitting becomes a more serious problem with larger grip length. Treated in this section is low temperature thermal load. However, the foregoing phenomenon is consistent with the one explained in section 5.1.3 that when subjected to thermal load, larger bolt variation occurs with larger grip length. The relationship between $\Delta\sigma_{rm}$ and L_f/d can be formulated using the least squares method:

$$\Delta\sigma_{rm} = 16.95\left(1 - e^{-0.3899\left(L_f/d\right)}\right) \tag{8.11}$$

Referring to JIS B 1451, the value of L_f/d in rigid flanged shaft couplings is rather small and roughly less than 5. Accordingly, it is expected that Eq. (8.11) can be used

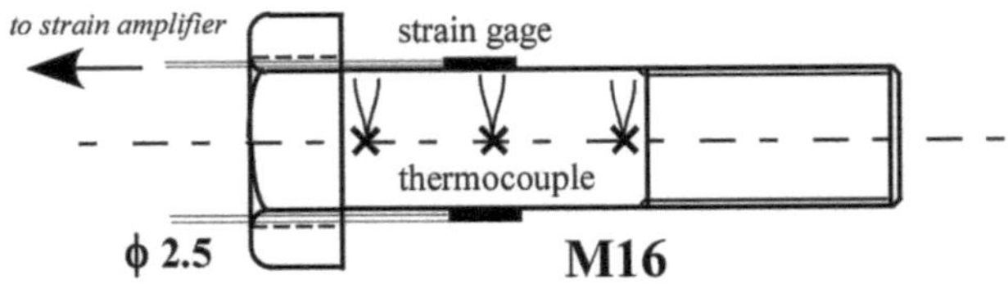

Fig. 8.41 Test specimen of M16 bolt for cooled fitting experiment.

not only for predicting the bolt force reduction ΔF_b due to thermal expansion, but also for enhancing the joint safety by adding ΔF_b to the original initial bolt force.

Fig. 8.41 shows a test specimen of M16 bolt. Attached to the cylindrical surface are a pair of strain gages and three thermocouples to measure the bolt force and bolt temperature. Immersing the test specimen in ethyl alcohol cooled by dry ice, its temperature drops to about −80 °C. Fig. 8.42 shows an example of the measured results on the relationship between axial bolt strain and bolt body temperature. The abscissa represents the temperature measured at the central portion of bolt body, depicted in Fig. 8.41. The bolt temperature is about −35 °C when starting the application of tightening torque. The amount of bolt strain relating to axial bolt stress is somewhat decreased while the bolt temperature rises from −35 °C to room temperature. Assuming that room temperature is 20 °C in this case, ΔT is 55 °C. Since $\Delta \sigma_{rm}$ ranges from about 1 to 2 MPa, according to Fig. 8.40, bolt stress may be reduced by as much as from 55 to 110 MPa. These values are obtained as the product of $\Delta \sigma_{rm}$ and ΔT.

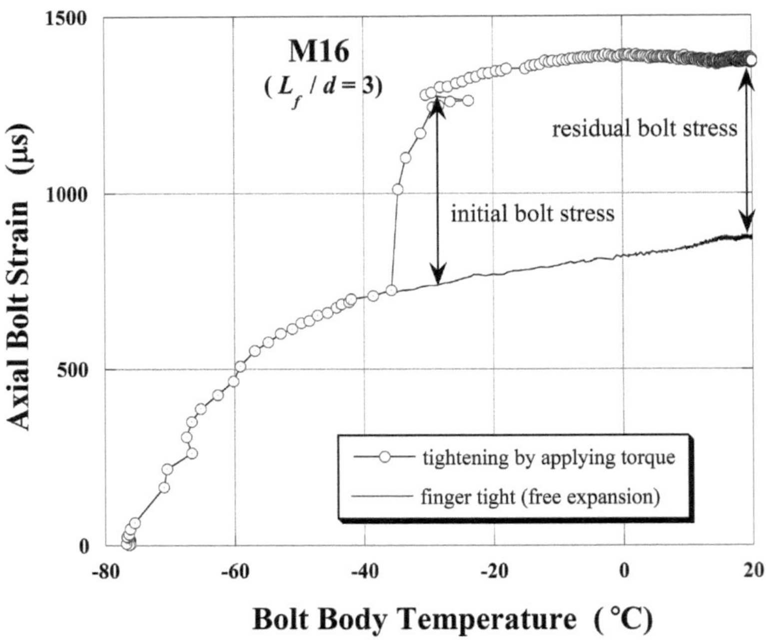

Fig. 8.42 Experimental results of the relationship between axial bolt strain and bolt body temperature.

The actual bolt stress of reamer bolts is considerably high; hence, there is little possibility that the bolt stress reduction calculated above will cause a fatal accident in shaft couplings. However, considering that the fatigue strength could be reduced due to insufficient bolt force, those values in bolt stress reduction are important when designing shaft couplings and determining the tightening procedure. The validity of the bolt stress reduction $\Delta\sigma_{rm}$ per unit temperature difference, calculated by Eq. (8.11), has been verified by experiments.

8.6.3 Proposition of tightening guidelines

Reamer bolts are commonly cooled by dry ice or liquid nitrogen. Hence, the reamer bolt temperature can be dropped to −80 °C by dry ice and −196 °C by liquid nitrogen. From the working efficiency, the use of liquid nitrogen is more desirable, because it produces a large shrinkage, which makes it easy to tighten shaft couplings with large interference. On the other hand, since the reamer bolt temperature at the start of tightening operation is very low, the temperature difference ΔT between reamer bolt and fastened plates becomes large when applying tightening torque. As a result, large bolt force reduction occurs at the completion of tightening operation. It should also be noted that the difference in working time yields a different temperature increase, thus producing a different amount of bolt force reduction.

From the foregoing reasons, it is fairly difficult to predict the bolt force reduction with high accuracy. As a countermeasure, a guideline is proposed here, based on the experimental results, that ΔT is set to be the difference between bolt head temperature, which is nearly equal to the mean temperature of bolt body, and the temperature of plate surface just next to the nut. The accuracy of the proposed guideline could be decreased according to the difference in the joint geometry and the reamer bolt temperature after cooling. Yet, it has an advantage that the amount of ΔT can be estimated by measuring the temperatures at the target points by use of a spot-type infrared thermography. If there is a margin in the joint strength, it is effective to apply the bolt force of $F_b+\Delta F_b$ to the target reamer bolt, in which F_b and ΔF_b represent the original bolt force and the anticipated bolt force reduction, respectively. The amount of ΔF_b can be obtained in the following procedure:

Step 1: $\Delta\sigma_{rm}$ is calculated by Eq. (8.11) for a given joint geometry, i.e., bolt nominal diameter d and grip length L_f.

Step 2: Bolt force reduction ΔF_b, which is predicted to occur in the process of cooled fitting, is calculated by the following equation, where the amount of ΔT is measured by a spot-type infrared thermography. Then, the target bolt force is raised by ΔF_b from the original value:

$$\Delta F_b = \Delta\sigma_{rm}\Delta T \times \frac{\pi d^2}{4} \tag{8.12}$$

The detailed tightening procedure is given in [164]. Incidentally, it has already been explained in section 3.2 how to apply an objective bolt force with practical accuracy by torque control method.

8.7 Sealing performance of oil seal plugs used for hydraulic equipment [166]

8.7.1 *Tightening characteristics and geometric errors of oil seal plugs*

Hydraulic equipment such as hydraulic motors usually consists of many parts. Among them, the parts called oil seal plugs, whose shape is similar to a bolt with low head height, are used in large numbers. Regarding the mechanical behavior of an oil seal plug, leakage from its head-bearing surface sometimes becomes a problem. The major cause of this is an increase in hydraulic oil pressure. Moreover, the oil pressure usually rises from zero to the maximum value in a relatively short time. It is considered, therefore, that the dynamic effect significantly influences the sealing performance.

Fig. 8.43A and B depicts the shape of oil seal plugs. There are two types of seal plugs, i.e., hexagon head and hexagon socket head plugs. In both types, the head height is comparatively lower than that of ordinary bolts. Generally, its dimensional accuracy is not so high; hence, a small clearance δ_{pl} can exist between the plug head and fastened plates, as depicted in Fig. 8.43C. Shown in Fig. 8.44 is the relationship between axial plug force and applied tightening torque, obtained by FE analysis. The parameter is δ_{pl}. Screw threads processed on the plug are parallel pipe thread of G1/2. The case of $\delta_{pl}=0$ is given by a straight line due to the linearity. In addition, when some clearance exists, $\delta_{pl}\neq 0$, the numerical results can approximately be expressed by a single straight line as well, regardless of the amount of δ_{pl}. Why does the effect of the difference in δ_{pl} not appear? That is because the plug head-bearing surface is brought into close contact with the surface of the main body when subjected to high plug clamping force, regardless of the amount of δ_{pl}.

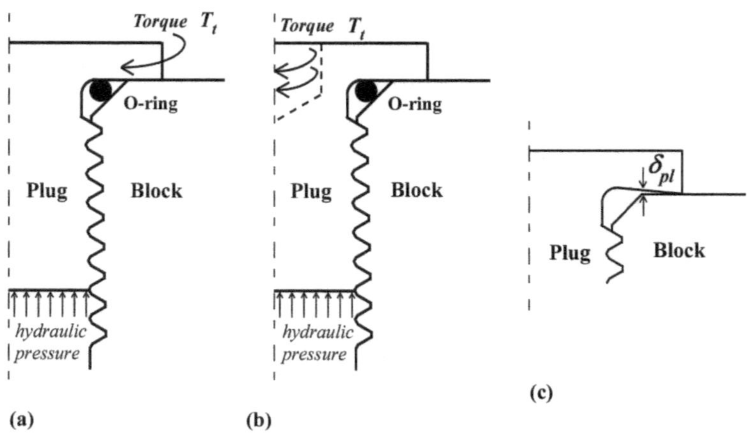

Fig. 8.43 Shape of oil seal plugs. (A) Hexagon head plug, (B) hexagon socket head plug, and (C) geometric error of hexagon head plug.

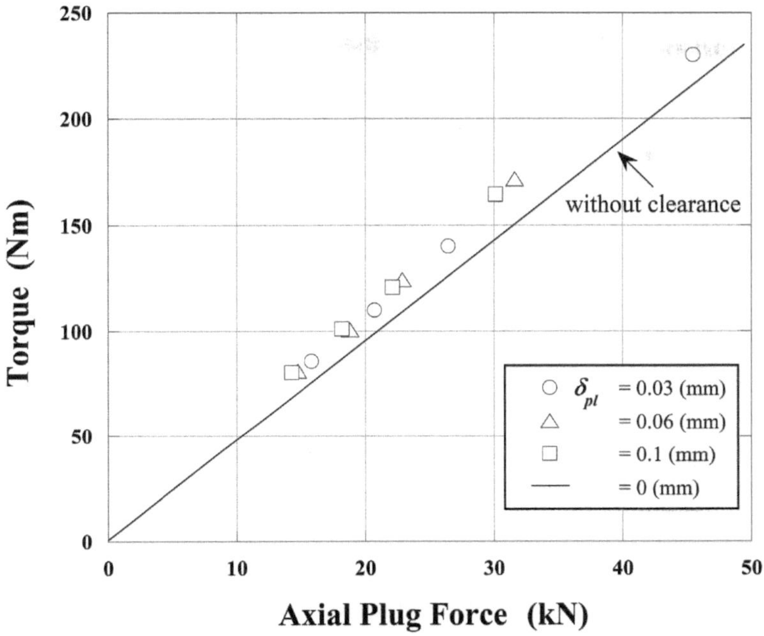

Fig. 8.44 Relationship between axial plug force and applied tightening torque.

Following the same manner as in the case of ordinary bolt-nut connections, nut factor K for an oil seal plug can be derived:

Case without clearance $\delta_{pl}=0$:

$$K = 1.029\mu + 0.0185 \text{ (hexagon head)}$$

$$K = 1.027\mu + 0.0186 \text{ (hexagon socket head)} \tag{8.13}$$

Case with clearance $\delta_{pl}\neq 0$:

$$K = 1.157\mu + 0.0171 \text{ (hexagon head)}$$

$$K = 1.018\mu + 0.0171 \text{ (hexagon socket head)} \tag{8.14}$$

Coefficients of friction μ are assumed to be identical on thread surface and plug head-bearing surface. Considering the possible processing errors in the plugs and main body, Eq. (8.14), which corresponds to $\delta_{pl}\neq 0$, is judged more practical. In the case without clearance, $\delta_{pl}=0$, there is little difference between hexagon head and hexagon socket head plugs. However, in Eq. (8.14), nut factor K is lower in hexagon socket head plug than in hexagon head plug. Substituting $\mu=0.15$ into Eq. (8.14) tentatively, nut factor K is calculated to be 0.190 for hexagon head plug and 0.170 for hexagon socket head plug. In any event, those values are somewhat smaller than K for ordinary bolt-nut connections.

8.7.2 Relationship between contact pressure distribution and sealing performance

Fig. 8.45 shows the contact pressure distributions on the bearing surface of a hexagon head plug of G1/2. They are evaluated by axisymmetric FEM, assuming the joint materials as being a linear elastic body. The axial plug force is 25.7 kN. Unlike ordinary bolt-nut connections, the range of contact zone is limited around the inner side of bearing surface. Therefore, the contact pressure steeply increases radially inwards. This phenomenon is opposite to the case of raised face flanges subjected to inner pressure, in which gasket contact pressure increases radially outwards. In the case of oil seal plugs, it is considered that this characteristic distribution pattern of the contact pressure works favorably for leakage prevention, because high oil pressure exerts on the inner edge of bearing surface. Although not shown in the figure, the contact length becomes relatively large in a small plug of G1/4; hence, the maximum contact pressure is decreased.

In Fig. 8.46, how the axial plug force is decreased with increasing hydraulic oil pressure is shown. Unlike the torque-plug force relationship in Fig. 8.44, the plug force shows complex behavior against the variation of the clearance δ_{pl} existing under the plug head. The axial plug force is decreased almost linearly as hydraulic oil pressure is raised. However, the occurrence of leakage cannot be predicted in the range of hydraulic oil pressure shown in the figure. Meanwhile, in actual hydraulic equipment, the oil pressure is frequently applied dynamically and it reaches the maximum value in

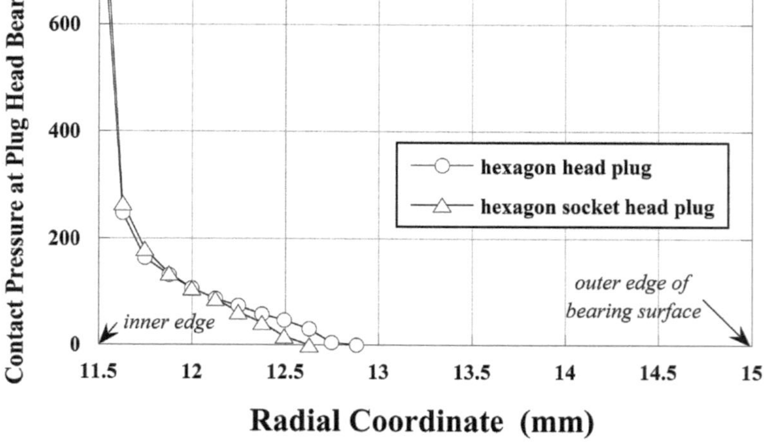

Fig. 8.45 Contact pressure distribution on the plug head-bearing surface.

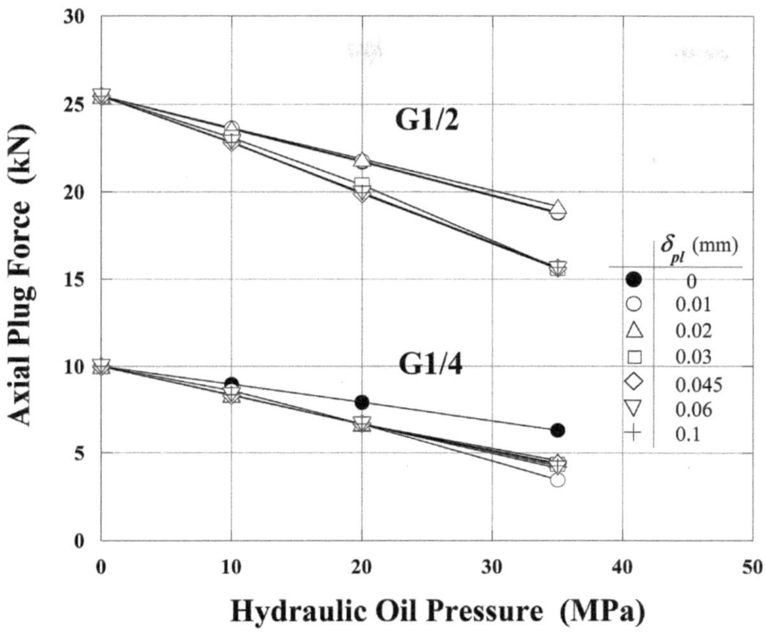

Fig. 8.46 Relationship between axial plug force and applied hydraulic oil pressure.

a very short time. In the next section, the dynamic effect of oil pressure on the plug force is evaluated by means of one-dimensional spring models.

8.7.3 Dynamic analysis of sealing performance using one-dimensional spring models

The equation of motion considering the dynamic effect is shown below, in which the effect of viscosity is ignored:

$$[M]\{\ddot{u}\} + [K]\{u\} = \{R\} \tag{8.15}$$

where $[M]$, $[K]$, and $\{R\}$ represent mass matrix, stiffness matrix, and load vector, and $\{\ddot{u}\}$ and $\{u\}$ denote acceleration and displacement vectors, respectively.

In Fig. 8.47, assuming the female threaded portion is a rigid body, an oil seal plug is divided into threaded portion and cylindrical one and they are expressed by a combination of one-dimensional spring models. The total length of the plug is 21 mm. Axial plug force is 25.7 kN and hydraulic oil pressure is 35 MPa. The oil pressure is applied stepwise corresponding to the most severe condition, as shown in the bottom right of the figure. Fig. 8.48 shows the variations of axial plug force with time, in which the stiffness of engaged threads is reduced simply by half to consider the effect of the stiffness of female threaded portion. Newmark β method is employed for time integration in the numerical analysis. It is found that the axial plug force significantly fluctuates

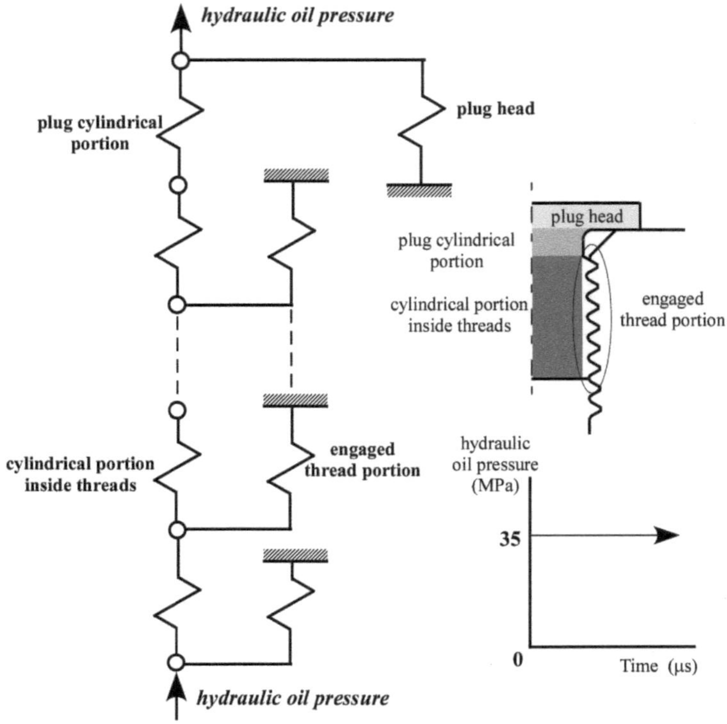

Fig. 8.47 One-dimensional spring model of a jointed portion clamped by oil seal plugs.

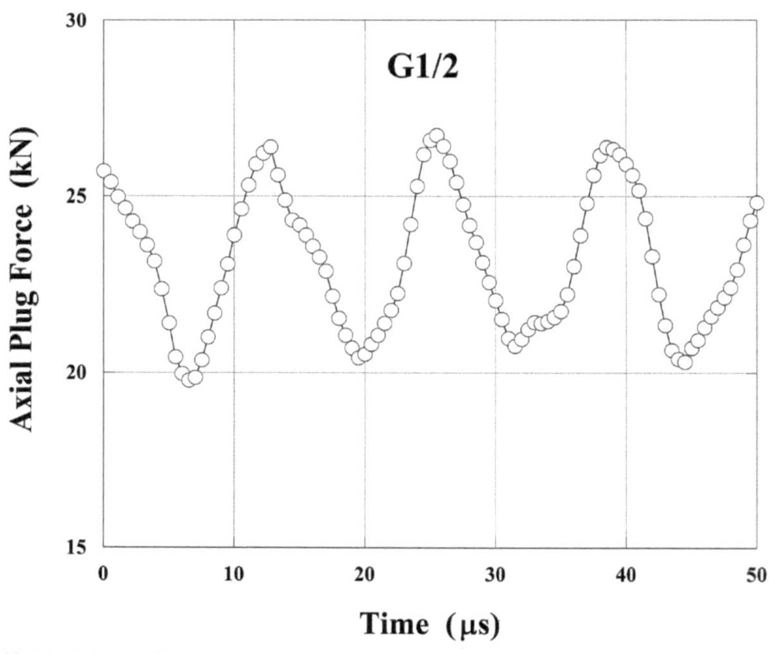

Fig. 8.48 Variations of axial plug force with time due to the dynamic effect of hydraulic oil pressure.

against the initial value because the hydraulic oil pressure exerts dynamically. Combining the result with the one presented in Fig. 8.46, it turns out that when the oil pressure is raised in a very short time, its dynamic effect increases the possibility of leakage occurrence. In order to evaluate the sealing performance of oil seal plugs with high accuracy, it is necessary to introduce two- or three-dimensional FE models. However, even with the simple numerical method proposed here, it is found that the dynamic effect of oil pressure significantly affects the sealing performance.

8.8 Mode analysis of bolted joints [56]

8.8.1 Relationship between axial bolt force and natural frequency

In machines and structures resonating subjected to vibrative external forces, the resonance can be stopped by tightening the clamping bolt harder. This phenomenon is explained by a rise in the natural frequency of jointed components with increasing bolt force. From the mechanical point of view, it is considered, e.g., that the mating surfaces, which were partially in contact under low bolt force, are brought into the perfect contact condition by raising the bolt force.

On the other hand, the foregoing phenomenon may occur without the existence of geometric errors at the interface. In that case, the phenomenon is explained as follows: contact pressure at the interface is increased by raising the bolt force; then interface stiffness is correspondingly increased with the progress of plastic deformation of microprojections existing on the surface. From the above considerations, it can be said that the evaluation of the natural frequency of a bolted joint, by taking account of the effect of interface stiffness, is particularly important when designing machines and structures to be robust against vibrative external forces.

Fig. 8.49 schematically shows the relationship between the natural frequency of the first mode and bolt force for a bolted joint consisting of two beams clamped by a single

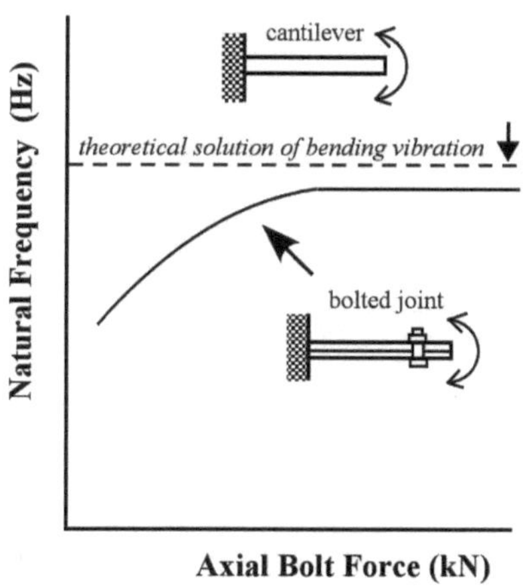

Fig. 8.49 Natural frequencies of bending vibration of a bolted joint and a cantilever.

bolt. For comparison, the natural frequency of a single beam is also shown, i.e., a cantilever having the same profile as the above two beams. It is assumed here that there is no difference in mass, due to the existence of a bolt-nut pair, between the two models. Natural frequencies of the bending vibration of cantilever are calculated by the following equation, corresponding to the single beam model stated above:

$$f = \frac{\lambda_i^2}{2\pi L^2}\sqrt{\frac{EI}{\rho_{dn}A}} \tag{8.16}$$

where λ_i is Eigen value of ith mode of vibration, L, I, and A represent length, second moment of area, and area of the objective cantilever, respectively, and E and ρ_{dn} are Young's modulus and density.

Natural frequencies of a bolted joint are increased with increasing bolt force, as shown in the figure. Then, as the contact pressure at the interface is further increased, the increasing rate starts to decrease and the natural frequencies approach a constant value. This phenomenon can be estimated by Ostrovskii's equation, given in eq. (2.33), which relates approach of interface ζ to contact pressure p_n.

Natural frequencies of the first mode of bending vibration of a cantilever can be measured by use of strain gages. Two beams are clamped by a pair of bolt and nut, and one end is fixed. Then, strain gages are attached to the surface of the test specimen near the fixed end. Applying bending vibration of the first mode to the test specimen, the strain gage readings fluctuate between tension and compression. Hence, the natural frequency can be calculated by the period of sign change. The relationship between natural frequencies of the first mode and bolt force measured by the above method is shown in Fig. 8.50 for a bolted joint consisting of two beams clamped by a pair of bolt and nut of M12 with coarse thread. Each beam measures 400 mm × 3.5 mm × 30 mm. The mounting position of a bolt-nut pair is selected to be at the center and the end of the cantilever. Surface roughness of the mating surfaces of the two beams is changed into two levels. In the case of small surface roughness, the sum of maximum height roughness of the mating surfaces, *Rzt* (μm), ranges between about 2 and 6 μm, and the value ranges between 18 and 33 μm in the case of large surface roughness.

The natural frequencies are increased with increasing bolt force, and they become almost constant beyond around the bolt force of 5 kN. The natural frequency is necessarily lower than that of a single beam indicated by a dashed line. In the case of zero bolt force, the natural frequency is about half the value of a single beam, which corresponds to the natural frequency of a beam whose height is half the clamped two beams. Focusing on the bolt positions, the natural frequencies for the center position are generally higher than those for the end position. The root cause is presumably due to the fact that the joint stiffness becomes higher when mounting the bolt in the center position. In addition, the natural frequencies are decreased with increasing surface roughness, and this tendency appears remarkably in the region of low bolt force.

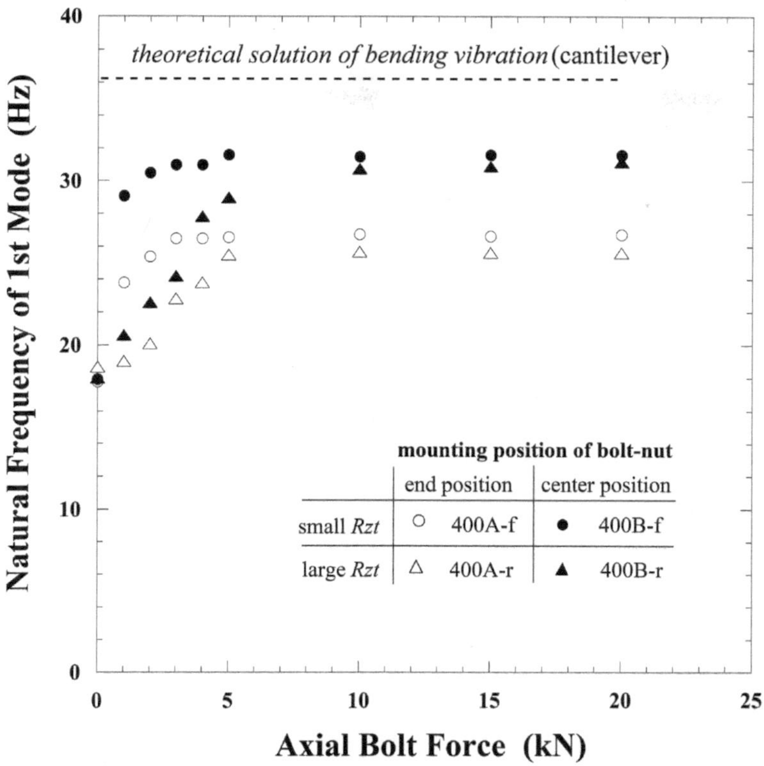

Fig. 8.50 Relationship between natural frequencies of first mode and bolt force.

8.8.2 Natural vibration analysis by taking account of Interface stiffness

In order to accurately analyze natural frequencies and natural modes of machines and structures having contact surfaces, the effect of interface stiffness needs to be considered. Natural vibration analysis by FEM is explained in detail, e.g., from [167–169]. Meanwhile, it has been shown in section 2.6.2 that once the magnitudes of surface roughness and contact pressure are given, the effect of interface stiffness can be quantitatively evaluated by spring constants in the normal and tangential directions, k_n and k_t.

Fig. 8.51 depicts a FE model of two plates clamped by a bolt-nut pair, in which the shapes of bolt and nut are not modeled. Instead, the effect of interface stiffness is considered using k_n and k_t by connecting the corresponding nodes at the interface consisting of the two plates. Mesh patterns are changed into three levels. The contact pressure at the interface is assumed to distribute following the idea of pressure cone, as explained in section 1.5.3. With respect to integration scheme in FE analysis, selective reduced integration (SRI) is employed [170]. If full integration (FI) is adopted, it

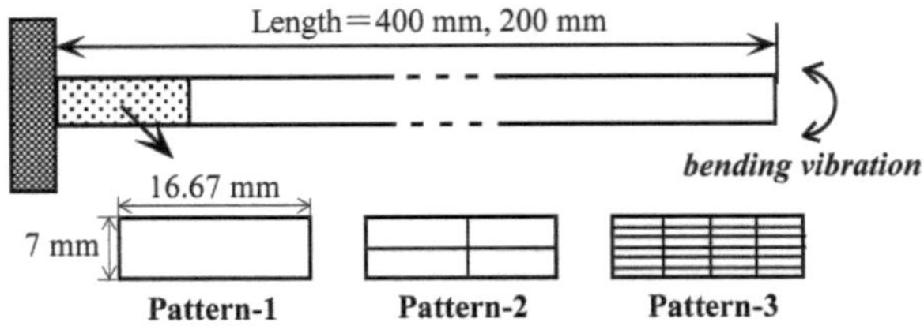

Fig. 8.51 Numerical models for natural vibration analysis.

is necessary to use a mesh pattern finer than Pattern-3. Use of SRI provides a practically sufficient accuracy even with the coarse mesh, Pattern-1.

In Fig. 8.52, the numerical results are compared to the experimental ones cited from Fig. 8.50, in which the analytical conditions are set to match the experimental ones. The values like 5 μm appearing in the figure indicate the sum of maximum height roughness of the mating surfaces *Rzt*. The difference between the numerical results and experimental ones is rather large in the range of small bolt force. However, from the viewpoint of modal analysis that takes account of the effect of surface roughness, the proposed numerical method is considered practically effective.

Fig. 8.53 shows the relationship between the ratio of the natural frequencies of a bolted joint to its equivalent single beam and the order of vibration mode. The parameters are beam length and *Rzt*. It shows that the natural frequencies of a bolted joint are

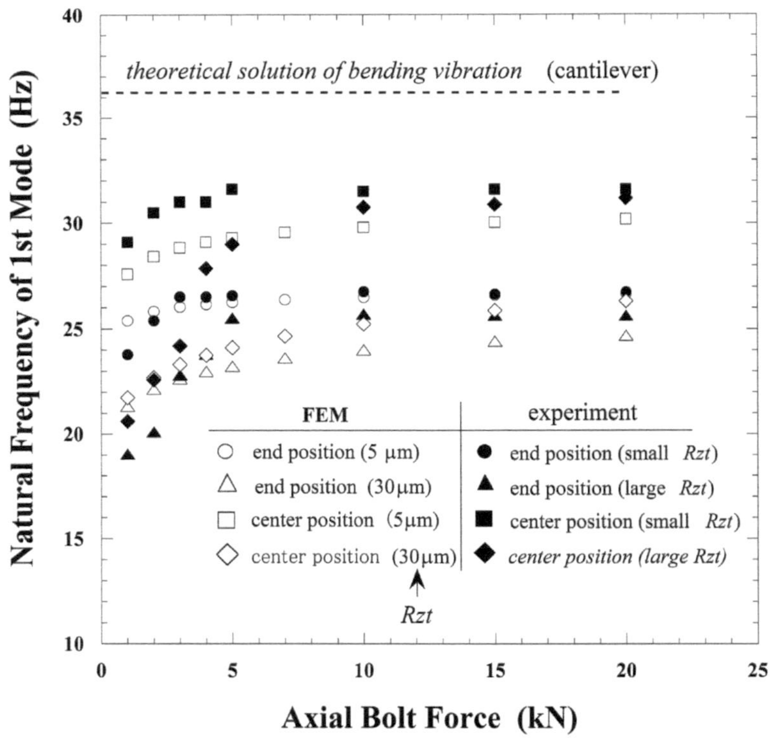

Fig. 8.52 Comparison of natural frequencies between numerical and experimental results.

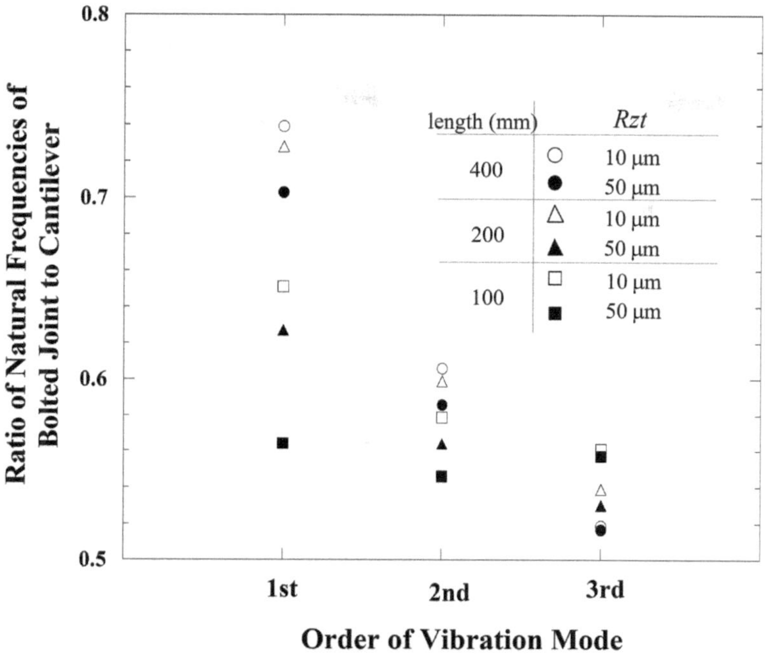

Fig. 8.53 Relationship between the ratio of natural frequencies and the order of vibration mode.

fundamentally lower than those of an equivalent single beam and this tendency becomes more remarkable as the order of vibration mode increases. As an application of the numerical method proposed here, modal analysis of bolted joints is a good candidate where the effect of interface stiffness becomes a problem.

8.9 Effective finite element analysis for bolted joints

8.9.1 Finite element models of bolted joints

Since the 1990s, numerical analyses represented by FEM have been widely used in the design stage of various industrial fields, due to the advent of high-speed personal computers and high-performance commercial software related to structural analysis, thermal fluid analysis, etc. These numerical analyses are also applied to the design of bolted joints and have achieved great results. On the other hand, even if introducing the latest technologies, a considerable number of complex problems related to the strength of actual bolted joints cannot be solved with sufficient accuracy. That is partly or mainly because of the complex geometry of threaded fasteners and the variety of loading conditions exerted on them.

In order to accurately reproduce the helical thread geometry, a fairly complicated modeling scheme is required, as explained in section 2.5. To analyze the mechanical and thermal behaviors of bolted joints with high accuracy, the introduction of helical thread FE model is effective. However, since it requires long computing time and high computing cost, an appropriate modeling scheme should be selected in accordance with the intended use, from the design point of view. When the evaluation target is the magnitude of bolt force, it is practically sufficient to divide engaged thread portion with rather coarse meshes. In some cases, it might be acceptable to omit the modeling of thread geometry. In the following, various modeling techniques for bolted joints are introduced from one-dimensional to three-dimensional models.

1) One-dimensional spring model and beam element

In section 4.1.3, the ratio of flank loads of engaged threads is evaluated by axisymmetric FEM. The amounts of ratio of flank loads are also calculated by one-dimensional spring model [13]. The more advanced applications introduced in this book are the numerical approaches to calculate shear load-bearing ratio in multibolted joints subjected to shear load in Section 8.4.4 and the bolt force variation with time when subjected to dynamic load in Section 8.7.3. Those examples can give some hints when applying the proposed technique to other purposes. Meanwhile, when dealing with a joint clamped by many bolts being subjected to the combination of tension, compression, bending, and twisting moments three-dimensionally, the numbers of nodes and elements of FE models can be decreased without a significant reduction in analytical accuracy as follows: engaged thread portion is replaced by an equivalent cylinder model with three-dimensional shape and bolt cylindrical portion is modeled by a series of beam elements [171], as shown in Fig. 8.54.

2) Axisymmetric model

Numerical analysis of the mechanical behavior of bolted joints by considering the engaged thread geometry was started in the 1970s using axisymmetric FEM [9,15]. After that, FE analysis came to be used to evaluate the effect of plastic deformation of threaded portions on the tightening characteristics [12,100,172] and calculate the bolt force variation and load factor when external force is applied annularly [173]. When the objective is not the calculation of stress concentration at thread root but the evaluation of ratio of flank loads or bolt force variation caused by external force, mesh divisions with the fineness shown in Fig. 8.55A and B can give numerical solutions with practical accuracy, in which circled numbers represent the element numbers

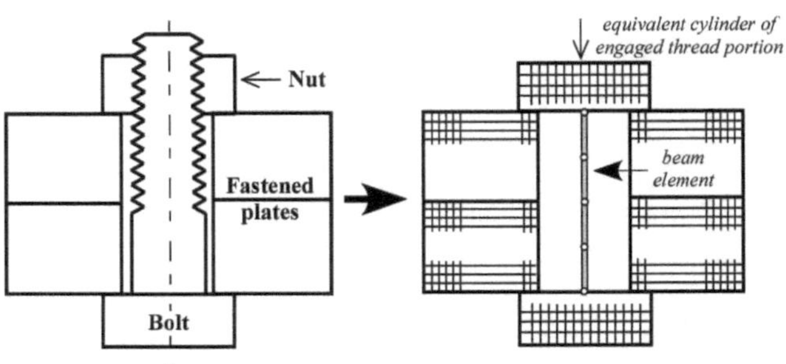

Fig. 8.54 Finite element models providing high computational efficiency.

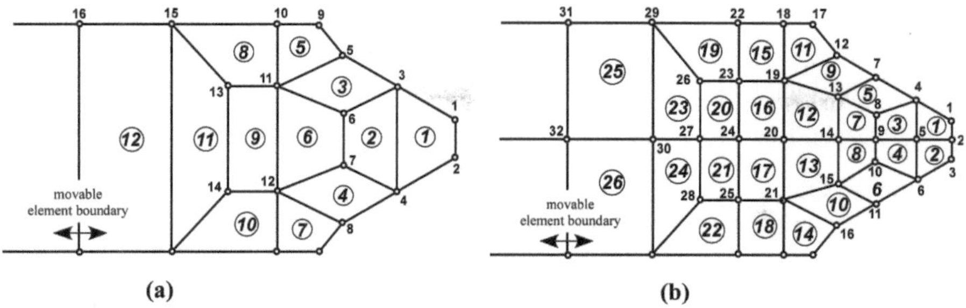

Fig. 8.55 Examples of mesh patterns for threaded portion. (A) Pattern-1 and (B) pattern-2.

of FE model. In each mesh pattern, three or four pairs of nodes are positioned on the pressure flank of male and female threads to be in contact. These mesh patterns are not fine enough to evaluate the stress concentration factor at thread root with high accuracy. However, they can be used for qualitatively examining the effect of tightening and loading conditions on the mechanical behavior of bolted joints. Furthermore, they are effective for quantitative evaluation at a certain level of accuracy.

It is difficult to distinctly show the relationship between the mesh fineness and analytical accuracy regarding the stress concentration factor at the thread root. In FE analysis, the variables to be directly calculated are nodal displacements, and the values of stress and strain are evaluated on Gauss points located inside each element. Gauss points are generally used for the purpose of numerical integration in element stiffness equations, which means that the stress values on the nodes placed along the thread root are calculated with the help of some sort of smoothing processing. Therefore, those values inevitably include a certain amount of numerical errors [33]. Focusing on linear triangular elements, it has been pointed out that stress concentration factors can be evaluated with almost two significant digits by making the dimension of the smallest element, which is placed along the notch, less than 1/10th of the notch root radius. Axisymmetric FE models, introduced above, have successfully been applied to analyze the tightening processes using hydraulic tensioners [80] and bolt heaters [81].

3) Pseudo three-dimensional model applicable to twisting load

To analyze the tightening process of torque control method, it is necessary to introduce three-dimensional analysis considering helical thread geometry. By devising the finite element formulation, however, the analyses can be performed by two-dimensional models [47]. The concrete procedure is explained below:

(1) The degree of freedom of each node in two-dimensional elements is set to be three to cope with r–θ–z coordinate system.
(2) For the nodes positioned along pressure flank of engaged threads, r–θ–z coordinate system is rotated by lead angle β and flank angle α'_1 so that the contact conditions can be evaluated in the plane perpendicular to thread surface.
(3) To simulate the process of applying tightening torque, nodal forces are applied in the circumferential direction to the nodes located on the outer periphery of nut top surface.
(4) As a result of numerical calculation, three components of displacements are obtained, and then six components of stress and strain are calculated.

This method is also applicable to elastic–plastic analysis of bolted joints. It has been clarified in fig. 4.12 how plastic deformations affect the tightening characteristics of torque control method [48].

4) Three-dimensional abacus model

External forces are generally exerted on bolted joints asymmetrically. In these cases, the introduction of three-dimensional models is necessary. However, unless the objective is to analyze the loosening phenomena, etc., influenced by helical thread geometry, three-dimensional abacus models, whose thread geometry is axisymmetric, can work effectively. Some examples solved by these models have already been introduced in this book. Meshes of engaged thread portion can be constructed by rotating axisymmetric models, shown in Fig. 8.55, around the bolt axis. As for the division number in the circumferential direction, for instance, the outer periphery of three-dimensional abacus model, shown in fig. 4.39, is divided into 48 elements.

5) Three-dimensional helical thread model

To analyze the loosening phenomena of screw threads, the use of FE models, in which helical thread geometry is accurately reproduced, is required. Helical thread models, constructed following the procedure explained in section 2.5, represent the geometry of screw threads precisely and fine elements can be placed around the region where high stress concentration occurs. Therefore, it is possible to evaluate stress and stress amplitude distributions along thread root with practical accuracy. As a result, it follows that helical thread models are effectively used to analyze the static and fatigue strengths of bolted joints with high accuracy and they are also applicable to simulate rotation loosening. On the other hand, when incorporating the helical thread models into a structure with complex geometry involves a certain amount of difficulty in constructing an FEM model of the whole structure.

8.9.2 Simple models for engaged thread portion

There are cases in which the stiffness of the engaged thread portion needs to be considered and the evaluation target is not the stress but the bolt force variation. The problems involved in this case are elastic interaction occurring in the tightening process of multibolted joints, discussed in section 3.7, evaluation of bolt force variation and bending moment generated in the bolts that clamp rigid flanged shaft couplings, explained in Section 8.5, etc. To reduce the computing time and cost in these cases, it is effective to replace the engaged thread portion with complex geometry by a simple cylinder. Then, the stiffness in the axial or tangential direction needs to be adjusted so as to conform with the actual one by changing the cylinder height. As a result, the proposed method can expediently be applied to a joint clamped by multiple bolts, since the required computing time can significantly be reduced. For example, the FE model shown in fig. 3.44 has been used to analyze elastic interaction occurring in the tightening process of a pipe flange connection. Regardless of the use of rather coarse meshes, it is found that the amounts of bolt forces can be evaluated with sufficient accuracy, referring to fig. 3.45A, in which the numerical results are compared to the experimental ones.

When analyzing contact problems by FEM, repeated calculations are commonly required to determine the contact conditions of stick, separation, and slip. Then, by forcibly putting the mating nodal points into stick condition with help of sophisticated functions of commercial software, repeated calculations can be avoided, which significantly shortens the computing time. More concretely, it is assumed that the pressure flank of engaged thread portion is in stick condition and the clearance flank is simply regarded as free surface. The aforementioned technique effectively works for the cases in which the phenomena of slip and separation occurring around the joint are not so significant.

8.9.3 Two-dimensional bolt model used for bolt force generation

There are cases in which the shapes of engaged threads and bolt head need not be modeled and bolt models with simple geometry are required instead, by which bolt force can be generated and changed when subjected to external force. For instance, in medium-speed diesel engines, the large end portion is commonly diagonally split, as shown in fig. 4.46, in order to firmly tighten the cap to the end of the connecting rod using crank pin bolts. In such types of connecting rods, fatigue failures are likely to occur in crank pin bolts and body-side female threads to be engaged with the bolt threads. It is reported that the fatigue failure also occurs around the interface consisting of the large end and the cap. For the purpose of withdrawing a piston in a periodic inspection, the large end of the connecting rod is diagonally split to make the rod width small, and the mating surfaces are processed to have zig-zag configuration, so-called serration, to support large shear forces due to piston movement produced by fuel combustion and inertia force. When the objective is the strength evaluation around the serrated region, a bolt can be modeled into a simple plate using two-dimensional elements and the analysis is performed as a plain strain problem. Hence, the stress distributions around the serrated region can be evaluated comparatively easily with practical accuracy [174].

Fig. 8.56 shows the FE model constructed to perform the numerical analysis. Nodal points located at both ends of the bolt model, depicted by black circles, are attached to the bolt head-bearing portion of the cap and the body-side female threaded portion of the connecting rod, which are also indicated by black circles. Hence, the mechanical behaviors of bolted joints in the tightening process and under external force can effectively be evaluated with less computational effort. In the proposed modeling method, a bolt with complex geometry is replaced by a simple plate using a small number of finite elements. Therefore, the bolt model stiffness necessarily becomes quite high. Then, in order to make the model stiffness match the actual one, selective reduced integration, introduced in Section 8.8.2, is effectively used at the stage of element matrix assembly in the finite element formulation.

Fig. 8.56 Finite element model of diagonally split connecting rod.

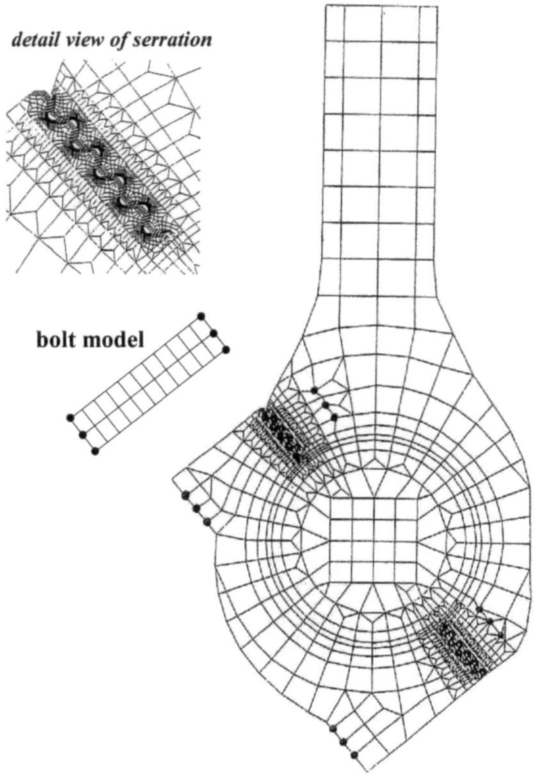

8.9.4 Improvement of computational efficiency utilizing geometric symmetry [175]

In many cases, bolted joints are composed of a repetition of structures with identical geometry. In such cases, a modeling scheme considering the geometric symmetry is effective to reduce computing time and cost. Fig. 8.57A and B depict three-dimensional models of pipe flange connections. When a pair of flanges is clamped with bolts and nuts, there exist engaged thread portions or bolt head portions on each flange surface. Meanwhile, the most important factor to be evaluated in pipe flange connections is gasket contact pressure. Accordingly, bolt-nut pairs are simply modeled into cylindrical shapes. In Fig. 8.57A and B, only one side of the pipe flange connection is modeled, hence, it follows that the symmetry is assumed against the central surface of gasket thickness. Engaged thread portions are modeled into cylindrical shapes as well as bolt head portions, and the cylinder height is set to be 0.27 times the nominal diameter d to make its axial stiffness match the actual one [87]. In the case of the number of clamping bolts being eight, mechanical

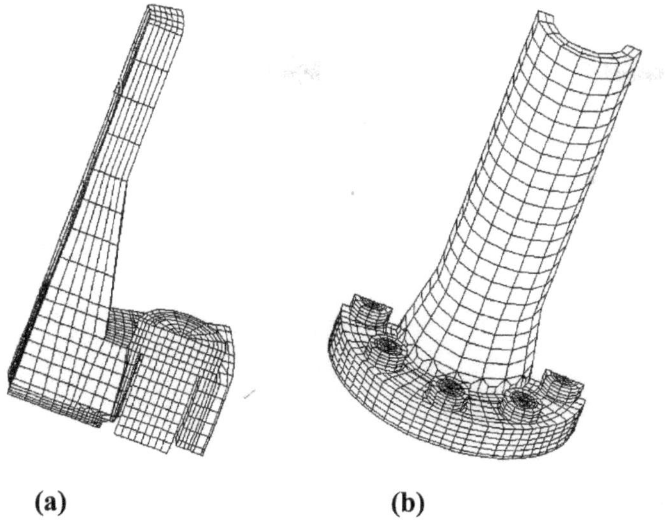

Fig. 8.57 Finite element models of pipe flange connections considering symmetry. (A) 1/32 model and (B) 1/4 model.

behaviors in the tightening process and under internal pressure can be analyzed by 1/32nd of the whole structure, shown in Fig. 8.57A, and the behavior under bending moment can be evaluated by the quarter model, shown in Fig. 8.57B. These modeling schemes are valid only when the symmetries in the joint shape and loading conditions are both fulfilled.

Meanwhile, if the symmetry is considered for each analysis, the number of models required necessarily increases. Supposing that computing time is the sum of modeling time and analysis time, it may be practical to select an FE model requiring shorter modeling time, even if the computing time is somewhat longer. In Fig. 8.57A, for example, one-half of the bolt is modeled, but there is an alternative method in which the bolt is wholly modeled and a 1/16th flange model is constructed. Furthermore, when analyzing the cases subjected to inner pressure and/or bending moment, there is an option to use the quarter model, shown in Fig. 8.57B, for all cases stated above. In the numerical analyses of multibolted joints in Section 8.4, one-half of the bolt body is modeled for some bolts, as shown in Fig. 8.25B, aiming at reducing the computing time. As a result, a quarter part of the whole joint is modeled by perfectly considering the symmetries in the joint shape and loading conditions.

In the analyses of rigid flanged shaft couplings in Section 8.5, a modeling scheme considering the periodic symmetry is introduced, because the joint clamped by multiple bolts is subjected to twisting moment with respect to the rotation center axis. In the analyses of stress amplitudes produced at the thread root of wheel bolts, discussed in Section 8.2.6, eight bolts are all modeled, since it is necessary to evaluate

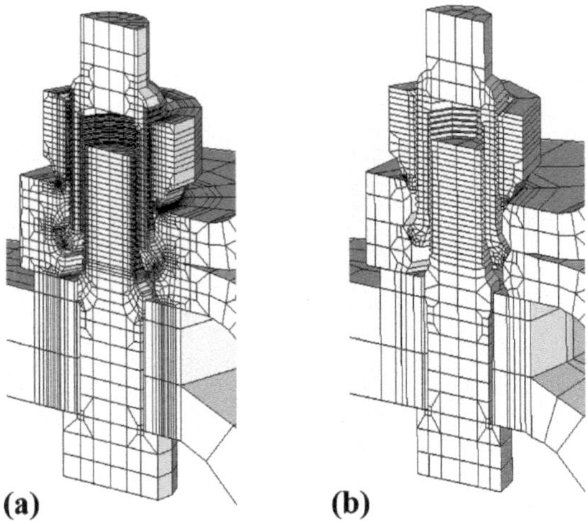

Fig. 8.58 (A) Fine and (B) coarse mesh patterns of the jointed portion of wheel bolts.

the bolt stress variation during one rotation of the wheel. Then, in order to enhance the computational efficiency, a fine mesh pattern is applied only to the bolt model used for stress evaluation at the thread root, and the remaining seven bolts are all divided with an identical mesh pattern using relatively coarse elements. Fig. 8.58A and B represents the fine and coarse mesh patterns of the jointed portion of wheel bolts, respectively.

References

[1] A. Yamamoto, Story of Screw Threads, Japanese Standards Association (JSA), 1990, pp. 19–32.
[2] O. Tamura, Knowledge of Screw Threads, Yokendo, 2008, pp. 1–7.
[3] F. Rötcher, Die Maschinenelemente (Erster Band), Julius Springer, 1927.
[4] J.N. Goodier, The distribution of load on the threads of screws, J. Appl. Mech. 7 (1) (1940) 10–16.
[5] D.G. Sopwith, The distribution of load in screw threads, Proc. Inst. Mech. Eng. 159 (45) (1948) 373–383.
[6] T. Sawa, K. Maruyama, On the deformations of the bolt head and nut in a bolted joint, Bull. JSME 19 (128) (1976) 203–211.
[7] H. Otaki, Stress distributions around bolt thread root in bolt-nut connections (1st report), Trans. Jpn. Soc. Mech. Eng. C 37 (303) (1971) 2197–2203.
[8] H. Otaki, Stress distributions around bolt thread root in bolt-nut connections (2nd report), Trans. Jpn. Soc. Mech. Eng. 38 (311) (1972) 1885–1894.
[9] K. Maruyama, Stress analysis of a bolt-nut joint by the finite element method and the copper-electroplating method (2nd report), Bull. JSME 17 (106) (1974) 442–450.
[10] J.L. Bretl, R.D. Cook, Modelling the load transfer in threaded connections by the finite element method, Int. J. Numer. Methods Eng. 14 (1979) 1359–1377.
[11] D.L. Miller, K.M. Marshek, M.R. Naji, Determination of load distribution in a threaded connection, Mech. Mach. Theory 18 (6) (1983) 421–430.
[12] M. Tanaka, H. Miyazawa, E. Asaba, K. Hongo, Application of the finite element method to bolt-nut joints, Bull. JSME 24 (192) (1981) 1064–1071.
[13] M. Tanaka, Considerations on static behavior of screw threads (2nd report), J. JFRI 12 (8) (1981) 185–191.
[14] M. Hetenyi, Photo elastic study of bolt and nut fastenings, Trans. ASME 10 (2) (1943) 93–100.
[15] K. Maruyama, Stress analysis of a bolt-nut joint by the finite element method and the copper-electroplating method (3rd report), Bull. JSME 19 (130) (1976) 360–368.
[16] M. Seika, S. Sasaki, K. Hosono, Measurement of stress concentrations in threaded connections, Bull. JSME 17 (111) (1974) 1151–1156.
[17] R.C. Juvinall, K.M. Marshek, Fundamentals of Machine Component Design, Wiley, 2000, pp. 395–461.
[18] N. Ohashi, Research and study on tapping screws (part-2), J. JFRI 30 (8) (1999) 230–235.
[19] K. Hongo, General remarks of screw threads, Design Draft. 15 (80) (1980) 281–287.
[20] T. Fukuoka, M. Nomura, Y. Takeda, U. Mori, Analysis of the tightening process and the cyclic stress amplitude of studs and tap bolts, in: ASME PVP Conference, 2011 (Paper No. 57118).
[21] S. Hashimura, Tightening characteristics of bolts made of non-ferrous materials, J. JFRI 45 (6) (2014) 166–172.
[22] JSME, The Modulus of Elasticity of Metals and Alloys, 1980.
[23] JSME, Data Book, Heat Transfer, fifth ed., 2009.
[24] JSME, Data Book, Heat Transfer, fourth ed., 1986.

[25] T. Fukuoka, M. Nomura, Y. Takasugi, Evaluation of thermal and mechanical behaviors of bolted joints made of titanium and titanium alloy and its application to robust joint design, in: ASME PVP Conference, 2013 (Paper No. 97156).
[26] T. Sakai, Introduction to Screw Thread Tightening Theory (enlarged ed.), Yokendo, 2010, p. 226.
[27] JFRI Publication Committee, Guide Book for Screw Thread Tightening (2004) 39–64.
[28] J. Oda, Y. Murotsu, Theory of Machine Design I, revised ed., Baifukan, 1999, pp. 48–51.
[29] JFRI Publication Committee, Systematic Calculation Procedure of High Strength Bolted Joints (VDI 2230 Blatt 1 (2003)), 2006, pp. 45–48.
[30] G. Yagawa, N. Miyazaki, Analysis of Thermal Stress, Creep and Heat Conduction by FEM, Saiens-Sha, 1985.
[31] T. Sakai, Introduction to Screw Thread Tightening Theory (enlarged ed.), Yokendo, 2010, pp. 63–65.
[32] J.H. Bickford, An Introduction to the Design and Behavior of Bolted Joints, third ed., Dekker, 1995, pp. 602–649.
[33] T. Fukuoka, Introduction to finite element analysis for marine engineers (part-3), J. JIME 49 (3) (2014) 381–386.
[34] T. Fukuoka, Analysis of the tightening process of bolted joint with a tensioner using spring elements, J. Press. Vessel Technol. 116 (1994) 443–448.
[35] T. Fukuoka, Analysis of the tightening process of bolted joint with a tensioner—effects of incorrect geometry at contact surface, Trans. Jpn. Soc. Mech. Eng. A 62 (602) (1996) 2372–2378.
[36] T. Fukuoka, Analysis of the tightening process of bolted joint with a tensioner-effects of interface stiffness, in: ASME DE Conference, 110, 2000, pp. 119–125.
[37] JFRI Publication Committee, Systematic Calculation Procedure of High Strength Bolted Joints (VDI 2230 Blatt 1 (2003)), 2006, pp. 20–31.
[38] T. Naruse, Y. Shibutani, Equivalent stiffness evaluation of clamped plates in bolted joints under loading, Trans. Jpn. Soc. Mech. Eng. A 75 (757) (2009) 1230–1238.
[39] M. Shibahara, Calculation method of spring constants of bolted joint, Design Draft. 15 (80) (1980) 307–313.
[40] K. Mitsunaga, Pressure distributions in fastened plates of bolted joint, Trans. Jpn. Soc. Mech. Eng. C 31 (231) (1965) 1750–1757.
[41] T. Fukuoka, M. Nomura, True cross sectional area of screw threads with helix and root radius geometries taken into consideration, J. Press. Vessel Technol. 131 (2009) (Paper No. 024501).
[42] T. Fukuoka, M. Nomura, Y. Maruo, Analytical expression of cross sectional geometry of various screw threads and finite element analysis of the mechanical behavior of multiple-thread screws, Trans. Jpn. Soc. Mech. Eng. C 79 (808) (2013) 5045–5059.
[43] K. Yanase, T. Fukuoka, Mathematical expressions of helical thread geometry and cross sectional areas of various shaped thread forms and finite element analysis, in: ASME PVP Conference, 2013 (Paper No. 97553).
[44] S. Izumi, T. Yokoyama, A. Iwasaki, S. Sakai, Three-dimensional finite element analysis on tightening and loosening mechanism of bolted joint, Trans. Jpn. Soc. Mech. Eng. A 71 (702) (2005) 204–212.
[45] T. Fukuoka, M. Nomura, D. Yasui, Numerical analysis of loosening phenomena of bolted joints under shear loads using helical thread models, J. JIME 55 (4) (2020) 522–528.
[46] T. Fukuoka, M. Nomura, D. Yasui, Effects of friction coefficient and various geometric parameters on loosening phenomena of bolted joints under shear loads, J. JIME 55 (5) (2020) 636–643.

[47] T. Fukuoka, N. Yamasaki, H. Kitagawa, M. Hamada, A stress analysis of threaded portions in fastening, Bull. JSME 28 (244) (1985) 2247–2253.

[48] T. Fukuoka, T. Takaki, Elastic plastic finite element analysis of bolted joint during tightening process, J. Mech. Des. ASME 125 (4) (2003) 823–830.

[49] T. Fukuoka, M. Nomura, Proposition of helical thread modeling with accurate geometry and finite element analysis, J. Press. Vessel Technol. 130 (2008) (Paper No. 011204).

[50] T. Fukuoka, Y. Hirai, Evaluation of mechanical behavior of taper pipe threads in the tightening process by finite element analysis and elementary theory of solid mechanics, in: ASME PVP Conference, 2017 (Paper No. 65061).

[51] R.H. Thornley, R. Connolly, M.M. Barash, F. Koenigsberger, The effect of surface topography upon the static stiffness of machine tool joints, Int. J. Mach. Tool Des. Res. 5 (1965) 57–74.

[52] V.I. Ostrovskii, The influence of machining methods on sideway contact stiffness, Mach. Tool., 36 (1) (1965) 17–19.

[53] A. Taniguchi, M. Tsutsumi, Y. Ito, Treatment of contact stiffness in structural analysis (1st report), Bull. JSME 27 (225) (1984) 601–607.

[54] T. Fukuoka, T. Takaki, Evaluations of the tightening process of bolted joint with elastic angle control method, in: ASME PVP Conference, vol. 478, 2004, pp. 11–18.

[55] N. Back, M. Burdekin, A. Cowley, Analysis of machine tool joints by the finite element method, in: 14th Int. MTDR Conf, 1973, pp. 529–537. Paper No. 210.

[56] T. Fukuoka, M. Nomura, N. Sugano, Experimental investigation and finite element analysis of the free vibration problem by taking account of interface stiffness, J. Environ. Eng. 4 (1) (2009) 101–111.

[57] V.N. Kirsanova, The shear compliance of flat joints, Mach. Tool., 38 (7) (1967) 30–34.

[58] J.H. Bickford, An Introduction to the Design and Behavior of Bolted Joints, third ed., Dekker, 1995, pp. 213–268.

[59] I. Yoshimoto (Ed.), Key Point in Designing Bolted Joints, Japanese Standards Association (JSA), 1992, pp. 159–196.

[60] A. Yamamoto, Practical guideline for plastic angle control method, J. JFRI 21 (11) (1990) 333–340.

[61] T. Makimae, H. Yamashita, Development of bolts used for diesel engines tightened by plastic angle control method, J. JFRI 18 (5) (1987) 129–138.

[62] Japan Institution of Marine Engineering, Research and Study on Tightening Guideline of Crucial Bolts for Diesel Engines, Research Committee Report No. 259, 1992, pp. 1–24.

[63] A. Yamamoto, Theory and Calculation for Tightening of Screw Threads, Yokendo, 1970, pp. 74–88.

[64] T. Fukuoka, T. Takaki, Mechanical behaviors of bolted joint in various clamping configurations, J. Press. Vessel Technol. 120 (1998) 226–231.

[65] T. Fukuoka, Theory and practice of bolt tightening process (part-I), J. JIME 46 (3) (2011) 406–412.

[66] JFRI Publication Committee, Guide Book for Screw Thread Tightening, 2004, pp. 77–100.

[67] K. Ikeda, H. Nakagawa, K. Mitsunaga, Study on tightening process of bolts, Trans. Jpn. Soc. Mech. Eng. C 36 (290) (1970) 1735–1744.

[68] T. Sakai, Friction coefficient of threaded fasteners, Trans. Jpn. Soc. Mech. Eng. C 43 (370) (1977) 2372–2382.

[69] T. Fukuoka, K. Nakano, Improvement of tightening accuracy of torque control method by taking into account geometric errors in threaded fasteners, in: ASME PVP Conference, 2016 (Paper No. 63083).

[70] T. Fukuoka, M. Nomura, Y. Fukuman, T. Kamihira, Y. Sugimoto, Evaluation of preload scatter of wheel bolts and safety assessment of large vehicles, Trans. Jpn. Soc. Mech. Eng. A 75 (759) (2009) 1577–1584.
[71] T. Fukuoka, M. Nomura, H. Kawabayashi, A new experimental approach for measuring friction coefficients of threaded fasteners focusing on the repetition of tightening operation and surface roughness, in: ASME PVP Conference, 2013 (Paper No. 97606).
[72] M. Hagiwara, Behavior of the thread torque after tightening, J. JSPE 50 (8) (1984) 1278–1282.
[73] K. Hayashi, Y. Kagami, H. Abe, Residual torsional deformation and residual torque due to screw fastening, Trans. Jpn. Soc. Mech. Eng. C 53 (489) (1987) 1096–1101.
[74] T. Fukuoka, T. Takaki, Mechanical behaviors of bolted joint during tightening using torque control, JSME Int. J. Ser. A 41 (2) (1998) 185–191.
[75] T. Fukuoka, K. Minami, S. Murakami, D. Yasui, Improving bolt tightening accuracy in torque control method, J. JIME 53 (4) (2018) 570–575.
[76] A.F.C. Brown, W. McClimont, Fatigue strength of five types of stud, Engineering 20 (1960) 430.
[77] H. Fessler, P.K. Jobson, Stresses in a bottoming stud assembly with chamfers at the ends of threads, J. Strain Anal. 18 (1) (1983) 15–22.
[78] A. Yamamoto, Principle and Design of Screw Threads, Yokendo, 1995, pp. 70–101.
[79] W. Fabry, Hydraulishes Shraubenanziehen an Reaktordeckeln, Konstruktion 23 (1) (1971) 13–19.
[80] T. Fukuoka, Finite element simulation of tightening process of bolted joint with a tensioner, J. Press. Vessel Technol. 114 (1992) 433–438.
[81] T. Fukuoka, Q. Xu, Finite element simulation of the tightening process of bolted joint with a bolt heater, J. Press. Vessel Technol. 124 (2002) 457–464.
[82] T. Fukuoka, T. Takaki, Simplified analysis of the tightening process of bolted joint with a bolt heater, in: ASME PVP Conference, 2005 (Paper No. 71168).
[83] T. Fukuoka, N. Yamasaki, H. Kitagawa, M. Hamada, Ratio of flank loads of screw threads in hollow bolt, Bull. JSME 29 (247) (1986) 265–272.
[84] T. Fukuoka, Application of bolt heater technique to tightening operation of large bolts placed horizontally, J. JIME 48 (6) (2013) 847–853.
[85] T. Fukuoka, K. Kageyama, H. Kusafuka, M. Hirao, New bolt tightening method with high accuracy utilizing real-time measurement of nut factor, J. JIME 53 (4) (2018) 589–594.
[86] T. Fukuoka, Theory and practice of bolt tightening process (part-III), J. JIME 46 (3) (2011) 419–423.
[87] T. Fukuoka, T. Takaki, Three-dimensional finite element analysis of pipe flange-effects of flange interface geometry, Trans. Jpn. Soc. Mech. Eng. A 64 (625) (1998) 2402–2407.
[88] T. Fukuoka, T. Takaki, Finite element simulation of bolt-up process of pipe flange connections, J. Press. Vessel Technol. 123 (2001) 282–287.
[89] T. Fukuoka, T. Takaki, Finite element simulation of bolt-up process of pipe flange connections with spiral wound gasket, J. Press. Vessel Technol. 125 (2003) 371–378.
[90] T. Takaki, T. Fukuoka, Three-dimensional finite element analysis of pipe flange connections—the case of using compressed asbestos sheet gasket, in: ASME PVP Conference, vol. 433, 2002, pp. 171–177.
[91] T. Fukuoka, T. Takaki, Finite element simulation of the disassembly process of pipe flange connections, in: ASME PVP Conference, vol. 433, 2002, pp. 139–145.

[92] T. Fukuoka, T. Takaki, Evaluations of bolt-up sequence of pipe flange using three-dimensional finite element analysis, in: ASME PVP Conference, vol. 382, 1999, pp. 87–93.
[93] T. Takaki, T. Fukuoka, Effective bolt-up procedure of pipe flange connections-finite element analyses and elastic interaction coefficient methods, Trans. Jpn. Soc. Mech. Eng. A 68 (668) (2002) 550–557.
[94] T. Fukuoka, M. Nomura, M. Sasai, Evaluation of specific mechanical behavior of fine screw threads by finite element analysis and experiments, in: ASME PVP Conference, 2014 (Paper No. 28241).
[95] A. Yamamoto, Theory and Calculation for Tightening of Screw Threads, Yokendo, 1970, pp. 39–54.
[96] T. Fukuoka, Mechanical properties of threaded regions in an eyebolt and an eyenut, JSME Int. J. Ser. A 34 (4) (1991) 505–511.
[97] M. Nishida, Stress Concentration, Morikita-Shuppan, 1967.
[98] S. Inada, T. Kawakita, Y. Honjo, Machine Design Procedure, Asakurashoten, 1983, pp. 5–11.
[99] T. Fukuoka, N. Yamasaki, H. Kitagawa, M. Hamada, Stress concentration at the root of a screw thread-effects of root radius and friction coefficient, Trans. Jpn. Soc. Mech. Eng. A 52 (481) (1986) 2201–2208.
[100] M. Tanaka, N. Henmi, H. Ishibashi, On the behavior of bolted joints tightened in plastic region, Trans. Jpn. Soc. Mech. Eng. C 66 (650) (2000) 3483–3488.
[101] T. Hattori, T. Nonaka, M. Taneda, Strength improvement of bolted joints by post-yield fastening, JHPI 23 (1) (1985) 7–13.
[102] E.A. Patterson, B. Kenny, Stress analysis of some nut-bolt connections with modifications to the external shape of the nut, J. Strain Anal. 22 (4) (1987) 187–193.
[103] T. Fukuoka, Evaluation of the method for lowering stress concentration at the thread root of bolted joints with modifications of nut shape, J. Press. Vessel Technol. 119 (1997) 1–9.
[104] JSME, Failure Cases and Its Analysis Techniques of Machines and Structures, 1984.
[105] Research Society of Loosening and Failures of Screw Threads, Reference Book of Loosening and Failures of Various Threaded Fasteners, Management Development Center, 1980.
[106] S. Tsunashima, Y. Simotsuma, Machine Design Theory, Ohmsha, 1973, pp. 10–16.
[107] H. Murakami, M. Ominami, Introduction to Fracture Mechanics, Ohmsha, 1979, pp. 121–161.
[108] F. Böhm, Progress in developing M.a.N. two-stroke and four-stroke diesel-engines, in: ISME TOKYO'78 Technical Papers, 1978, pp. 157–169.
[109] A. Yamamoto, Principle and Design of Screw Threads, Yokendo, 1995, pp. 147–178.
[110] K.H. Kloss, W. Schneider, Fatigue strength of bolted joints clamped with high bolt force, J. JFRI 21 (10) (1990) 306–318.
[111] T. Sawa, K. Maruyama, Y. Maekawa, The force ratio of bolted joints—the case where clamped parts are connecting rods, Bull. JSME 20 (150) (1977) 1623–1630.
[112] O. Zhang, J.A. Poirier, New analytical model of bolted joints, J. Mech. Des. ASME 126 (3) (2004) 721–728.
[113] O. Zhang, Discussions on behavior of bolted joints in tension, J. Mech. Des. ASME 127 (2) (2005) 506–510.
[114] I. Yoshimoto, A hypothesis on fatigue strength of screw threads, J. JSPE 49 (6) (1983) 801–803.
[115] M. Tanaka, The stress analysis of bolted joint with model using spring-beam elements—a case of single bolted joint, J. JSPE 51 (12) (1985) 2265–2270.

[116] M. Tanaka, K. Hongo, The stress analysis of bolted joint with model using spring-beam elements—a case of bolted joint received eccentric load, J. JSPE 52 (4) (1986) 655–660.
[117] M. Tanaka, K. Hongo, The stress analysis of bolted joint with model using spring-beam elements—application to the force-deformation diagram of bolted joint, J. JSPE 54 (9) (1988) 1747–1752.
[118] T. Fukuoka, H. Taniuchi, Evaluation of stress amplitude of bolted joints by FE analysis and joint diagram and a simple strategy for its reduction, in: ASME PVP Conference, 2015 (Paper No. 45084).
[119] T. Fukuoka, M. Nomura, T. Fuchikami, Finite element analysis of the cyclic stress amplitude of threaded fasteners using helical thread models, J. Press. Vessel Technol. 133 (2011) (Paper No. 061201).
[120] Y. Takeuchi, Thermal Stress, Nisshin-Shuppan, 1971, pp. 1–16.
[121] T. Fukuoka, M. Nomura, Evaluation of thermal contact resistance at the interface of dissimilar materials, J. Press. Vessel Technol. 135 (2013) (Paper No. 021403).
[122] T. Fukuoka, Q. Xu, Evaluations of thermal contact resistance in an atmospheric environment, in: ASME PVP Conference, 382, 1999, pp. 145–151.
[123] A.M. Clausing, Heat transfer at the Interface of dissimilar materials—the influence of thermal strain, Int. J. Heat Mass Transf. 9 (1966) 791–801.
[124] T. Fukuoka, Finite element analysis of the thermal and mechanical behaviors of a bolted joint, J. Press. Vessel Technol. 127 (2005) 402–407.
[125] T. Fukuoka, M. Nomura, K. Shino, Analysis of heat flow around bolted joints and variations of axial bolt force, J. Press. Vessel Technol. 131 (2009) (Paper No. 061203).
[126] L. Guessous, G.C. Barber, Q. Zou, S.A. Nassar, A numerical investigation of bolt underhead temperature evolution under various fastening conditions, Tribol. Trans. 51 (2008) 494–503.
[127] M. Muraki, Tribology by Illustration, Nikkan Kogyo Shimbun, 2007, pp. 48–53.
[128] Y. Yamamoto, S. Kaneda, Tribology, Rikogakusha, 1998, pp. 48–56.
[129] M. Muraki, Tribology by Illustration, Nikkan Kogyo Shimbun, 2007, pp. 27–37.
[130] T. Sakai, Introduction to Screw Thread Tightening Theory (enlarged ed.), Yokendo, 2010, pp. 44–72.
[131] A. Yamamoto, Principle and Design of Screw Threads, Yokendo, 1995, pp. 102–146.
[132] A. Yamamoto, Story of Screw Threads, Japanese Standards Association (JSA), 1990, pp. 73–87.
[133] T. Sakai, Loosening of bolts (2nd report), Trans. Jpn. Soc. Mech. Eng. C 44 (377) (1978) 288–292.
[134] J.N. Goodier, R.J. Sweeny, Loosening by vibration of threaded fasteners, Mech. Eng. 67 (1945) 798–802.
[135] S. Sato, S. Hosokawa, A. Yamamoto, Studies on loosening mechanism of bolt nut units (2nd report), J. JSPE 51 (8) (1985) 1540–1546.
[136] K. Hongo, Loosening of bolt and nut fastening, Trans. Jpn. Soc. Mech. Eng. A 30 (215) (1964) 934–939.
[137] K. Koga, Considerations on loosening of screw threads due to impact load, Trans. Jpn. Soc. Mech. Eng. C 35 (273) (1969) 1104–1111.
[138] H. Isono, K. Koga, A study on loosening of thread fastenings by repeated impacts, J. JSPE 51 (12) (1985) 2247–2252.
[139] S. Kasei, P. Sun, Fundamental considerations on thread-loosening due to transverse impact, Trans. Jpn. Soc. Mech. Eng. C 60 (570) (1994) 625–631.
[140] O. Tamura, Knowledge of Screw Threads, Yokendo, 2008, pp. 91–96.

[141] I. Yoshimoto (Ed.), Key Point in Designing Bolted Joints, Japanese Standards Association (JSA), 1992, pp. 197–229.
[142] M. Tanaka, H. Miyazawa, Finite element analysis of the tightening process using double nut, Trans. Jpn. Soc. Mech. Eng. C 47 (417) (1981) 592–601.
[143] A. Yamamoto, Theory and Calculation for Tightening of Screw Threads, Yokendo, 1970, pp. 89–101.
[144] T. Fukuoka, M. Nomura, K. Minami, Evaluation of various methods to inspect bolted joint integrity by torque measurement, J. JIME 56 (2) (2021) 257–264.
[145] T. Fukuoka, M. Nomura, Y. Hata, T. Nishikawa, Development of test equipment for measuring compression characteristics of sheet gaskets at elevated temperature, in: ASME PVP Conference, 2007 (Paper No. 26382).
[146] T. Fukuoka, M. Nomura, T. Nishikawa, Analysis of thermal and mechanical behavior of pipe flange connections by taking account of gasket compression characteristics at elevated temperature, J. Press. Vessel Technol. 134 (2012) (Paper No. 021202).
[147] T. Fukuoka, M. Nomura, Y. Uemori, K. Sato, Bolt preload reduction of pipe flange connections after plant shutdown, in: ASME PVP Conference, 2012 (Paper No. 78054).
[148] T. Fukuoka, Evaluation of thermal and mechanical behaviors of pipe flange connections for low temperature fluids by numerical analysis and experiments, in: ASME PVP Conference, 2016 (Paper No. 63212).
[149] T. Fukuoka, M. Nomura, A. Nakada, Evaluation of sealing performance of pipe flange connections for low temperature fluids by numerical analysis, J. JIME 54 (5) (2019) 758–764.
[150] Japan National Traffic Safety and Environment Laboratory, Research Report on Wheel Falling-off Accidents of Large Vehicles Due to the Rupture of Wheel Bolts, 2004.
[151] Japan Railway Construction, Transport and Technology Agency, Interim Report of Basic Research and Development Program in Transport Field, 2008.
[152] T. Fukuoka, M. Nomura, M. Kizawa, Y. Fukuman, Evaluation of the tightening process of wheel bolts for large vehicles by finite element analysis, Trans. Jpn. Soc. Mech. Eng. C 75 (750) (2009) 446–453.
[153] T. Fukuoka, M. Nomura, T. Kamihira, K. Kitano, Evaluation of the stress amplitude of wheel bolts for large vehicles by full-scale test, Trans. Jpn. Soc. Mech. Eng. C 76 (772) (2010) 3768–3775.
[154] T. Fukuoka, M. Nomura, T. Kamihira, Finite element analysis of the cyclic stress amplitude of wheel bolts for large vehicles, Trans. Jpn. Soc. Mech. Eng. C 77 (782) (2011) 3840–3849.
[155] T. Fukuoka, M. Nomura, T. Kamihira, Development of tightening apparatus with multiple shafts and torque control mechanism for wheel bolts of large vehicles, Trans. Jpn. Soc. Mech. Eng. C 76 (768) (2010) 1970–1977.
[156] S. Timoshenko, D.H. Young, (Translated by Maezawa). Elements of Strength of Materials, revised ed., Corona Pub. Co. Ltd, 1972, pp. 40–46.
[157] T. Sawa, Why the Roller Coaster Accident Occurred, Monozukuri, Nikkei, 2007, pp. 36–39.
[158] T. Fukuoka, Finite element analysis of the mechanical behavior of multibolted joints subjected to shear loads, J. Press. Vessel Technol. 140 (2018) (Paper No. 051201).
[159] J.H. Bickford, An Introduction to the Design and Behavior of Bolted Joints, third ed., Dekker, 1995, pp. 504–526.
[160] T. Fukuoka, M. Nomura, M. Yamashita, Evaluation of strength and load capacity of reamer bolts by numerical analysis, in: ASME PVP Conference, 2008 (Paper No. 61226).

[161] K. Okayama, T. Fukuoka, Three-dimensional finite element analysis of rigid flanged shaft coupling subjected to twisting moment, in: ASME PVP Conference, 2013 (Paper No. 97482).
[162] Research Committee of Propeller Shaft System of MESJ, Standards for finish processing of propeller shaft, J. MESJ 23 (4) (1988) 273–280.
[163] J.R. Mancuso, Couplings and Joints, Dekker, 1999, pp. 23–29.
[164] T. Fukuoka, M. Nomura, T. Sakamoto, Evaluation of tightening process of reamer bolts by cooled fitting, J. JIME 43 (5) (2008) 773–779.
[165] S. Ishihara, Marine Propeller and Shaft System, Seizando, 2002, pp. 1–3.
[166] T. Fukuoka, Finite element simulation of mechanical behaviors and seal performance of oil seal plug, Trans. Jpn. Soc. Mech. Eng. A 64 (625) (1998) 2395–2401.
[167] H. Togawa, Finite Element Analysis of Vibration Problems, Saiensu-sha, 1975.
[168] K.J. Bathe, E.L. Wilson, (Translated by F. Kikuchi). Numerical Methods in Finite Element Analysis, Kagaku Gijutsu Shuppan, 1979, pp. 407–587.
[169] G. Yagawa, Y. Aoyama, Eigenvalue Analysis by Finite Element Method, Morikita Pub. Co. Ltd., 2001.
[170] N. Kikuchi, Finite Element Methods in Mechanics, Cambridge University Press, 1986, pp. 218–221.
[171] T. Fukuoka, Introduction to finite element analysis for marine engineers (part-1), J. JIME 49 (1) (2014) 111–118.
[172] T. Sunamoto, Elastic-plastic deflections and strains of thread ridges in screw thread connections, Trans. Jpn. Soc. Mech. Eng. C 45 (399) (1979) 1287–1298.
[173] M. Tanaka, K. Hongo, E. Asaba, Finite element analysis of bolted joint under external loads, Trans. Jpn. Soc. Mech. Eng. C 47 (418) (1981) 766–775.
[174] T. Fukuoka, Y. Senoh, Finite element simulation of the strength of the big end of diagonally split connecting rod, Trans. Jpn. Soc. Mech. Eng. A 64 (617) (1998) 104–110.
[175] T. Fukuoka, Introduction to finite element analysis for marine engineers (part-2), J. JIME 49 (2) (2014) 239–246.

Index

Note: Page numbers followed by *f* indicate figures and *t* indicate tables.

W

Y

MIX
Papier | Fördert gute Waldnutzung
FSC® C083411

Kolibri 360 GmbH
Ferdinand-Jühlke-Straße 7
99095 Erfurt, Deutschland
produktsicherheit@kolibri360.de